Lecture Notes in Physics

Volume 886

Founding Editors

W. Beiglböck
J. Ehlers
K. Hepp
H. Weidenmüller

Editorial Board

B.-G. Englert, Singapore, Singapore
P. Hänggi, Augsburg, Germany
W. Hillebrandt, Garching, Germany
M. Hjorth-Jensen, Oslo, Norway
R.A.L. Jones, Sheffield, UK
M. Lewenstein, Barcelona, Spain
H. von Löhneysen, Karlsruhe, Germany
M.S. Longair, Cambridge, UK
J.-F. Pinton, Lyon, France
J.-M. Raimond, Paris, France
A. Rubio, Donostia, San Sebastian, Spain
M. Salmhofer, Heidelberg, Germany
S. Theisen, Potsdam, Germany
D. Vollhardt, Augsburg, Germany
J.D. Wells, Geneva, Switzerland

For further volumes:
http://www.springer.com/series/5304

The Lecture Notes in Physics

The series Lecture Notes in Physics (LNP), founded in 1969, reports new developments in physics research and teaching-quickly and informally, but with a high quality and the explicit aim to summarize and communicate current knowledge in an accessible way. Books published in this series are conceived as bridging material between advanced graduate textbooks and the forefront of research and to serve three purposes:

- to be a compact and modern up-to-date source of reference on a well-defined topic
- to serve as an accessible introduction to the field to postgraduate students and nonspecialist researchers from related areas
- to be a source of advanced teaching material for specialized seminars, courses and schools

Both monographs and multi-author volumes will be considered for publication. Edited volumes should, however, consist of a very limited number of contributions only. Proceedings will not be considered for LNP.

Volumes published in LNP are disseminated both in print and in electronic formats, the electronic archive being available at springerlink.com. The series content is indexed, abstracted and referenced by many abstracting and information services, bibliographic networks, subscription agencies, library networks, and consortia.

Proposals should be sent to a member of the Editorial Board, or directly to the managing editor at Springer:

Christian Caron
Springer Heidelberg
Physics Editorial Department I
Tiergartenstrasse 17
69121 Heidelberg/Germany
christian.caron@springer.com

Tilman Plehn

Lectures on LHC Physics

Second Edition

Springer

Tilman Plehn
Institut für Theoretische Physik
University of Heidelberg
Heidelberg
Germany

ISSN 0075-8450   ISSN 1616-6361 (electronic)
Lecture Notes in Physics
ISBN 978-3-319-05941-9   ISBN 978-3-319-05942-6 (eBook)
DOI 10.1007/978-3-319-05942-6
Springer Cham Heidelberg New York Dordrecht London

Library of Congress Control Number: 2014941877

© Springer International Publishing Switzerland 2012, 2015
This work is subject to copyright. All rights are reserved by the Publisher, whether the whole or part of the material is concerned, specifically the rights of translation, reprinting, reuse of illustrations, recitation, broadcasting, reproduction on microfilms or in any other physical way, and transmission or information storage and retrieval, electronic adaptation, computer software, or by similar or dissimilar methodology now known or hereafter developed. Exempted from this legal reservation are brief excerpts in connection with reviews or scholarly analysis or material supplied specifically for the purpose of being entered and executed on a computer system, for exclusive use by the purchaser of the work. Duplication of this publication or parts thereof is permitted only under the provisions of the Copyright Law of the Publisher's location, in its current version, and permission for use must always be obtained from Springer. Permissions for use may be obtained through RightsLink at the Copyright Clearance Center. Violations are liable to prosecution under the respective Copyright Law.
The use of general descriptive names, registered names, trademarks, service marks, etc. in this publication does not imply, even in the absence of a specific statement, that such names are exempt from the relevant protective laws and regulations and therefore free for general use.
While the advice and information in this book are believed to be true and accurate at the date of publication, neither the authors nor the editors nor the publisher can accept any legal responsibility for any errors or omissions that may be made. The publisher makes no warranty, express or implied, with respect to the material contained herein.

Printed on acid-free paper

Springer is part of Springer Science+Business Media (www.springer.com)

for Thomas Binoth

Preface

These notes are based on lectures at Heidelberg University between Summer 2009 and Winter 2013/2014, written up in coffee shops around the world. Obviously, in the Fall of 2012 they were heavily adapted to the new and exciting experimental realities. It felt great to rewrite the Higgs chapter from a careful description of possible experimental signals into a description of an actual experimental research program. I promise I will do it again once the LHC discovers physics beyond the Standard Model.

To those familiar with the German system, it will be obvious that the target audience of the lecture are students who know field theory and are starting to work on their master thesis; carefully studying these notes should put you into a position to start actual research in LHC physics. The way I prefer to learn physics is by calculating things on a piece of paper or on the blackboard. This is why the notes look the way they look. Because this is not a text book there is less text in the notes than actual talk during the lecture. So when reading these notes, take a break here and there, get a coffee and think about the physics behind the calculation you just followed.

The text is divided into three main parts:

- In the first part, I focus on Higgs physics and collider searches. To understand what we are looking for I start with the most minimalistic and not renormalizable models describing massive gauge bosons. I then slowly advance to the usual fundamental Higgs scalar we are really searching for. At the end of this part what everybody should understand is the usual set of ATLAS or CMS graphs shown in Figure 1.10, where many colored lines represent different search channels and their discovery potential. Many QCD issues affecting Higgs searches I will skip in the Higgs part and postpone to the ...
- ... QCD part. Here, I am taking at least one step back and study the theory which describes Higgs production and everything else at the LHC. Two core discussions shape this part: first, I derive the DGLAP equation by constructing the splitting kernels. This leads us to the parton shower and to the physical interpretation of resumming different logarithms in the QCD perturbation series.

Second, there are two modern approaches combining parton shower and matrix element descriptions of jet radiation, which I introduce at least on the level of simplified models. Throughout the QCD discussion I avoid the more historically interesting deep inelastic scattering process and instead rely on the Drell-Yan process or its inverted R ratio process for motivation and illustration. Because the first two parts of the lecture notes are really advanced quantum field theory, something is missing: there are ...

- ... many aspects of LHC physics we need to take into account once we look at experimental LHC results. Some of them, like old fashioned jets and fat jets, helicity amplitudes, or missing transverse energy I cover in the third part. This part will expand in the online version over the coming years while I will keep these lecture notes up to date with actual discussions of LHC data.

What is almost entirely missing is an introduction to searches for new physics completing the Standard Model of particle physics beyond the weak scale. Covering this topic appropriately would at least double the length of these notes. For the structure of such models and their signatures I instead refer to our review article [14] and in particular to its second chapter.

Last, but not least, the literature listed at the end of each part is not meant to cite original or relevant research papers. Instead, I collected a set of review papers or advanced lecture notes supplementing these lecture notes in different directions. Going through some of these mostly introductory papers will be instructive and fun once the basics have been covered by these lecture notes.

Heidelberg, Germany Tilman Plehn
February 2014

Acknowledgements

The list of people I would like to thank is long and still growing: starting with Peter Zerwas, Wim Beenakker, Roland Höpker, Michael Spira and Michael Krämer I would like to thank all the people who taught me theoretical physics and phenomenology over many years. This later on included Tao Han, Dieter Zeppenfeld, Uli Baur, and Thomas Binoth. The latter two hopefully got to watch the Higgs discovery from their clouds up there. Since moving to Heidelberg, it has been great fun to benefit from the physics knowledge in our happy institute on Philosophenweg. To all of these people I would like to say: I am very sorry, but what you are holding in your hands is the best I could do.

Of all the great experimentalists who have taught me LHC physics I would like to single out Dirk Zerwas and thank him for his continuous insight into experimental physics, from our first semester at Heidelberg to the last day of writing these notes. Another experimentalist, Kyle Cranmer, taught me enough statistics to avoid major disasters, which I am very grateful for. On the other Heidelberg hill, I would like to thank Manfred Lindner for his many comments on my black board notes when his class started right after mine. For anything to do with Higgs couplings and statistics I would like to thank Michael Rauch for many years of fun collaboration and for helping me with the text. As a long-term collaborator I am missing Dave Rainwater who should have stayed in LHC physics and who would now be the leading Higgs expert in the USA.

The QCD part of this lecture is based on my 2008 TASI lectures, and I would like to thank Tom DeGrand and Ben Allanach for their comments on the TASI notes and Sally Dawson for her encouragement to put these notes on the web. For this longer version I am indebted to Steffen Schumann for helping me out on many QCD questions over the past years, to Jan Pawlowski for uncountable cups of coffee on field theory and QCD, to Fabio Maltoni and Johan Alwall for teaching me jet merging, to Michelangelo Mangano for many detailed comments, and to Peter Schichtel for helping me sort out many QCD topics. Alert readers like David Lopez-Val, Sebastian Bock, Florian Görtz, Michael Spannowsky, Martin Weber, Anja Butter, Malte Buschmann, and Manuel Scinta followed my appeal to give me lists of typos, and Manuela Wirschke carefully read the notes removing language mistakes—thank you very much to everyone who helped make this writeup more correct and more readable.

Most importantly, I would like to thank all the people who have convinced me that theoretical physics even including QCD is fun—at least most of the time. Not surprisingly this includes many US colleagues from our TASI year 1997.

Contents

1 Higgs Physics 1
    1.1   Electroweak Symmetry Breaking 2
           1.1.1   What Masses? 2
           1.1.2   Massive Photon 5
           1.1.3   Standard Model Doublets 7
           1.1.4   Sigma Model 12
           1.1.5   Higgs Boson 18
           1.1.6   Custodial Symmetry 21
    1.2   The Standard Model 29
           1.2.1   Higgs Potential to Dimension Six 30
           1.2.2   Mexican Hat 36
           1.2.3   Unitarity 38
           1.2.4   Renormalization Group Analysis 43
           1.2.5   Top–Higgs Renormalization Group 48
           1.2.6   Two Higgs Doublets and Supersymmetry 50
           1.2.7   Coleman–Weinberg Potential 59
    1.3   Higgs Decays and Signatures 68
    1.4   Higgs Discovery 69
    1.5   Higgs Production in Gluon Fusion 77
           1.5.1   Effective Gluon–Higgs Coupling 78
           1.5.2   Low–Energy Theorem 84
           1.5.3   Effective Photon-Higgs Coupling 86
           1.5.4   Signatures 88
    1.6   Higgs Production in Weak Boson Fusion 93
           1.6.1   Production Kinematics 94
           1.6.2   Jet Ratios and Central Jet Veto 97
           1.6.3   Decay Kinematics and Signatures 103
    1.7   Associated Higgs Production 105

	1.8	Beyond Higgs Discovery	106
		1.8.1 Coupling Measurement	107
		1.8.2 Higgs Quantum Numbers	116
		1.8.3 Higgs Self Coupling	120
	1.9	Alternatives and Extensions	123
		1.9.1 Technicolor	123
		1.9.2 Hierarchy Problem and the Little Higgs	132
		1.9.3 Little Higgs Models	140
	1.10	Higgs Inflation	148
	References		155
2	**QCD**		157
	2.1	Drell–Yan Process	157
		2.1.1 Gauge Boson Production	158
		2.1.2 Massive Intermediate States	163
		2.1.3 Parton Densities	169
		2.1.4 Hadron Collider Kinematics	171
		2.1.5 Phase Space Integration	175
	2.2	Ultraviolet Divergences	179
		2.2.1 Counter Terms	180
		2.2.2 Running Strong Coupling	185
		2.2.3 Resumming Scaling Logarithms	191
	2.3	Infrared Divergences	195
		2.3.1 Single Jet Radiation	195
		2.3.2 Parton Splitting	197
		2.3.3 DGLAP Equation	210
		2.3.4 Parton Densities	218
		2.3.5 Resumming Collinear Logarithms	222
	2.4	Scales in LHC Processes	227
	2.5	Parton Shower	231
		2.5.1 Sudakov Form Factor	231
		2.5.2 Multiple Gluon Radiation	237
		2.5.3 Catani–Seymour Dipoles	243
		2.5.4 Ordered Emission	248
	2.6	Multi–Jet Events	254
		2.6.1 Jet Radiation Patterns	254
		2.6.2 CKKW and MLM Schemes	265
	2.7	Next–to–Leading Orders and Parton Shower	274
		2.7.1 Next–to–Leading Order in QCD	275
		2.7.2 MC@NLO Method	280
		2.7.3 POWHEG Method	283
	References		288
3	**LHC Phenomenology**		291
	3.1	Jets and Fat Jets	292
		3.1.1 Jet Algorithms	293
		3.1.2 Fat Jets	295

3.2	Helicity Amplitudes		298
3.3	Missing Transverse Energy		303
	3.3.1	Measuring Missing Energy	304
	3.3.2	Missing Energy in the Standard Model	306
	3.3.3	Missing Energy and New Physics	308
3.4	Uncertainties		317
References			324

Index 325

Chapter 1
Higgs Physics

Understanding the nature of electroweak symmetry breaking—or slightly more specifically deciphering the Higgs mechanism—is the main goal of the ATLAS and CMS experiments at the LHC. Observing some kind of Higgs boson and studying its properties involves many experimental and theoretical issues focused around understanding hadron collider data and QCD predictions to unprecedented precision. The latter will be the main topic of the second half of this lecture.

On the other hand, before we discuss the details of Higgs signatures, backgrounds, and related QCD aspects we should start with a discussion of electroweak symmetry breaking. Higgs physics at the LHC means much more than just finding a light fundamental Higgs boson as predicted by the Standard Model of particle physics. As a matter of fact, the discovery of a light Higgs boson was announced on July 4th, 2012, and we will briefly discuss it in Sect. 1.4.

In our theory derivation in Sect. 1.1 we prefer to follow an effective theory approach. This means we do not start by writing down the Higgs potential and deriving the measured gauge boson and fermion masses. Instead, we step by step include gauge boson and fermion masses in our gauge theories, see what this means for the field content, and show how we can embed this mechanism in a renormalizable fundamental gauge theory. Only this last step will lead us to the Standard Model and the Higgs potential. In Sect. 1.2 we will return to the usual path and discuss the properties of the renormalizable Standard Model including high energy scales. This includes new physics effects in terms of higher-dimensional operators in Sect. 1.2.1, an extended supersymmetric Higgs sector in Sect. 1.2.6, and general effects of new particles in the Higgs potential in Sect. 1.2.7.

In Sect. 1.3 we will start discussing Higgs physics at colliders, leading us to the Higgs discovery papers presented in Sect. 1.4. Higgs production in gluon fusion, weak boson fusion, and in association with a gauge boson will be in the focus of Sects. 1.5–1.7, with a special focus on QCD issues linked to jet radiation in Sect. 1.6.2. The LHC experiments have shown that they cannot only discover a Higgs resonance, but also study many Higgs properties, some of which we discuss in Sect. 1.8.

In our approach to the Higgs mechanism it is clear that a fundamental Higgs particle is not the only way to break the electroweak symmetry. We therefore discuss alternative embeddings in strongly interacting physics in Sect. 1.9. This part will also include a very brief introduction to the hierarchy problem. Finally, in Sect. 1.10 we will touch on a slightly more speculative link between Higgs physics and inflation.

1.1 Electroweak Symmetry Breaking

As a starting point, let us briefly remind ourselves what the Higgs boson really is all about and sketch the Standard Model Lagrangian with mass terms for gauge bosons and fermions. As a matter of fact, in a first step in Sect. 1.1.2 we will try to make a photon massive without introducing a physical Higgs field. Even for the $SU(2)$ gauge theory of the electroweak Standard Model we might get away without a fundamental Higgs boson, as we will show in Sect. 1.1.3. Then, we will worry about quantum fluctuations of the relevant degrees of freedom, which leads us to the usual picture of the Higgs potential, the Higgs boson, and the symmetries related to its implementation. This approach is not only interesting because it allows us to understand the history of the Higgs boson and identify the key steps in this triumph of quantum field theory, it also corresponds to a modern effective field theory picture of this sector. Such an effective field theory approach will make it easy to ask the relevant questions about the experimental results, guiding us towards the experimental confirmation of the Higgs mechanism as part of the Standard Model of elementary particles.

1.1.1 What Masses?

The relevance of the experimental Higgs discovery cannot be over–stated. The fact that local gauge theories describe the fundamental interactions of particles has been rewarded with a whole list of experimental and theoretical Nobel prizes. What appears a straightforward construction to us now has seen many challenges by alternative approaches, some justified and some not all that justified. The greatest feature of such a gauge theory lead to the Nobel prize given to Gerald 't Hooft and Martinus Veltman: the absence of a cutoff scale. Mathematically we call this validity to arbitrarily high energy scales renormalizability, physically it means that the Standard Model describing the interactions of quarks and leptons is truly fundamental. There is the usual argument about the Planck scale as an unavoidable cutoff and the apparent non–renormalizability of gravity, but the final word on that is still missing.

1.1 Electroweak Symmetry Breaking

What is important to notice is that the Nobel prize for 't Hooft and Veltman would have had to be exchanged for a Fields medal without the experimental discovery of the Higgs boson. Massive local gauge theories are not fundamental without the Higgs mechanism, i.e. without spontaneous symmetry breaking in a relativistic framework with doublet fields and hence predicting the existence of a narrow and relatively light Higgs boson. The Higgs discovery is literally the keystone to a fundamental theory of elementary particles—without it the whole construction would collapse.

When people say that the Higgs mechanism is responsible for the masses of (all) particles they actually mean very specific masses. These are not the proton or neutron masses which are responsible for the masses of people, furniture, or the moon. In a model describing the fundamental interactions, the mass of the weak gauge bosons as the exchange particles of the weak nuclear force has structural impact. Let us briefly remind ourselves of the long history of understanding these masses.

The first person to raise the question about the massive structure of the weak gauge boson, without knowing about it, was Enrico Fermi. In 1934 he wrote a paper on *Versuch einer Theorie der β-Strahlen*, proposing an interaction responsible for the neutron decay into a proton, an electron, and an antineutrino. He proposed an interaction Lagrangian which we nowadays write as

$$\mathscr{L} \supset G_F \left(\overline{\psi}_1 * \psi_2\right) \left(\overline{\psi}_3 * \psi_4\right) . \tag{1.1}$$

The four external fermions, in our case quarks and leptons, are described by spinors ψ. The star denotes the appropriate scalar, vector, or tensor structure of the interaction current, which we will leave open at this stage. On the one hand we know that spinors have mass dimension 3/2, on the other hand the Lagrangian density has to integrate to the action and therefore has to have mass dimension four. This means that in the low energy limit the Fermi coupling constant has to have mass dimension $G_F \sim 1/\Lambda^2$ with an appropriate mass scale Λ. We now know that in the proper theory, where the Fermi interaction should include an exchange particle, this scale Λ should be linked to the mass of this exchange particle, the W boson.

The key to understanding such a dimensionful coupling in terms of massive exchange particles was published by Hideki Yukawa in 1935 under the title *On the interaction of elementary particles*. He links the mass of exchange particles to the potential they generate after Fourier transform,

$$V(r) = -\frac{e}{r} \qquad \text{massless particle exchange}$$
$$V(r) = -g^2 \frac{e^{-mr}}{r} \qquad \text{massive particle exchange with } m. \tag{1.2}$$

Yukawa did not actually talk about the weak nuclear force at the quark level. His model was based on fundamental protons and neutrons, and his exchange particles were pions. But his argument applies perfectly to Fermi's theory at the quark level. Using Eq. (1.2) we can link the mass of the exchange particle, in units of $c = 1$

and $\hbar = 1$, to the reach of the interaction. For radii above $1/m$ the massive Yukawa potential is suppressed exponentially. For the weak nuclear force this is the structure we need, because the force which links quarks for example to protons and neutrons is incredibly hard to observable at larger distances.

What is still missing in our argument is the link between a coupling strength with mass dimension and massive exchange particles. Since the 1920s many physicists had been using a theory with a quantized electromagnetic field to compute processes involving charged particles, like electrons, and photons. The proper, renormalizable quantum field theory of charged particles and the photon as the exchange particle was proposed by Sin-Itiro Tomonaga in 1942. Julian Schwinger independently developed the same theory, for example in his papers *Quantum electrodynamics I A covariant formulation* and *On quantum electrodynamics and the magnetic moment of the electron*, both published in 1948. The development of quantum electrodynamics as the theory of massless photon exchange was from the beginning driven by experimental observations. For example the calculation of the Lamb shift was a key argument to convince physicists that such a theory was not only beautiful, but also useful or 'correct' by experimental standards. The extension of QED to a non–abelian $SU(2)$ gauge group was proposed by Sheldon Glashow, Julian Schwinger's student, in 1961, but without any hint of the Higgs mechanism.

Combining these three aspects gives us a clear picture of what people in the 1950s knew as the basis to solve the puzzle of the weak nuclear force: interactions between fundamental particles are very successfully described by massless photon exchange; interactions with a finite geometric range correspond to a massive exchange particle; and in the low energy limit such a massive particle exchange can reproduce Fermi's theory.

The main question we will discuss in these lecture notes is how to construct a QED–like quantum theory of massive exchange particles, the $W^\pm$ and Z^0 bosons. Usually, the first reason we give to why the photon is massless is the local $U(1)$ gauge invariance which essentially defines the QED Lagrangian. However, we will see in Sect. 1.1.2 that making the photon massive requires much more fundamental changes to the theory. This problem, linked to work by Yoichiro Nambu and specifically to Goldstone's theorem, was what Peter Higgs and his contemporaries solved for Lorentz-invariant gauge theories in 1964. The idea of spontaneous symmetry breaking was well established in solid state physics, going back to the work by Landau and Ginzburg, by Bardeen, Cooper, Schrieffer, and by Anderson on superconductivity. However, these systems did not have a Higgs state. Historically, Walter Gilbert triggered Peter Higgs' first paper in 1964 by making the wrong statement that spontaneous symmetry breaking would fail for Lorentz-invariant theories. This fundamental mistake did not keep him from receiving a Nobel Prize in 1980, but for chemistry. Statements of this kind were very popular at the time, based on more and more rigorous proofs of Goldstone's theorem. Needless to say, the Higgs discovery is a good indication that all of them are wrong for local gauge theories.

The actual Higgs particle only features in Peter Higgs' second paper in 1964. The very clear prediction of the new particle in this paper is supposedly due to Yoichiro

Nambu, the journal referee for the paper. The same mechanism of spontaneous symmetry breaking in high energy physics was, independently of and even slightly before Peter Higgs' papers, proposed by Francois Englert and Robert Brout. It is probably fair to assume that Robert Brout, had he not passed away in 2011, would have been the third Nobel Laureate in Physics, 2013. Still in 1964 the group of Gerald Guralnik, Carl Hagen, and Thomas Kibble published a more detailed and rigorous field theoretical study of the Higgs mechanism. In 1966 Peter Higgs wrote a third paper, in which he worked out many details of the Higgs mechanism and the Higgs boson, including scattering rates and decay widths. Still without linking the Higgs mechanism to the weak force, this can be considered the first phenomenological study of the Higgs boson.

Combining the Higgs mechanism with QED and applying this combination to the weak interaction is the birth of the Standard Model of elementary particles. Steven Weinberg proposed *A model of leptons* in 1967, for the first time including fermion masses generated by the Higgs mechanism. Together with Abdus Salam's paper on *Weak and Electromagnetic Interactions* from 1968 the Standard Model, as we know it today, was now complete. However, most physics aspects which we link to the Higgs boson nowadays, did not feature at the time. One reason is that the actual proof of renormalizability by Gerald 't Hooft and Martinus Veltman still had to be given in 1972, so asking questions about the high energy behavior of the Standard Model was lacking the formal basis. Because in these lecture notes we are to some degree following the historic or effective field theory approach the discussion of the renormalizable field theory and its ultraviolet behavior will have to wait until Sect. 1.2.3.

If we want to follow this original logic of the Higgs mechanism and the prediction of the new particle we now know that to do. First, we need to understand what really keeps the photon from acquiring even a tiny mass. This will allow us to construct a gauge theory Lagrangian for massive weak bosons. Finally, we will see how the same mechanism will allow us to include massive fermions in the electroweak theory.

1.1.2 Massive Photon

Even though this is not the physical problem we are interested in, we start by breaking electrodynamics and giving a mass to the (massless) photon of our usual locally $U(1)_Q$-symmetric Lagrangian. To its kinetic $F \cdot F$ term we would like to add a photon mass term $m^2 A^2/2$, which we know is forbidden by the gauge symmetry. We will see that adding such a photon mass term to the Lagrangian requires a bit of work, but it not a very hard problem. The key idea it to also add an innocent looking real (uncharged) scalar field without a mass and without a coupling to the photon, but with a *scalar–photon mixing* term and a non–trivial gauge transformation. The result is called the Boulware-Gilbert model

$$\mathcal{L} = -\frac{1}{4} F_{\mu\nu} F^{\mu\nu} + \frac{1}{2} e^2 f^2 A_\mu^2 + \frac{1}{2} (\partial_\mu \phi)^2 - e f A_\mu \partial^\mu \phi$$
$$= -\frac{1}{4} F_{\mu\nu} F^{\mu\nu} + \frac{1}{2} e^2 f^2 \left(A_\mu - \frac{1}{ef} \partial_\mu \phi \right)^2. \quad (1.3)$$

where f is a common mass scale for the photon mass and the mixing. It ensures that all terms in the Lagrangian have mass dimension four—remembering that bosonic fields like A_μ and ϕ have mass dimension one. The additional factor e will become the usual electric charge, but at this stage it is a dimensionless number without any specific relevance in this interaction-less Lagrangian. Because all terms in Eq. (1.3) have mass dimension four and there are not inverse powers of mass our theory should be renormalizable.

We can define a simultaneous gauge transformation of both fields in the Lagrangian

$$A_\mu \longrightarrow A_\mu + \frac{1}{ef} \partial_\mu \chi \qquad \phi \longrightarrow \phi + \chi, \quad (1.4)$$

under which the Lagrangian is indeed invariant: the kinetic term for the photon we leave untouched, so it will be gauge invariant just as it was before. The simultaneous gauge invariance is then defined to keep the second term in Eq. (1.3) invariant. If we now re-define the photon field as $B_\mu = A_\mu - \partial_\mu \phi / (ef)$ we need to compare the new and the old kinetic terms

$$F_{\mu\nu}\Big|_B = \partial_\mu B_\nu - \partial_\nu B_\mu = \partial_\mu \left(A_\nu - \frac{1}{ef} \partial_\nu \phi \right) - \partial_\nu \left(A_\mu - \frac{1}{ef} \partial_\mu \phi \right)$$
$$= \partial_\mu A_\nu - \partial_\nu A_\mu = F_{\mu\nu}\Big|_A, \quad (1.5)$$

and then rewrite the Lagrangian of Eq. (1.3) as

$$\boxed{\mathcal{L} = -\frac{1}{4} F_{\mu\nu} F^{\mu\nu} + \frac{1}{2} e^2 f^2 B_\mu^2 = -\frac{1}{4} F_{\mu\nu} F^{\mu\nu} + \frac{1}{2} m_B^2 B_\mu^2.} \quad (1.6)$$

This Lagrangian effectively describes a *massive photon field* B_μ, which has absorbed the real scalar ϕ as its additional longitudinal component. This is because a massless gauge boson A_μ has only two on–shell degrees of freedom, a left handed and a right handed polarization, while the massive B_μ has an additional longitudinal polarization degree of freedom. Without any fundamental Higgs boson appearing, the massive photon has 'eaten' the real scalar field ϕ. Of course, the new field B_μ is not simply a photon with a mass term, because this is still forbidden by gauge invariance. Our way out is to split the massive photon field into the transverse degrees of freedom A_μ and the longitudinal mode ϕ with their different gauge transformations given by Eq. (1.4).

What kind of properties does this field ϕ need to have, so that we can use it to provide a photon mass? From the combined gauge transformation in Eq. (1.4) we immediately see that any additional purely scalar term in the Lagrangian, like a scalar potential $V(\phi)$, needs to be symmetric under the linear shift $\phi \to \phi + \chi$, to not spoil gauge invariance. This means that we cannot write down polynomial terms ϕ^n, like a mass or a self coupling of ϕ. An interaction term ϕAA would not be possible, either. Only *derivative interactions* proportional to $\partial \phi$ which are attached to other (conserved) currents are allowed. In that case we can absorb the shift by χ into a total derivative in the Lagrangian.

This example illustrates a few vital properties of *Nambu–Goldstone bosons* (NGB). Such massless physical states appear in many areas of physics and are described by *Goldstone's theorem*. It applies to global continuous symmetries of the Lagrangian which are violated by a non–symmetric vacuum state, a mechanism called spontaneous symmetry breaking. Based on Lorentz invariance and states with a positively definite norm we can then prove:

If a *global symmetry* group is spontaneously broken into a group of lower rank, its broken generators correspond to physical Goldstone modes. These scalar fields transform non–linearly under the larger and linearly under the smaller group. This way they are massless and cannot form a potential, because the non–linear transformation only allows derivative terms in the Lagrangian.

One common modification of this situation is an explicit breaking of the smaller symmetry group. In that case the Nambu-Goldstone bosons become pseudo–Goldstones and acquire a mass of the size of this hard-breaking term.

Before Peter Higgs and his colleagues proposed their mechanism of electroweak symmetry breaking they were caught between two major no-go theorems. First, they needed an additional degree of freedom to make massless gauge bosons massive. Secondly, the spontaneous breaking of a gauge symmetry supposedly predicted massless scalar states which were clearly ruled out experimentally. These two problems solve each other once we properly treat the special case that the spontaneously broken symmetry is a *local gauge symmetry*. It turns out that the Goldstone theorem does not apply, because a local gauge theory cannot be Lorentz invariant and only have positively defined states simultaneously. Instead of becoming massless scalars the Goldstone modes are then 'eaten' by the additional degrees of freedom of the massive gauge bosons. This defines the incredibly elegant Higgs mechanism. The gauge boson mass is given by the vacuum expectation value breaking the larger symmetry. A massive additional scalar degree of freedom, the *Higgs boson*, appears if there are more Goldstone modes than degrees of freedom for the massive gauge bosons.

1.1.3 Standard Model Doublets

One of the complications of the Standard Model is its $SU(2)$ doublet structure. In the last section we have chosen not to introduce a charged $SU(2)$ doublet, which is why there are no degrees of freedom left after the photon gets its mass. This means that our toy model is not going to be well suited to provide the three degrees of freedom needed to make $SU(2)$ gauge bosons massive. What it illustrates is only how by introducing a neutral scalar particle without an interaction but with a mixing term we make gauge bosons heavy, in spite of gauge invariance.

Fermion fields have mass dimension 3/2, so we know how mass and interaction terms in the renormalizable dimension-4 Lagrangian have to look. For example, the interaction of fermions with gauge bosons is most easily written in terms of *covariant derivatives*. The terms

$$\mathscr{L}_{D4} = \overline{Q}_L i \slashed{D} Q_L + \overline{Q}_R i \slashed{D} Q_R + \overline{L}_L i \slashed{D} L_L + \overline{L}_R i \slashed{D} L_R - \frac{1}{4} F_{\mu\nu} F^{\mu\nu} \dots \quad (1.7)$$

describe electromagnetic interactions introducing a covariant derivative $D_\mu = \partial_\mu + ieqA_\mu$ with the photon field also appearing in the field strength tensor $F_{\mu\nu} = \partial_\mu A_\nu - \partial_\nu A_\mu$. The same form works for the weak $SU(2)$ interactions, except that the weak interaction knows about the chirality of the fermion fields, so we have to distinguish $\slashed{D}_L$ and $\slashed{D}_R$. The covariant derivatives we write in terms of the $SU(2)$ basis matrices or *Pauli matrices* $\tau_{1,2,3}$ or $\tau_{+,-,3}$, with $\tau_\pm = (\tau_1 \pm i \tau_2)/2$.

$$D_{L\mu} = \partial_\mu + ig' \left(q - \frac{\tau_3}{2}\right) B_\mu + ig \sum_{a=1,2,3} W_\mu^a \frac{\tau_a}{2}$$

$$= \partial_\mu + ieqA_\mu + ig_Z \left(-qs_w^2 + \frac{\tau_3}{2}\right) Z_\mu + i\frac{g}{2} \left(\tau_1 W_\mu^1 + \tau_2 W_\mu^2\right)$$

$$\equiv \partial_\mu + ieqA_\mu + ig_Z \left(-qs_w^2 + \frac{\tau_3}{2}\right) Z_\mu + i\frac{g}{\sqrt{2}} \left(\tau_+ W_\mu^+ + \tau_- W_\mu^-\right)$$

$$D_{R\mu} = D_{L\mu}\Big|_{\tau=0}$$

$$\tau_+ = \begin{pmatrix} 0 & 1 \\ 0 & 0 \end{pmatrix} \quad \tau_- = \begin{pmatrix} 0 & 0 \\ 1 & 0 \end{pmatrix}$$

$$\tau_1 = \begin{pmatrix} 0 & 1 \\ 1 & 0 \end{pmatrix} \quad \tau_2 = \begin{pmatrix} 0 & -i \\ i & 0 \end{pmatrix} \quad \tau_3 = \begin{pmatrix} 1 & 0 \\ 0 & -1 \end{pmatrix}, \quad (1.8)$$

The explicit sum in the first line we will omit in the rest of this lecture. All indices appearing twice include an implicit sum. The fields B_μ and W_μ^a are the new gauge bosons. In the second line we re-write the covariant derivative in the photon A_μ and the Z boson mass eigenstates. What is not obvious from this argument is that we can actually write the ratio g'/g in terms of a rotation angle, which implicitly assumes that we can rotate the B and W^3 fields into the physical mass-eigenstate photon and Z fields

$$\begin{pmatrix} A_\mu \\ Z_\mu \end{pmatrix} = \begin{pmatrix} c_w & s_w \\ -s_w & c_w \end{pmatrix} \begin{pmatrix} B_\mu \\ W_\mu^3 \end{pmatrix}. \quad (1.9)$$

1.1 Electroweak Symmetry Breaking

The details of this rotation do not matter for the Higgs sector. The normalization of the charged gauge fields we will fix later. At this level the two weak couplings g and g_Z do not necessarily coincide, but we will get back to this issue in Sect. 1.1.6.

Before we generalize the Boulware-Gilbert model to the weak gauge symmetry of the Standard Model it is instructive to review the form of the mass term for massive gauge bosons following from Eq. (1.8). In particular, there will appear a relative factor 2 between the two bases of the Pauli matrices, i.e. in terms of $W^{1,2}$ and $W^\pm$, which often causes confusion. For later use we also need a sum rule for the $SU(2)$ generators or *Pauli matrices* $\tau_{1,2,3}$ as written out in Eq. (1.8). They satisfy the relation $\tau_a \tau_b = \delta_{ab} + i\epsilon_{abc}\tau_c$ or the commutator relation $[\tau_a, \tau_b] = 2i\epsilon_{abc}\tau_c$. Summing over indices we see that

$$\sum_{a,b} \tau^a \tau^b = \sum_{a,b} \left(\delta^{ab} + i\epsilon^{abc}\tau_c\right) = \sum \delta^{ab} + i \sum_{a \neq b} \epsilon^{abc} \tau_c$$

$$= \sum \delta^{ab} + i \sum_{a<b} \left(\epsilon^{abc} + \epsilon^{bac}\right) \tau_c = \sum \delta^{ab} . \quad (1.10)$$

The basis of three Pauli matrices we can write in terms of $\tau_{1,2,3}$ as well as in terms of $\tau_{+,-,3}$. The latter correspond to two charged and one neutral vector bosons. While the usual basis is written in terms of complex numbers, the second set of generators reflects the fact that for $SU(2)$ as for any $SU(N)$ group we can find a set of real generators of the adjoint representation. When we switch between the two bases we only have to make sure we get the standard normalization of all fields as shown in Eq. (1.8),

$$\sqrt{2}\left(\tau_+ W_\mu^+ + \tau_- W_\mu^-\right) = \sqrt{2} \begin{pmatrix} 0 & W_\mu^+ \\ 0 & 0 \end{pmatrix} + \sqrt{2} \begin{pmatrix} 0 & 0 \\ W_\mu^- & 0 \end{pmatrix}$$

$$\stackrel{!}{=} \tau_1 W_\mu^1 + \tau_2 W_\mu^2 = \begin{pmatrix} 0 & W_\mu^1 \\ W_\mu^1 & 0 \end{pmatrix} + \begin{pmatrix} 0 & -iW_\mu^2 \\ iW_\mu^2 & 0 \end{pmatrix}$$

$$\Leftrightarrow \quad W_\mu^+ = \frac{1}{\sqrt{2}}\left(W_\mu^1 - iW_\mu^2\right)$$

$$W_\mu^- = \frac{1}{\sqrt{2}}\left(W_\mu^1 + iW_\mu^2\right) . \quad (1.11)$$

To track these factors of 2 in the definitions of the weak gauge field we have a close look at the dimension-2 mass term for charged and neutral gauge bosons

$$\mathcal{L}_{D2} = -\frac{m_W^2}{2}\left(W^{1,\mu}W_\mu^1 + W^{2,\mu}W_\mu^2\right) - \frac{m_Z^2}{2} Z^\mu Z_\mu$$

$$= -m_W^2 W^{+,\mu} W_\mu^- - \frac{m_Z^2}{2} Z^\mu Z_\mu . \quad (1.12)$$

The relative factor 2 in front of the W mass appears because the Z field is neutral and the W field is charged. This difference also appear for neutral and charged scalars discussed in field theory. In our conventions it corresponds to the factors $1/\sqrt{2}$ in the $SU(2)$ generators $\tau_\pm$.

Of course, in the complete Standard Model Lagrangian there are many additional terms involving the massive gauge bosons, e.g. kinetic terms of all kinds, but they do not affect our discussion of $U(1)_Y$ and $SU(2)_L$ gauge invariance.

Guessing the form of the fermion masses the one thing we have to ensure is that we combine the left handed and right handed doublets (Q_L, L_L) and singlets (Q_R, L_R) properly:

$$\mathscr{L}_{D3} = -\overline{Q}_L m_Q Q_R - \overline{L}_L m_L L_R + \ldots \qquad (1.13)$$

This form strictly speaking requires a doublet structure of the Higgs–Goldstone fields, which we will briefly comment on later. For now we ignore this notational complication. Following our labeling scheme by mass dimension fermion masses will be included as $\mathscr{L}_{D3}$. *Dirac mass* terms simply link $SU(2)$ doublet fields for leptons and quarks with right handed singlets and give mass to all fermions in the Standard Model. This helicity structure of mass terms we can easily derive by introducing left handed and right handed projectors

$$\psi_L = \frac{\mathbb{1} - \gamma_5}{2} \psi \equiv \mathbb{P}_L \psi \qquad \psi_R = \frac{\mathbb{1} + \gamma_5}{2} \psi \equiv \mathbb{P}_R \psi \,, \qquad (1.14)$$

where ψ is a generic Dirac spinor and $\mathbb{P}_{L,R}$ are projectors in this 4×4 Dirac space. At this stage we do not need the explicit form of the gamma matrices which we will introduce in Eq. (2.109). The mass term for a Dirac fermion reads

$$\begin{aligned}
\overline{\psi} \mathbb{1} \psi &= \overline{\psi} \left(\mathbb{P}_L + \mathbb{P}_R\right) \psi \\
&= \overline{\psi} \left(\mathbb{P}_L^2 + \mathbb{P}_R^2\right) \psi \\
&= \psi^\dagger \gamma_0 \left(\mathbb{P}_L^2 + \mathbb{P}_R^2\right) \psi & \text{with} \quad & \overline{\psi} = \psi^\dagger \gamma^0 \\
&= \psi^\dagger \left(\mathbb{P}_R \gamma^0 \mathbb{P}_L + \mathbb{P}_L \gamma^0 \mathbb{P}_R\right) \psi & \text{with} \quad & \{\gamma_5, \gamma_\mu\} = 0 \\
&= (\mathbb{P}_R \psi)^\dagger \gamma^0 (\mathbb{P}_L \psi) + (\mathbb{P}_L \psi)^\dagger \gamma^0 (\mathbb{P}_R \psi) & \text{with} \quad & \gamma_5^\dagger = \gamma_5, \mathbb{P}_{L,R}^\dagger = \mathbb{P}_{L,R} \\
&= (\overline{\mathbb{P}_R \psi}) \mathbb{1} (\mathbb{P}_L \psi) + (\overline{\mathbb{P}_L \psi}) \mathbb{1} (\mathbb{P}_R \psi) \\
&= \overline{\psi}_R \mathbb{1} \psi_L + \overline{\psi}_L \mathbb{1} \psi_R \,.
\end{aligned} \qquad (1.15)$$

The kinetic term stays diagonal

$$\begin{aligned}
\overline{\psi} \partial \psi &= \overline{\psi} \partial \left(\mathbb{P}_L^2 + \mathbb{P}_R^2\right) \psi \\
&= \overline{\psi} \left(\mathbb{P}_R \partial \mathbb{P}_L + \mathbb{P}_L \partial \mathbb{P}_R\right) \psi
\end{aligned}$$

1.1 Electroweak Symmetry Breaking

$$= (\overline{\mathbb{P}_L \psi}) \partial (\mathbb{P}_L \psi) + (\overline{\mathbb{P}_R \psi}) \partial (\mathbb{P}_R \psi)$$
$$= \overline{\psi}_L \partial \psi_L + \overline{\psi}_R \partial \psi_R . \tag{1.16}$$

In general, these mass terms can be matrices in generation space, which implies that we might have to rotate the fermion fields from an interaction basis into the mass basis, where these mass matrices are diagonal. Flavor physics dealing with such 3×3 mass matrices is its own field of physics with its own reviews and lecture notes, so we will omit this complication here. For our discussion of electroweak symmetry breaking it is sufficient to study one fermion generation at a time.

The well known problem with the mass terms in Eq. (1.13) is that they are not gauge invariant. To understand this issue of *fermion masses* we check the local weak $SU(2)_L$ transformation

$$U(x) = \exp\left(i\alpha^a(x)\frac{\tau_a}{2}\right) \equiv e^{i(\alpha \cdot \tau)/2} , \tag{1.17}$$

which only transforms the left handed fermion fields and leaves the right handed fields untouched

$$L_L \xrightarrow{U} UL_L \qquad Q_L \xrightarrow{U} UQ_L$$
$$L_R \xrightarrow{U} L_R \qquad Q_R \xrightarrow{U} Q_R . \tag{1.18}$$

It is obvious that there is no way we can make left–right mixing fermion mass terms as shown in Eq. (1.13) invariant under this left handed $SU(2)_L$ gauge transformation, where one of the fermion field picks up a factor U and the other is unchanged,

$$\overline{Q}_L m_Q Q_R \xrightarrow{U} \overline{Q}_L U^{-1} m_Q Q_R \neq \overline{Q}_L m_Q Q_R . \tag{1.19}$$

In analogy to the massive photon case, to write a gauge–invariant Lagrangian for massive fermions we have to add something else to our minimal Standard Model Lagrangian. Note that this addition does not have to be a fundamental scalar Higgs field, dependent on how picky we are with the properties of our new Lagrangian beyond the issue of gauge invariance.

To see what we need to add let us also look at the local $U(1)$ transformations involved. We start with a slightly complicated-looking way of writing the abelian *hypercharge* $U(1)_Y$ and *electric charge* $U(1)_Q$ transformations, making it more obvious how they mix with the neutral component of $SU(2)_L$ to give the electric charge.

Let us start with the neutral component of the $SU(2)_L$ transformation $V = \exp(i\beta\tau_3/2)$. Acting on a field with an $SU(2)_L$ charge this it not a usual $U(1)$ transformation. What we can do is combine it with another, appropriately chosen transformation. This hypercharge transformation is proportional to the unit matrix and hence commutes with all other matrices

$$\exp(i\beta q)\, V^\dagger = \exp(i\beta q)\, \exp\left(-\frac{i}{2}\beta\tau_3\right) \qquad \text{with} \quad V = U(x)\Big|_{\tau_3} = \exp\left(\frac{i}{2}\beta\tau_3\right)$$

$$= \exp\left(i\beta\frac{y\mathbb{1}+\tau_3}{2}\right)\exp\left(-\frac{i}{2}\beta\tau_3\right) \quad \text{with} \quad \boxed{q \equiv \frac{y\mathbb{1}+\tau_3}{2}} \; y_Q = \frac{1}{3} \quad y_L = -1$$

$$= \exp\left(i\frac{\beta}{2}y\mathbb{1}\right). \tag{1.20}$$

The relation between the charge q, the hypercharge y, and the isospin τ_3 is called the Gell-Mann–Nishijima formula. The indices Q and L denote quark and lepton doublets. Acting on a left handed field the factor τ_3 above is replaced by its eigenvalue ± 1 for up–type and down–type fermions. The $U(1)_Y$ charges or quantum numbers y are the quark and lepton hypercharges of the Standard Model. As required by the above argument, properly combined with the isospin they give the correct electric charges $q_{Q,L}$. Since τ_3 and the unit matrix commute with each other the combined exponentials have no additional factor a la Baker–Campbell–Hausdorff $e^A e^B = e^{A+B} e^{[A,B]/2}$. In analogy to Eq. (1.18) left handed and right handed quark and lepton fields transform under this $U(1)_Y$ symmetry as

$$L_L \to \exp(i\beta q_L) V^\dagger L_L = \exp\left(i\frac{\beta}{2}y_L\mathbb{1}\right) L_L \quad Q_L \to \exp(i\beta q_Q) V^\dagger Q_L = \exp\left(i\frac{\beta}{2}y_Q\mathbb{1}\right) Q_L$$

$$L_R \to \exp(i\beta q_L) L_R \qquad\qquad\qquad Q_R \to \exp(i\beta q_Q) Q_R . \tag{1.21}$$

Under a combined $SU(2)_L$ and $U(1)_Y$ transformation the left handed fermions see the hypercharge, while the right handed fermions only see the electric charge. Just as for the $SU(2)_L$ transformation U we do not have to compute anything to see that such different transformations of the left handed and right handed fermion fields do not allow for a Dirac mass term.

1.1.4 Sigma Model

One way of solving this problem is to introduce an additional field $\Sigma(x)$. This field will in some way play the role of the real scalar field we used for the photon mass generation. Its physical properties will become clear piece by piece from the way it appears in the Lagrangian and from the required gauge invariance. The equation of motion for the Σ field will also have to follow from the way we introduce it in the Lagrangian.

Following the last section, we first introduce Σ into the *fermion mass* term. This will tell us what it takes to make this mass term gauge invariant under the weak $SU(2)_L$ transformation defined in Eq. (1.18)

$$\overline{Q}_L \Sigma m_Q Q_R \xrightarrow{U} \overline{Q}_L U^{-1} \Sigma^{(U)} m_Q Q_R \stackrel{!}{=} \overline{Q}_L \Sigma m_Q Q_R \;\Leftrightarrow\; \Sigma \to \Sigma^{(U)} = U\Sigma .$$
$$\tag{1.22}$$

1.1 Electroweak Symmetry Breaking

If the result should be a dimension-4 Lagrangian the mass dimension of Σ has to be $m^0 = 1$. The same we can do for the $U(1)_Y$ transformation V as described in Eq. (1.21)

$$\overline{Q}_L \Sigma m_Q Q_R \xrightarrow{V} \overline{Q}_L \exp\left(-i\frac{\beta}{2} y \mathbb{1}\right) \Sigma^{(V)} m_Q \exp(i\beta q) Q_R$$

$$= \overline{Q}_L \Sigma^{(V)} \exp\left(-i\frac{\beta}{2} y \mathbb{1}\right) \exp(i\beta q) m_Q Q_R \qquad \exp\left(i\frac{\beta}{2} y \mathbb{1}\right) \text{ always commuting}$$

$$= \overline{Q}_L \Sigma^{(V)} V m_Q Q_R$$

$$\stackrel{!}{=} \overline{Q}_L \Sigma m_Q Q_R \qquad \Leftrightarrow \qquad \Sigma \to \Sigma^{(V)} = \Sigma V^\dagger \,. \tag{1.23}$$

Combining this with Eq. (1.22) gives us the *transformation property* we need

$$\boxed{\Sigma \to U \Sigma V^\dagger} \,. \tag{1.24}$$

For any Σ with this property the $\mathscr{L}_{D3}$ part of the Lagrangian has the required $U(1)_Y \times SU(2)_L$ symmetry, independent of what this Σ field really means. From the way it transforms we see that Σ is a 2×2 matrix with mass dimension zero. In other words, including a Σ field in the fermion mass terms gives a $U(1)_Y$ and $SU(2)_L$-invariant Lagrangian, without saying anything about possible representations of Σ in terms of physical fields

$$\boxed{\mathscr{L}_{D3} = -\overline{Q}_L \Sigma m_Q Q_R - \overline{L}_L \Sigma m_L L_R + \text{h.c.} + \ldots} \tag{1.25}$$

Fixing the appropriate transformations of the Σ field allows us to include fermion masses without any further complication.

In a second step, we deal with the *gauge boson masses*. We start with the left handed covariant derivative already used in Eq. (1.8)

$$D_{L\mu} = \partial_\mu + ig'\left(q - \frac{\tau_3}{2}\right) B_\mu + ig W^a_\mu \frac{\tau_a}{2} = \partial_\mu + ig' \frac{y}{2} B_\mu + ig W^a_\mu \frac{\tau_a}{2} \,. \tag{1.26}$$

Instead of deriving the gauge transformation of Σ let us start with a well-chosen ansatz and work backwards step by step, to check that we indeed arrive at the correct masses. First, we consistently require that the covariant derivative acting on the Σ field in the gauge-symmetric Lagrangian reads

$$D_\mu \Sigma = \partial_\mu \Sigma + ig' \Sigma B_\mu \frac{y}{2}\bigg|_{q=0} + ig W^a_\mu \frac{\tau_a}{2} \Sigma = \partial_\mu \Sigma - ig' \Sigma B_\mu \frac{\tau_3}{2} + ig W^a_\mu \frac{\tau_a}{2} \Sigma, \tag{1.27}$$

If we introduce the abbreviations $V_\mu \equiv \Sigma(D_\mu \Sigma)^\dagger$ and $T \equiv \Sigma \tau_3 \Sigma^\dagger$ we claim we can write the gauge boson mass terms as

$$\boxed{\mathscr{L}_{D2} = \frac{v^2}{4} \text{Tr}[V_\mu V^\mu] + \Delta\rho \frac{v^2}{8} \text{Tr}[TV_\mu] \, \text{Tr}[TV^\mu]} \,. \tag{1.28}$$

The trace acts on the 2×2 $SU(2)$ matrices. The parameter $\Delta\rho$ is conventional and will be the focus of Sect. 1.1.6. We will show below that this form is gauge invariant and gives the correct gauge boson masses.

Another structural question is what additional terms of mass dimension four we can write down using the dimensionless field Σ and which are gauge invariant. Our first attempt of a building block

$$\Sigma^\dagger \Sigma \xrightarrow{U,V} (U\Sigma V^\dagger)^\dagger (U\Sigma V^\dagger) = V\Sigma^\dagger U^\dagger U\Sigma V^\dagger = V\Sigma^\dagger \Sigma V^\dagger \neq \Sigma^\dagger \Sigma \tag{1.29}$$

is forbidden by $SU(2)_L$ gauge invariance according to Eq. (1.23). On the other hand, a circularly symmetric trace $\text{Tr}(\Sigma^\dagger \Sigma) \to \text{Tr}(V\Sigma^\dagger \Sigma V^\dagger) = \text{Tr}(\Sigma^\dagger \Sigma)$ changes this into a gauge invariant combination, which allows for the additional *potential terms*, meaning terms with no derivatives

$$\boxed{\mathscr{L}_\Sigma = -\frac{\mu^2 v^2}{4} \text{Tr}(\Sigma^\dagger \Sigma) - \frac{\lambda v^4}{16} \left(\text{Tr}(\Sigma^\dagger \Sigma)\right)^2 + \cdots} \,, \tag{1.30}$$

with properly chosen prefactors μ, v, λ. This fourth term finalizes our construction of the relevant weak Lagrangian

$$\mathscr{L} = \mathscr{L}_{D2} + \mathscr{L}_{D3} + \mathscr{L}_{D4} + \mathscr{L}_\Sigma \,, \tag{1.31}$$

organized by mass dimension.

As rule of thumb we will later notice that once we express the potential of Eq. (1.30) in terms of the usual Higgs doublet $|\phi|^2$, the prefactors will just be μ and λ. The parameter μ and the factor v appearing with every power of Σ have mass dimension one, while λ has mass dimension zero. Higher–dimensional terms in a dimension-4 Lagrangian are possible as long as we limit ourselves to powers of $\text{Tr}(\Sigma^\dagger \Sigma)$. However, they lead to higher powers in v which we will see makes them higher–dimensional operators in our complete quantum theory.

To check that Eq. (1.28) gives the correct masses in the Standard Model we start with $\text{Tr}(\Sigma^\dagger \Sigma)$ and assume it acquires a finite (expectation) value after we properly deal with Σ. The definitely simplest way to achieve this is to assume

$$\boxed{\Sigma(x) = \mathbb{1}} \,. \tag{1.32}$$

1.1 Electroweak Symmetry Breaking

This choice is called *unitary gauge*. It looks like a dirty trick to first introduce $\Sigma(x) = \mathbb{1}$ and then use this field for a gauge invariant implementation of gauge boson masses. Clearly, a constant does not exhibit the correct transformation property under the U and V symmetries, but we can always work in a specific gauge and only later check the physical predictions for gauge invariance. The way the sigma field breaks our gauge symmetry we can schematically see from

$$\Sigma \to U\Sigma V^\dagger = U\mathbb{1}V^\dagger = UV^\dagger \stackrel{!}{=} \mathbb{1} , \qquad (1.33)$$

which requires $U = V$ to be the remaining $U(1)$ gauge symmetry after including the Σ field. Certainly, $\Sigma = \mathbb{1}$ gives the correct fermion masses in $\mathscr{L}_{D3}$ and makes the potential $\mathscr{L}_\Sigma$ an irrelevant constant. What we need to check is $\mathscr{L}_{D2}$ which is supposed to reproduce the correct gauge boson masses. Using the covariant derivative from Eq. (1.27) acting on a constant field we can compute the auxiliary field V_μ in unitary gauge

$$\begin{aligned}
V_\mu &= \Sigma(D_\mu \Sigma)^\dagger = \mathbb{1}(D_\mu \Sigma)^\dagger \\
&= -igW^a_\mu \frac{\tau_a}{2} + ig'B_\mu \frac{\tau_3}{2} \\
&= -igW^+_\mu \frac{\tau_+}{\sqrt{2}} - igW^-_\mu \frac{\tau_-}{\sqrt{2}} - igW^3_\mu \frac{\tau_3}{2} + ig'B_\mu \frac{\tau_3}{2} \\
&= -i\frac{g}{\sqrt{2}}\left(W^+_\mu \tau_+ + W^-_\mu \tau_-\right) - ig_Z Z_\mu \frac{\tau_3}{2} ,
\end{aligned} \qquad (1.34)$$

with $Z_\mu = c_w W^3_\mu - s_w B_\mu$ and the two coupling constants $g_Z = g/c_w$ and $g' = gs_w/c_w$ as defined in Eq. (1.9). This gives us the first of the two terms in the gauge boson mass Lagrangian $\mathscr{L}_{D2}$

$$\begin{aligned}
\text{Tr}[V_\mu V^\mu] &= -2\frac{g^2}{2}W^+_\mu W^{-\mu} \text{Tr}(\tau_+ \tau_-) - \frac{g^2_Z}{4}Z_\mu Z^\mu \text{Tr}(\tau^2_3) \\
&= -g^2 W^+_\mu W^{-\mu} - \frac{g^2_Z}{2} Z_\mu Z^\mu ,
\end{aligned} \qquad (1.35)$$

using $\tau^2_\pm = 0$, $\text{Tr}(\tau_3 \tau_\pm) = 0$, $\text{Tr}(\tau_\pm \tau_\mp) = 1$, and $\text{Tr}(\tau^2_3) = \text{Tr}\mathbb{1} = 2$. The second mass term in $\mathscr{L}_{D2}$ proportional to $\Delta\rho$ is equally simple in unitary gauge

$$T = \Sigma\tau_3\Sigma^\dagger = \tau_3$$

$$\Rightarrow \qquad \text{Tr}(TV_\mu) = \text{Tr}\left(-ig_Z Z_\mu \frac{\tau^2_3}{2}\right) = -ig_Z Z_\mu$$

$$\Rightarrow \qquad \text{Tr}(TV_\mu)\,\text{Tr}(TV^\mu) = -g^2_Z Z_\mu Z^\mu . \qquad (1.36)$$

Inserting both terms into Eq. (1.28) yields the complete gauge boson mass term

$$\mathcal{L}_{D2} = \frac{v^2}{4}\left(-g^2 W_\mu^+ W^{-\mu} - \frac{g_Z^2}{2}Z_\mu Z^\mu\right) + \Delta\rho\frac{v^2}{8}\left(-g_Z^2 Z_\mu Z^\mu\right)$$

$$= -\frac{v^2 g^2}{4}W_\mu^+ W^{-\mu} - \frac{v^2 g_Z^2}{8}Z_\mu Z^\mu - \Delta\rho\frac{v^2 g_Z^2}{8}Z_\mu Z^\mu$$

$$= -\frac{v^2 g^2}{4}W_\mu^+ W^{-\mu} - \frac{v^2 g_Z^2}{8}(1+\Delta\rho)\,Z_\mu Z^\mu\,. \tag{1.37}$$

Identifying the masses with the form given in Eq. (1.12) and assuming universality of *neutral and charged current* interactions ($\Delta\rho = 0$) we find

$$\boxed{m_W = \frac{gv}{2}} \qquad \boxed{m_Z = \sqrt{1+\Delta\rho}\,\frac{g_Z v}{2} \stackrel{\Delta\rho=0}{=} \frac{g_Z v}{2} = \frac{gv}{2c_w}}. \tag{1.38}$$

The role of a possible additional and unwanted Z-mass contribution $\Delta\rho$ we will discuss in Sect. 1.1.6 on custodial symmetry. Given that we know the heavy gauge boson masses ($m_W \sim 80\,\text{GeV}$) and the weak coupling ($g \sim 0.7$) from experiment, these relations experimentally tell us $v \sim 246\,\text{GeV}$.

Let us just recapitulate what we did until now—using this Σ field with its specific transformation properties and its finite constant value $\Sigma = \mathbb{1}$ in unitary gauge we have made the fermions and electroweak gauge boson massive. Choosing this constant finite field value for Σ is not the only and not the minimal assumption needed to make the gauge bosons heavy, but it leads to the most compact Lagrangian. From the photon mass example, however, we know that there must be more to this mechanism. We should for example be able to see the additional degrees of freedom of the longitudinal gauge boson modes if we step away from unitary gauge.

If a finite expectation value of the terms in the potential $\mathcal{L}_\Sigma$ should be linked to electroweak symmetry breaking and the gauge boson masses we can guess that the *minimal assumption* leading to finite gauge boson masses is $\langle\text{Tr}(\Sigma^\dagger(x)\Sigma(x))\rangle \neq 0$ in the vacuum. Every parameterization of Σ with this property will lead to the same massive gauge bosons, so they are all physically equivalent—as they should be given that they are only different gauge choices. In the canonical normalization we write

$$\boxed{\frac{1}{2}\langle\text{Tr}(\Sigma^\dagger(x)\Sigma(x))\rangle = 1} \qquad \forall x\,, \tag{1.39}$$

which instead of our previous $\Sigma(x) = \mathbb{1}$ we can also fulfill through

$$\Sigma^\dagger(x)\Sigma(x) = \mathbb{1} \qquad \forall x\,. \tag{1.40}$$

1.1 Electroweak Symmetry Breaking

This means that $\Sigma(x)$ is a unitary matrix which like any 2×2 unitary matrix can be expressed in terms of the Pauli matrices. This solution still forbids fluctuations which in the original condition equation (1.39) on the expectation value only vanish on average. However, in contrast to $\Sigma(x) = \mathbb{1}$ it allows a non-trivial x dependence. A unitary matrix Σ with the appropriate normalization can for example be written as a simple linear combination of the basis elements, i.e. in the *linear representation*

$$\Sigma(x) = \frac{1}{\sqrt{1 + \frac{w_a w_a}{v^2}}} \left(\mathbb{1} - \frac{i}{v} \vec{w}(x) \right) \quad \text{with} \quad \vec{w}(x) = w_a(x) \tau_a \, , \quad (1.41)$$

where $\vec{w}(x)$ has mass dimension one which is absorbed by the mass scale v. These fields are a set of scalar *Nambu-Goldstone modes*. From the photon mass example for Goldstone's theorem we know that they will become the missing degrees of freedom for the three now massive gauge bosons $W^\pm$ and Z. The normalization scale v fixes the relevant energy scale of our Lagrangian.

Another way of parameterizing the unitary field Σ in terms of the Pauli matrices is

$$\boxed{\Sigma(x) = \exp\left(-\frac{i}{v} \vec{w}(x) \right)} \quad \text{with} \quad \vec{w}(x) = w_a(x) \tau_a \, , \quad (1.42)$$

Because the relation between Σ and $\vec{w}$ is not linear, this is referred to as a *non-linear representation* of the Σ field. Using the commutation properties of the Pauli matrices we can expand Σ as

$$\Sigma = \mathbb{1} - \frac{i}{v} \vec{w} + \frac{1}{2} \frac{(-1)}{v^2} w_a \tau_a w_b \tau_b + \frac{1}{6} \frac{i}{v^3} w_a \tau_a w_b \tau_b w_c \tau_c + \mathcal{O}(w^4)$$

$$= \mathbb{1} - \frac{i}{v} \vec{w} - \frac{1}{2v^2} w_a w_a \mathbb{1} + \frac{i}{6v^3} w_a w_a \vec{w} + \mathcal{O}(w^4) \quad \text{using Eq. (1.10)}$$

$$= \left(1 - \frac{1}{2v^2} w_a w_a + \mathcal{O}(w^4) \right) \mathbb{1} - \frac{i}{v} \left(1 - \frac{1}{6v^2} w_a w_a + \mathcal{O}(w^4) \right) \vec{w} \, . \quad (1.43)$$

From this expression we can for example read off Feynman rules for the longitudinal gauge fields $\vec{w}$, which we will use later. The different ways of writing the Σ field in terms of the Pauli matrices of course cannot have any impact on the physics.

Before we move on and introduce a physical Higgs boson we briefly discuss different gauge choices and the appearance of Goldstone modes. If we break the full electroweak gauge symmetry $SU(2)_L \times U(1)_Y \to U(1)_Q$ we expect three Goldstone bosons which become part of the weak gauge bosons and promote those from massless gauge bosons (with two degrees of freedom each) to massive gauge bosons (with three degrees of freedom each). This is the point of view of the unitary gauge, in which we never see Goldstone modes.

In the general renormalizable R_ξ gauge we can actually see these Goldstone modes appear separately in the *gauge boson propagators*

$$\Delta^{\mu\nu}_{VV}(q) = \frac{-i}{q^2 - m_V^2 + i\epsilon}\left[g^{\mu\nu} + (\xi-1)\frac{q^\mu q^\nu}{q^2 - \xi m_V^2}\right]$$

$$= \begin{cases} \dfrac{-i}{q^2 - m_V^2 + i\epsilon}\left[g^{\mu\nu} - \dfrac{q^\mu q^\nu}{m_V^2}\right] & \text{unitary gauge } \xi \to \infty \\ \dfrac{-i}{q^2 - m_V^2 + i\epsilon}\, g^{\mu\nu} & \text{Feynman gauge } \xi = 1 \\ \dfrac{-i}{q^2 - m_V^2 + i\epsilon}\left[g^{\mu\nu} - \dfrac{q^\mu q^\nu}{q^2}\right] & \text{Landau gauge } \xi = 0\,. \end{cases} \quad (1.44)$$

Obviously, these gauge choices are physically equivalent. However, something has to compensate, for example, for the fact that in Feynman gauge the whole Goldstone term vanishes and the polarization sum looks like a massless gauge boson, while in unitary gauge we can see the effect of these modes directly. The key is the Goldstone propagator, with its additional propagating scalar degrees of freedom

$$\Delta_{VV}(q^2) = \frac{-i}{q^2 - \xi m_V^2 + i\epsilon}\,, \quad (1.45)$$

for both heavy gauge bosons ($V = Z, W^+$). The Goldstone mass $\sqrt{\xi}m_V$ depends on the gauge: in unitary gauge the infinitely heavy Goldstones do not propagate ($\Delta_{VV}(q^2) \to 0$), while in Feynman gauge and in Landau gauge we have to include them as particles. From the form of the Goldstone propagators we can guess that they will indeed cancel the second term of the gauge boson propagators.

These different gauges have different Feynman rules and Green's functions, even a different particle content, so for a given problem one or the other might be the most efficient to use in computations or proofs. For example, the proof of renormalizability was first formulated in unitary gauge. Loop calculations might be more efficient in Feynman gauge, because of the simplified propagator structure, while many QCD processes benefit from an explicit projection on the physical external gluons. Tree level helicity amplitudes are usually computed in unitary gauge, etc.

1.1.5 Higgs Boson

At this stage we have defined a perfectly fine electroweak theory with massive gauge bosons. All we need is a finite vacuum expectation value for Σ, which means this field *spontaneously breaks* the electroweak symmetry not by explicit terms in the Lagrangian but via the vacuum. The origin of this finite vacuum expectation value

1.1 Electroweak Symmetry Breaking

is not specified. This aspect that the Higgs mechanism does not actually specify where the vacuum expectation value v comes from is emphasized by Peter Higgs in his original paper. If we are interested in physics at or above the electroweak energy scale $E \sim v$ some kind of ultraviolet completion of this Σ model should tell us what the Σ field's properties as a quantum object are.

If we consider our Σ model itself the fundamental theory and promote the Σ field to a quantum field like all other Standard Model fields, we need to allow for *quantum fluctuations* of $\text{Tr}(\Sigma^\dagger \Sigma)$ around the vacuum value $\text{Tr}(\Sigma^\dagger \Sigma) = 2$. Omitting the Goldstone modes we can parameterize these new degrees of freedom as a real scalar field

$$\boxed{\Sigma \to \left(1 + \frac{H}{v}\right)\Sigma}, \tag{1.46}$$

as long as this physical field H has a vanishing vacuum expectation value and therefore

$$\frac{1}{2}\langle \text{Tr}(\Sigma^\dagger \Sigma)\rangle = \left\langle \left(1 + \frac{H}{v}\right)^2\right\rangle = 1 \quad \Leftrightarrow \quad \langle H \rangle = 0. \tag{1.47}$$

This real Higgs field is the fourth direction in the basis choice for the unitary matrix Σ for example shown in Eq. (1.41), where only w_a are originally promoted to quantum fields.

The factor in front of the fluctuation term H/v is not fixed until we properly define the physical Higgs field and make sure that its kinetic term does not come with an unexpected prefactor. On the other hand, if we assume that the neutral Goldstone mode w_3 has the correct normalization, the Higgs field should be added to Σ such that it matches this Goldstone, as we will see later in this section and then in more detail in Sect. 1.2.

The non–dynamical limit of this Higgs ansatz is indeed our sigma model in unitary gauge $\Sigma^\dagger \Sigma = \mathbb{1}$, equivalent to $H = 0$. Interpreting the fluctuations around the non–trivial vacuum as a *physical Higgs field* is the usual Higgs mechanism.

For this new Higgs field the Lagrangian $\mathscr{L}_\Sigma$ defines a potential following the original form of Eq. (1.30)

$$\mathscr{L}_\Sigma = -\frac{\mu^2 v^2}{2}\left(1 + \frac{H}{v}\right)^2 - \frac{\lambda v^4}{4}\left(1 + \frac{H}{v}\right)^4 + \ldots \tag{1.48}$$

The dots stand for higher–dimensional terms which might or might not be there. We will have a look at them in Sect. 1.2.1. Some of them are not forbidden by any symmetry, but they are not realized at tree level in the Standard Model either. The minimum of this potential occurs at $H = 0$, but this potential is not actually needed to give mass to gauge bosons and fermions. Therefore, we postpone a detailed study of the Higgs potential to Sect. 1.2.1.

Let us recall one last time how we got to the Higgs mechanism from a static gauge invariant theory, the Σ model. From an effective field theory point of view we can introduce the Goldstone modes and with them gauge boson masses without introducing a fundamental Higgs scalar. All we need is the finite vacuum expectation value for Σ to spontaneously break electroweak symmetry. For this symmetry breaking we do not care about quantum fluctuations of the Σ field, which means we do not distinguish between the invariant $\text{Tr}(\Sigma^\dagger \Sigma)$ and its expectation value. Any properties of the Σ field as a quantum field are left to the ultraviolet completion, which has to decide for example if Σ is a fundamental or composite field. This way, the Higgs field could just be one step in a ladder built out of effective theories. Such a non–fundamental Higgs field is the basis for so-called strongly interacting light Higgs models where the Higgs field is a light composite field with a different canonical normalization as compared to a fundamental scalar.

Counting degrees of freedom we should be able to write Σ as a complex doublet with four degrees of freedom, three of which are eaten Goldstones and one is the fundamental Higgs scalar. On the pure Goldstone side we can choose for example between the linear representation of Eq. (1.41) and the non–linear representation of Eq. (1.42). If we extend the linear representation and for now ignore the normalization we find

$$\Sigma = \left(1 + \frac{H}{v}\right)\mathbb{1} - \frac{i}{v}\vec{w} = \frac{1}{v}\begin{pmatrix} v + H - iw_3 & -w_2 - iw_1 \\ w_2 - iw_1 & v + H + iw_3 \end{pmatrix} = \frac{\sqrt{2}}{v}(\tilde{\phi}\phi). \tag{1.49}$$

The last step is just another way to write the 2×2 matrix as a bi-doublet in terms of the two $SU(2)_L$ doublets containing the physical Higgs field and the Goldstone modes for the massive vector bosons W and Z,

$$\phi = \frac{1}{\sqrt{2}}\begin{pmatrix} -w_2 - iw_1 \\ v + H + iw_3 \end{pmatrix} \qquad \tilde{\phi} = -i\tau_2\,\phi^* = \frac{1}{\sqrt{2}}\begin{pmatrix} v + H - iw_3 \\ w_2 - iw_1 \end{pmatrix}. \tag{1.50}$$

This description of the Higgs field as part of the $SU(2)_L$ doublet in the *linear representation* has a profound effect on the form of the Lagrangian: we can only include the Higgs field in $SU(2)_L$-invariant ways, for example using the combination $\phi^\dagger \phi$. In the presence of a new mass scale Λ the structure and mass dimension of the doublet field ϕ will help us organize the most general electroweak and Higgs Lagrangians, for example allowing for additional terms $\phi^\dagger \phi/\Lambda^2$. In contrast, the *non–linear representation* of Eq. (1.42) cannot be cast into such a $SU(2)$ invariant form, and the Higgs field appears as a singlet under the weak gauge symmetry. In an extended Lagrangian we can simply add a general power series in H/v or H/Λ to any gauge operator.

The vacuum expectation value v appearing in the ϕ and $\tilde{\phi}$ doublets corresponds to $\langle \Sigma \rangle = \mathbb{1}$. In this form the normalization of the two real scalars w_3 and H is

1.1 Electroweak Symmetry Breaking

indeed the same, so their kinetic terms will be of the same form. The over–all factors $1/\sqrt{2}$ in the definition of the doublets are purely conventional and sometimes lead to confusion when some people define $v = 246\,\text{GeV}$ while others prefer $v = 174\,\text{GeV}$. The latter choice is a little less common but has the numerological advantage of $v \sim m_t$. For the fermion sector equation (1.13) this bi-doublet structure is important, because it means that we give mass to up–type fermions and down–type fermions not with the same field ϕ, but with ϕ and $\tilde{\phi}$.

Following Eq. (1.49) we can for example derive the couplings of the physical Higgs boson to the massive W and Z gauge bosons from Eq. (1.37) with custodial symmetry,

$$\begin{aligned}\mathscr{L}_{D2} &= -\frac{(v+H)^2 g^2}{4} W^+_\mu W^{-\mu} - \frac{(v+H)^2 g_Z^2}{8} (1+\Delta\rho) Z_\mu Z^\mu \\ &\supset -\frac{2vHg^2}{4} W^+_\mu W^{-\mu} - \frac{2vHg_Z^2}{8} (1+\Delta\rho) Z_\mu Z^\mu \\ &= -g m_W H W^+_\mu W^{-\mu} - \frac{g_Z m_Z}{2} (1+\Delta\rho) H Z_\mu Z^\mu\,. \end{aligned} \quad (1.51)$$

The same we can do for each fermion, where Eq. (1.13) in the diagonal limit and with the appropriate normalization of the Yukawa coupling $y_f = \sqrt{2} m_f/v$ becomes

$$\mathscr{L}_{D3} \to -y_f \frac{(v+H)}{\sqrt{2}} \overline{\psi}_f \psi_f \supset -\frac{y_f}{\sqrt{2}} H \overline{\psi}_f \psi_f\,. \quad (1.52)$$

The couplings of the scalar Higgs boson are completely determined by the masses of the particles it is coupling to. This includes the unwanted correction $\Delta\rho$ to the Z mass.

Apart from problems arising when we ask for higher precision and quantum corrections, the effective sigma model clearly breaks down at large enough energies which can excite the fluctuations of the sigma field and for example produce a Higgs boson. This is the *job of the LHC*, which is designed and built to take us into an energy range where we can observe the structure of electroweak symmetry breaking beyond the effective theory and the Goldstone modes. The observation of a light and narrow Higgs resonance roughly compatible with its Standard Model definition in Eq. (1.46) is only a first step into this direction.

1.1.6 Custodial Symmetry

Analyzing the appearance of $\Delta\rho$ in Eqs. (1.28) and (1.38) we will see that not only higher energies, but also higher precision leads to a breakdown of the effective sigma model. At some point we start seeing how the relative size of the W and

Z masses are affected by quantum fluctuations of the sigma field, i.e. the three Goldstone modes and the Higgs boson itself. Diagrammatically, we can compute these quantum effects by evaluating Higgs contributions to the one-loop form of the W and Z propagators.

From the construction in Sect. 1.1.4 we know that electroweak symmetry breaking by a sigma field or Higgs doublet links the couplings of neutral and charged currents firmly to the masses of the W and Z bosons. On the other hand, the general renormalizable Lagrangian for the gauge boson masses in Eq. (1.28) involves two terms, both symmetric under $SU(2)_L \times U(1)_Y$ and hence allowed in the electroweak Standard Model. The mass values coming from $\text{Tr}[V_\mu V^\mu]$ give m_W and m_Z proportional to $g \equiv g_W$ and g_Z. The second term involving $(\text{Tr}[TV_\mu])^2$ only contributes to m_Z.

The relative size of the two gauge boson masses can be expressed in terms of the weak mixing angle θ_w, together with the assumption that G_F or g universally govern charged current ($W^\pm$) and neutral-current (W^3) interactions. At tree level this experimentally very well tested relation corresponds to $\Delta\rho = 0$ or

$$\frac{m_W^2}{m_Z^2} = \frac{g^2}{g_Z^2} = \cos^2\theta_w \equiv c_w^2 . \tag{1.53}$$

In general, we can introduce a free parameter ρ which breaks this relation

$$\boxed{g_Z^2 \to g_Z^2\, \rho} \qquad m_Z \to m_Z\, \sqrt{\rho} = m_Z\, \sqrt{1+\Delta\rho}\,, \tag{1.54}$$

which from measurements is very strongly constrained to be unity. It is defined to correspond to our theoretically known allowed deviation $\Delta\rho$. In $\mathscr{L}_{D2}$ the Z-mass term precisely predicts this deviation. To bring our Lagrangian into agreement with measurements we better find a reason to constrain $\Delta\rho$ to zero, and the $SU(2)_L \times U(1)_Y$ gauge symmetry unfortunately does not do the job.

Looking ahead, we will find that in the Standard Model $\rho = 1$ is actually violated at the one-loop level. This means we are looking for an *approximate symmetry* of the Standard Model. What we can hope for is that this symmetry is at least a good symmetry in the $SU(2)_L$ gauge sector and slightly broken elsewhere. One possibility along those lines is to replace the $SU(2)_L \times U(1)_Y$ symmetry with a larger $SU(2)_L \times SU(2)_R$ symmetry. At this stage this extended symmetry does not have to be a local gauge symmetry, a global version of $SU(2)_L$ combined with a global $SU(2)_R$ is sufficient. This global symmetry would have to act like

$$\Sigma \to U \Sigma V^\dagger \qquad U \in SU(2)_L \qquad V \in SU(2)_R$$
$$\text{Tr}(\Sigma^\dagger \Sigma) \to \text{Tr}\left(V \Sigma^\dagger U^\dagger U \Sigma V^\dagger\right) = \text{Tr}(\Sigma^\dagger \Sigma)\,. \tag{1.55}$$

In this setup, the three components of W^μ form a triplet under $SU(2)_L$ and a singlet under $SU(2)_R$. If we cannot extract τ_3 as a special generator of $SU(2)_L$ and

1.1 Electroweak Symmetry Breaking

combine it with the $U(1)_Y$ hypercharge the W and Z masses have to be identical, corresponding to $c_w = 1$ at tree level.

In the gauge boson and fermion mass terms computed in unitary gauge the Σ field becomes identical to its vacuum expectation value $\mathbb{1}$. The combined global $SU(2)_L$ transformations act on the symmetry breaking vacuum expectation value the same way as shown in Eq. (1.24),

$$\langle \Sigma \rangle \to \langle U \Sigma V^\dagger \rangle = \langle U \mathbb{1} V^\dagger \rangle = U V^\dagger \stackrel{!}{=} \mathbb{1} . \tag{1.56}$$

The last step, i.e. the symmetry requirement for the Lagrangian can only be satisfied if we require $U = V$. In other words, the vacuum expectation value for Σ breaks $SU(2)_L \times SU(2)_R$ to the *diagonal subgroup* $SU(2)_{L+R}$. The technical term is precisely defined this way—the two $SU(2)$ symmetries reduce to one remaining symmetry which can be written as $U = V$. Depending on if we look at the global symmetry structure in the unbroken or broken phase the *custodial symmetry* group either refers to $SU(2)_R$ or $SU(2)_{L+R}$.

Even beyond tree level the global $SU(2)_L \times SU(2)_R$ symmetry structure can protect the relation $\rho = 1$ between the gauge boson masses shown in Eq. (1.53). From Eq. (1.55) we immediately see that it allows all terms in the Higgs potential $\mathscr{L}_\Sigma$, but it changes the picture not only for gauge boson but also for fermion masses. If fermions reside in $SU(2)_L$ as well as $SU(2)_R$ doublets we cannot implement any difference between up–type and down–type fermions in the Lagrangian. The custodial symmetry is only intact in the limit for example of identical third generation fermion masses $m_b = m_t$.

The measured masses $m_t \gg m_b$ change the protected tree level value $\rho = 1$: self energy loops in the W propagator involve a mixture of the bottom and top quark, while the Z propagator includes pure bottom and top loops. Skipping the loop calculation we quote their different contributions to the gauge boson masses as

$$\begin{aligned}
\Delta\rho &\supset \frac{3G_F}{8\sqrt{2}\pi^2} \left(m_t^2 + m_b^2 - 2\frac{m_t^2 m_b^2}{m_t^2 - m_b^2} \log \frac{m_t^2}{m_b^2} \right) \\
&= \frac{3G_F}{8\sqrt{2}\pi^2} \left(2m_b^2 + m_b^2\delta - 2m_b^2 \frac{1+\delta}{\delta} \log(1+\delta) \right) \quad \text{defining } m_t^2 = m_b^2(1+\delta) \\
&= \frac{3G_F}{8\sqrt{2}\pi^2} \left(2m_b^2 + m_b^2\delta - 2m_b^2 \left(\frac{1}{\delta}+1\right)\left(\delta - \frac{\delta^2}{2} + \frac{\delta^3}{3} + \mathcal{O}(\delta^4)\right) \right) \\
&= \frac{3G_F}{8\sqrt{2}\pi^2} m_b^2 \left(2 + \delta - 2 - 2\delta + \delta + \delta^2 - \frac{2}{3}\delta^2 + \mathcal{O}(\delta^3) \right) \\
&= \frac{3G_F}{8\sqrt{2}\pi^2} m_b^2 \left(\frac{1}{3}\delta^2 + \mathcal{O}(\delta^3) \right) \\
&= \frac{G_F m_W^2}{8\sqrt{2}\pi^2} \left(\frac{(m_t^2 - m_b^2)^2}{m_W^2 m_b^2} + \cdots \right) .
\end{aligned} \tag{1.57}$$

In the Taylor series above the assumption of δ being small is of course not realistic, but the result is nevertheless instructive: the shift vanishes very rapidly towards the chirally symmetric limit $m_t \sim m_b$. The sign of the contribution of a chiral fermion doublet to $\Delta\rho$ is always positive. In terms of realistic Standard Model mass ratios it scales like

$$\Delta\rho \supset \frac{3G_F}{8\sqrt{2}\pi^2}m_t^2\left(1 - 2\frac{m_b^2}{m_t^2}\log\frac{m_t^2}{m_b^2}\right) = \frac{3G_F m_W^2}{8\sqrt{2}\pi^2}\frac{m_t^2}{m_W^2}\left(1 + \mathcal{O}\left(\frac{m_b^2}{m_t^2}\right)\right), \tag{1.58}$$

remembering that the Fermi coupling constant has a mass dimension fixed by $G_F \propto 1/m_W^2$.

We have already argued that hypercharge or electric charge break custodial symmetry. From the form of the covariant derivative $D_\mu \Sigma$ including a single τ_3 we can guess that the $SU(2)_R$ symmetry will not allow B field interactions which are proportional to $s_w \sim \sqrt{1/4}$. A second contribution to the ρ parameter therefore arises from Higgs loops in the presence of $g' \neq 0$

$$\boxed{\Delta\rho \supset -\frac{11G_F m_Z^2 s_w^2}{24\sqrt{2}\pi^2}\log\frac{m_H^2}{m_Z^2}.} \tag{1.59}$$

The loop diagrams responsible for the contribution are simply virtual Higgs exchanges in the W and Z self energies, which not only have a different factorizing couplings, but also different W and Z masses inside the loop. These masses inside loop diagrams appear as logarithms. The sign of this contribution implies that larger Higgs masses give increasingly negative contributions to the ρ parameter.

There is another parameterization of the same effect, namely the T parameter. It is part of an effective theory parameterization of deviations from the tree level relations between gauge boson masses, mixing angles, and neutral and charged current couplings. If we allow for deviations from the Standard Model gauge sector induced by vacuum polarization corrections $\Pi(p^2)$ and their momentum derivatives $\Pi'(p^2)$ we can write down additional Lagrangian terms

$$\Delta\mathcal{L} = -\frac{\Pi'_{\gamma\gamma}}{4}\hat{F}_{\mu\nu}\hat{F}^{\mu\nu} - \frac{\Pi'_{WW}}{2}\hat{W}_{\mu\nu}\hat{W}^{\mu\nu} - \frac{\Pi'_{ZZ}}{4}\hat{Z}_{\mu\nu}\hat{Z}^{\mu\nu}$$
$$- \frac{\Pi'_{\gamma Z}}{4}\hat{F}_{\mu\nu}\hat{Z}^{\mu\nu} - \Pi_{WW}\hat{m}_W^2 \hat{W}_\mu^+ \hat{W}^{-\mu} - \frac{\Pi_{ZZ}}{2}\hat{m}_Z^2 \hat{Z}_\mu \hat{Z}^\mu. \tag{1.60}$$

The field strengths $\hat{F}_{\mu\nu}, \hat{W}_{\mu\nu}, \hat{Z}_{\mu\nu}$ are based the fields $\hat{A}_\mu, \hat{W}_\mu, \hat{Z}_\mu$. The hats are necessary, because the kinetic terms and hence the fields do not (yet) have the canonical normalization. To compute the proper field normalization we assume that all Π' are small, so we can express the hatted gauge fields in terms of the properly normalized fields as

1.1 Electroweak Symmetry Breaking

$$\hat{A}_\mu = \left(1 - \frac{\Pi'_{\gamma\gamma}}{2}\right) A_\mu + \Pi'_{\gamma Z} Z_\mu \qquad \hat{W}_\mu = \left(1 - \frac{\Pi'_{WW}}{2}\right) W_\mu$$

$$\hat{Z}_\mu = \left(1 - \frac{\Pi'_{ZZ}}{2}\right) Z_\mu. \tag{1.61}$$

To check this ansatz, we can for example extract all terms proportional to the photon–Z mixing $\Pi'_{\gamma Z}$ arising from Eq. (1.61) and find

$$\begin{aligned}
-\frac{1}{4} \hat{F}_{\mu\nu} \hat{F}^{\mu\nu}\Big|_{\gamma Z} &= -\frac{1}{4} \left(\partial_\mu \hat{A}_\nu - \partial_\nu \hat{A}_\mu\right)\left(\partial_\mu \hat{A}_\nu - \partial_\nu \hat{A}_\mu\right)\Big|_{\gamma Z} \\
&= -\frac{1}{4} \left(\partial_\mu (A + \Pi'_{\gamma Z} Z)_\nu - \partial_\nu (A + \Pi'_{\gamma Z} Z)_\mu\right) \\
&\quad \left(\partial_\mu (A + \Pi'_{\gamma Z} Z)_\nu - \partial_\nu (A + \Pi'_{\gamma Z} Z)_\mu\right)\Big|_{\gamma Z} \\
&= -\frac{\Pi'_{\gamma Z}}{4} \left(\partial_\mu A_\nu - \partial_\nu A_\mu\right)\left(\partial_\mu Z_\nu - \partial_\nu Z_\mu\right) \\
&\quad - \frac{\Pi'_{\gamma Z}}{4} \left(\partial_\mu Z_\nu - \partial_\nu Z_\mu\right)\left(\partial_\mu A_\nu - \partial_\nu A_\mu\right) + \mathcal{O}(\Pi'^2) \\
&= -\frac{\Pi'_{\gamma Z}}{2} \left(\partial_\mu Z_\nu - \partial_\nu Z_\mu\right)\left(\partial_\mu A_\nu - \partial_\nu A_\mu\right) + \mathcal{O}(\Pi'^2) \\
&= -\frac{\Pi'_{\gamma Z}}{2} Z_{\mu\nu} F^{\mu\nu} + \mathcal{O}(\Pi'^2) = -\frac{\Pi'_{\gamma Z}}{2} \hat{Z}_{\mu\nu} \hat{F}^{\mu\nu} + \mathcal{O}(\Pi'^2).
\end{aligned} \tag{1.62}$$

The field shift in Eq. (1.61) indeed absorbs the explicit new contribution in Eq. (1.60). Assuming that this works for all gauge field combinations the Lagrangian including the loop–induced $\Delta\mathscr{L}$ gets the canonical form

$$\mathscr{L} \supset -\frac{1}{4} F_{\mu\nu} F^{\mu\nu} - \frac{1}{2} W_{\mu\nu} W^{\mu\nu} - \frac{1}{4} Z_{\mu\nu} Z^{\mu\nu}$$
$$- (1 + \Pi_{WW} - \Pi'_{WW}) \hat{m}_W^2 W_\mu^+ W^{-\mu} - \frac{1}{2}(1 + \Pi_{ZZ} + \Pi'_{ZZ}) \hat{m}_Z^2 Z_\mu Z^\mu. \tag{1.63}$$

The physical Z mass now has to be $m_Z^2 = (1 + \Pi_{ZZ} + \Pi'_{ZZ}) \hat{m}_Z^2$. Just as in the usual Lagrangian we can link the two gauge boson masses through the (hatted) weak mixing angle $\hat{m}_W = \hat{c}_w \hat{m}_Z$. In terms of this mixing angle we can compute the muon decay constant, the result of which we quote as

$$\frac{\hat{s}_w^2}{s_w^2} = 1 + \frac{c_w^2}{c_w^2 - s_w^2} \left(\Pi'_{\gamma\gamma} - \Pi'_{ZZ} - \Pi_{WW} + \Pi_{ZZ}\right)$$

or $\quad\dfrac{\hat{c}_w^2}{c_w^2} = 1 - \dfrac{s_w^2}{c_w^2 - s_w^2} \left(\Pi'_{\gamma\gamma} - \Pi'_{ZZ} - \Pi_{WW} + \Pi_{ZZ}\right) .$ \quad (1.64)

With this result for example the complete W–mass term in Eq. (1.63) reads

$$\mathscr{L} \supset -(1 + \Pi_{WW} - \Pi'_{WW}) \, \hat{c}_w^2 \, \hat{m}_Z^2 \, W_\mu^+ W^{-\mu}$$

$$= -(1 + \Pi_{WW} - \Pi'_{WW}) \left[1 - \dfrac{s_w^2}{c_w^2 - s_w^2} \left(\Pi'_{\gamma\gamma} - \Pi'_{ZZ} - \Pi_{WW} + \Pi_{ZZ}\right)\right]$$

$$c_w^2 (1 - \Pi_{ZZ} + \Pi'_{ZZ}) \, m_Z^2 \, W_\mu^+ W^{-\mu}$$

$$= -\left[1 - \Pi'_{WW} + \Pi'_{ZZ} + \Pi_{WW} - \Pi_{ZZ} - \dfrac{s_w^2}{c_w^2 - s_w^2} \right.$$

$$\left. \left(\Pi'_{\gamma\gamma} - \Pi'_{ZZ} - \Pi_{WW} + \Pi_{ZZ}\right)\right] m_Z^2 \, W_\mu^+ W^{-\mu}$$

$$= -\left[1 - \dfrac{\alpha S}{2(c_w^2 - s_w^2)} + \dfrac{c_w^2 \alpha T}{c_w^2 - s_w^2} + \dfrac{\alpha U}{4 s_w^2}\right] m_Z^2 \, W_\mu^+ W^{-\mu} . \quad (1.65)$$

In the last step we have defined three typical combinations of the different correction factors as

$$S = \dfrac{4 s_w^2 c_w^2}{\alpha} \left(-\Pi'_{\gamma\gamma} + \Pi'_{ZZ} - \Pi'_{\gamma Z} \dfrac{c_w^2 - s_w^2}{c_w s_w}\right)$$

$$T = \dfrac{1}{\alpha} \left(\Pi_{WW} - \Pi_{ZZ}\right)$$

$$U = \dfrac{4 s_w^4}{\alpha} \left(\Pi'_{\gamma\gamma} - \dfrac{\Pi'_{WW}}{s_w^2} + \Pi'_{ZZ} \dfrac{c_w^2}{s_w^2} - 2\Pi'_{\gamma Z} \dfrac{2 c_w}{s_w}\right) . \quad (1.66)$$

Two of these so-called *Peskin–Takeuchi parameters* can be understood fairly easily: the S-parameter corresponds to a shift of the Z mass. This is not completely obvious because it seems to also involve photon terms. We have to remember that the weak mixing angle is defined such that the photon is massless, while all mass terms are absorbed into the Z boson. The T parameter obviously compares contributions to the W and Z masses. The third parameter U is less important for most models.

To get an idea how additional fermions contribute to S and T we quote the contributions from the heavy fermion doublet:

$$\Delta S = \dfrac{N_c}{6\pi} \left(1 - 2Y \log \dfrac{m_t^2}{m_b^2}\right)$$

$$\Delta T = \dfrac{N_c}{4\pi s_w^2 c_w^2 m_Z^2} \left(m_t^2 + m_b^2 - \dfrac{2 m_t^2 m_b^2}{m_t^2 - m_b^2} \log \dfrac{m_t^2}{m_b^2}\right), \quad (1.67)$$

1.1 Electroweak Symmetry Breaking

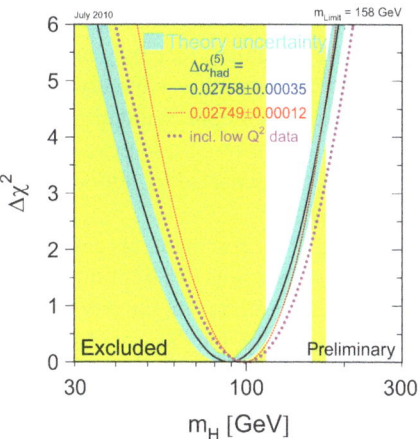

Fig. 1.1 Allowed range of Higgs masses in the Standard Model after taking into account electroweak precision data, most notably the ρ parameter contribution from the Higgs itself, Eq. (1.59) (Figure from the LEP electroweak working group, with updates available under http://lepewwg.web.cern.ch/LEPEWWG)

with $Y = 1/6$ for quarks and $Y = -1/2$ for leptons. While the parameter S has nothing to do with our custodial symmetry, ρ and $T \sim \Delta\rho/\alpha$ are closely linked. Their main difference is the reference point, where $\rho = 1$ refers to its tree level value and $T = 0$ is often chosen for some kind of light Higgs mass and including the Standard Model top-bottom corrections. A similar third set of parameters going back to Altarelli and Barbieri consists of $\epsilon_{1,2,3}$, where the leading effect on the custodial symmetry can be translated via $\epsilon_1 = \alpha T$.

Typical experimental constraints form an ellipse in the S vs T plane along the diagonal. They are usually quoted as ΔT with respect to a reference Higgs mass. Compared to a 125 GeV Standard Model Higgs boson the measured values range around $T \sim 0.1$ and $S \sim 0.05$. Additional contributions $\Delta T \sim 0.1$ tend to be within the experimental errors, much larger contributions are in contradiction with experiment.

There are two reasons to discuss these loop contributions breaking the custodial symmetry in the Standard Model. First, $\Delta\rho$ is experimentally very strongly constrained by *electroweak precision measurements*, which means that alternative models for electroweak symmetry breaking usually include the same kind of approximate custodial symmetry by construction. As a matter of fact, this constraint is responsible for the death of technicolor models, which describe the Higgs boson as a bound state under a new QCD-like interaction and which we will discuss in Sect. 1.9.1.

Even more importantly, in the Standard Model we can measure the symmetry violations from the heavy quarks and from the Higgs sector shown in Eqs. (1.57) and (1.59) in electroweak precision measurements. Even though the Higgs contributions depend on the Higgs mass only logarithmically, we can then derive an *upper bound on the Higgs mass* of the order of $\mathcal{O}(200\,\text{GeV})$, as shown in Fig. 1.1. This strongly suggests that if we are faced (mostly) with the Standard Model at the weak scale the Tevatron and at the LHC will be looking for a fairly light Higgs boson—or something that very much looks like a light fundamental Higgs boson. This is the reason why in the absence of other hints for new physics at the LHC the discovery of a light Standard–Model–like Higgs boson is not unexpected. Any significant deviation of a Higgs boson from the Standard Model prediction would have to be compensated by additional yet unobserved particles in the relevant self energy diagrams.

Turning this argument around, we should firmly keep in mind that the ρ parameter only points to a light fundamental Higgs boson if we assume the Standard Model Higgs mechanism. For any other model it might point to something similar to a light Higgs field, but does not have to be fundamental. Including additional fields in a model can even turn around this prediction and prefer a heavy Higgs state. By now, studying electroweak precision data given the measured Higgs mass is one of the most sensitive consistency tests of the Standard Model.

The symmetry breaking pattern we describe in this section is a nice example to check the predictions from the Goldstone theorem quoted in Sect. 1.1.2, so our last question is: how do physical modes which we introduce as $\Sigma(x) = \exp(-i\vec{w}/v)$ transform under the different broken and unbroken global $SU(2)$ symmetries which make up the custodial symmetry and can we construct a model of electroweak symmetry breaking around the custodial symmetry? This brings us back to the example of the photon mass, where we first saw Goldstone's theorem at work.

Under the usual $SU(2)_L$ symmetry we know the transformation reads $\Sigma \to U\Sigma$ with $U = \exp(i\alpha \cdot \tau/2)$. The transformation properties of the Goldstone modes $\vec{w}$ follow from the infinitesimal transformations

$$\mathbb{1} - i\frac{w \cdot \tau}{v} \to \left(\mathbb{1} + i\frac{\alpha \cdot \tau}{2}\right)\left(\mathbb{1} - i\frac{w \cdot \tau}{v}\right)$$

$$= \mathbb{1} + \frac{i}{v}\left(-w \cdot \tau + \frac{v}{2}\alpha \cdot \tau\right) + \cdots$$

$$\stackrel{!}{=} \mathbb{1} - i\frac{w' \cdot \tau}{v} \qquad \text{implying} \quad \boxed{w \to w' = w - \frac{v}{2}\alpha}, \qquad (1.68)$$

so U is a *non–linear transformation*, since w'_a is not proportional to w_a. The same independent structure we find for the $SU(2)_R$ transformation. This model of electroweak symmetry breaking we call a non–linear sigma model. In our discussion of the Goldstone theorem we already quoted its most important feature: when we construct a Lagrangian this non–linear symmetry transformation forbids mass terms, gauge interactions, Yukawa couplings, and quadratic potential terms for these modes in Σ. As discussed in Sect. 1.1.2 only derivative terms like the kinetic term and derivative couplings are allowed under the $SU(2)_L$ and $SU(2)_R$ symmetries.

Similarly, we can evaluate the transformation of these physical modes under the remaining diagonal symmetry group $SU(2)_{L+R}$ with $\Sigma \to U\Sigma U^\dagger$ and instead find

$$\mathbb{1} - i\frac{w \cdot \tau}{v} \to \left(\mathbb{1} + i\frac{\alpha \cdot \tau}{2}\right)\left(\mathbb{1} - i\frac{w \cdot \tau}{v}\right)\left(\mathbb{1} - i\frac{\alpha \cdot \tau}{2}\right)$$

$$= \left(\mathbb{1} + i\frac{\alpha \cdot \tau}{2}\right)\left(\left[\left(\mathbb{1} - i\frac{w \cdot \tau}{v}\right), \left(\mathbb{1} - i\frac{\alpha \cdot \tau}{2}\right)\right]\right.$$
$$\left. + \left(\mathbb{1} - i\frac{\alpha \cdot \tau}{2}\right)\left(\mathbb{1} - i\frac{w \cdot \tau}{v}\right)\right)$$

$$= \left(\mathbb{1} + i\frac{\alpha \cdot \tau}{2}\right)\left[-i\frac{w \cdot \tau}{v}, -i\frac{\alpha \cdot \tau}{2}\right] + \left(\mathbb{1} - i\frac{w \cdot \tau}{v}\right) + \cdots$$

$$= \left(\mathbb{1} + i\frac{\alpha \cdot \tau}{2}\right)\frac{1}{2v} 2i\tau(\alpha \times w) + \left(\mathbb{1} - i\frac{w \cdot \tau}{v}\right) + \cdots$$

$$= \mathbb{1} - i\frac{w \cdot \tau}{v} + i\frac{\tau(\alpha \times w)}{v} + \cdots$$

implying $\boxed{w_i \to w'_i = w_i - \varepsilon_{ijk}\alpha_j w_k}$, (1.69)

which is a *linear transformation*. In the fourth line we use the commutator

$[\tau_a, \tau_b] = 2i\varepsilon_{abc}\tau_c \quad \Rightarrow \quad (\alpha \cdot \tau)(w \cdot \tau) = \alpha \cdot w + i\tau(\alpha \times w) \quad$ using Eq. (1.10)

$\Rightarrow \quad [(\alpha \cdot \tau), (w \cdot \tau)] = 2i\tau(\alpha \times w)$. (1.70)

In other words, when we transform the physical modes corresponding to the broken generators in Σ by the larger symmetry $SU(2)_L \times SU(2)_R$ we find a non–linear transformation, while the approximate symmetry $SU(2)_{L+R}$ leads to a linear transformation. This is precisely what Goldstone's theorem predicts for the spontaneous breaking of a global electroweak symmetry.

1.2 The Standard Model

Before we discuss all the ways we can look for a Higgs Boson and go through the Higgs discovery in the summer of 2012 we briefly review the Higgs mechanism in the Standard Model. This will link the somewhat non–standard effective theory approach we used until now to the standard textbook arguments. In the last sections we have seen that there does not really need to be such a fundamental scalar, but electroweak precision data tells us whatever it is the Higgs should look very similar to a light fundamental scalar, unless we see some seriously new states and effects around the weak scale.

To make it a little more interesting and since we are already in the mood of not taking the Standard Model Higgs sector too literally, in Sect. 1.2.1 we include higher–dimensional operators on top of the usual *renormalizable* dimension-4 operators in the Higgs potential. Such operators generally occur in effective theories based on ultraviolet completions of our Standard Model, but their effects are often small or assumed to be small.

Once we want to analyze the behavior of the Higgs sector over a wide range of energy scales, like we will do in Sects. 1.2.3 and 1.2.4, we need to take the Standard Model seriously and in turn find constraints on the structure and the validity of our Standard Model with a fundamental Higgs boson.

1.2.1 Higgs Potential to Dimension Six

In the *renormalizable Standard Model* all terms in the Lagrangian are defined to be of mass dimension four, like $m_f \bar{\psi}\psi$ or $\bar{\psi}\partial_\mu\psi$ or $F_{\mu\nu}F^{\mu\nu}$. This mass dimension is easy to read off if we remember that for example scalar fields or vector-boson fields contribute mass dimension one while fermion spinors carry mass dimension 3/2. The same renormalizability assumption we usually make for the Higgs potential, even though from the previous discussion it is clear that higher–dimensional terms— stemming from higher powers of $\text{Tr}(\Sigma^\dagger \Sigma)$—can and should exist.

Starting from the Higgs doublets introduced in Eq. (1.50) and for now ignoring the Goldstone modes the simplified Higgs–only doublet

$$\phi = \frac{1}{\sqrt{2}} \begin{pmatrix} -w_2 - iw_1 \\ v + H + iw_3 \end{pmatrix} \sim \frac{1}{\sqrt{2}} \begin{pmatrix} 0 \\ v + H \end{pmatrix} \quad (1.71)$$

leaves us with only two renormalizable potential terms in Eq. (1.48), now written in terms of the Higgs doublet and with μ^2 and λ as prefactors

$$-\mathscr{L}_\Sigma = V_{\text{SM}} = \mu^2 |\phi|^2 + \lambda |\phi|^4 + \text{const}. \quad (1.72)$$

To emphasize that renormalizability is a strong and not necessarily very justified theoretical assumption in LHC Higgs physics, we allow for more operators in the Higgs potential. Following the discussion in Sect. 1.1.5 we will use the linear representation in terms of the doublet ϕ to organize the extended Lagrangian by mass dimensions. If we expand the possible mass dimensions and the operator basis, there are exactly *two gauge–invariant operators* of dimension six we can write down in terms of the Higgs doublet $|\phi|^2$, i.e. before electroweak symmetry breaking

$$\mathscr{O}_1 = \frac{1}{2}\partial_\mu(\phi^\dagger\phi)\,\partial^\mu(\phi^\dagger\phi) \qquad \mathscr{O}_2 = -\frac{1}{3}(\phi^\dagger\phi)^3. \quad (1.73)$$

There exists one more possible operator $(D_\mu\phi)^\dagger\phi\ \phi^\dagger(D^\mu\phi)$, but it violates custodial symmetry, so we ignore it in our analysis. The prefactors in the Lagrangian are conventional, because to construct a Lagrangian we have to multiply these operators with general coefficients of mass dimension minus two, parameterized in terms of an unknown mass scale Λ

$$\mathscr{L}_{D6} = \sum_{i=1}^{2} \frac{f_i}{\Lambda^2} \mathscr{O}_i. \quad (1.74)$$

1.2 The Standard Model

As long as the typical energy scale E in the numerator in our matrix element is small ($E \ll \Lambda$), the corrections from the additional operators are small as well.

Before we compute the Higgs potential including $\mathcal{O}_2$ we look at the effects of the dimension-6 operator $\mathcal{O}_1$. It contributes to the kinetic term of the Higgs field in the Lagrangian, before or after symmetry breaking

$$\begin{aligned}\mathcal{O}_1 &= \frac{1}{2}\partial_\mu(\phi^\dagger\phi)\,\partial^\mu(\phi^\dagger\phi) \\ &= \frac{1}{2}\partial_\mu\left(\frac{(\hat{H}+v)^2}{2}\right)\partial^\mu\left(\frac{(\hat{H}+v)^2}{2}\right) \\ &= \frac{1}{2}(\hat{H}+v)^2\,\partial_\mu\hat{H}\,\partial^\mu\hat{H}\,.\end{aligned} \qquad (1.75)$$

We use the symbol $\hat{H}$ for the Higgs field as part of ϕ. From the similar case of the gauge fields in Eq. (1.60) we can guess that there will be a difference between $\hat{H}$ and the physical Higgs field H at the end of the day. The contribution from $\mathcal{O}_1$ leaves us with a combined kinetic term

$$\mathcal{L}_{\text{kin}} = \frac{1}{2}\partial_\mu\hat{H}\,\partial^\mu\hat{H}\left(1+\frac{f_1 v^2}{\Lambda^2}\right) \stackrel{!}{=} \frac{1}{2}\partial_\mu H\,\partial^\mu H \quad \Leftrightarrow \quad \boxed{H = \sqrt{1+\frac{f_1 v^2}{\Lambda^2}}\,\hat{H}}\,. \qquad (1.76)$$

This is a simple rescaling to define the *canonical kinetic term* in the Lagrangian, corresponding to a finite wave function renormalization which ensures that the residuum of the Higgs propagator is one. This kind of condition is well known from the LSZ equation and the proper definition of outgoing states. It means we have to eventually replace $\hat{H}$ with H in the entire Higgs sector. In most cases such a wave function renormalization will not lead to observable physics effects, because it can be absorbed for example in coupling renormalization. However, in this case the setup of the electroweak sector does not give us enough freedom to absorb this scaling factor, so it will appear in the observable couplings similar to a form factor in strongly interacting models.

Taking into account the additional dimension-6 operator $\mathcal{O}_2$ we can write the Higgs potential as

$$\boxed{V = \mu^2|\phi|^2 + \lambda|\phi|^4 + \frac{f_2}{3\Lambda^2}|\phi|^6}\,. \qquad (1.77)$$

The positive sign in the last term of the potential V ensures that for $f_2 > 0$ the potential is bounded from below for large field values ϕ. The non–trivial minimum at $\phi \neq 0$ is given by

$$\frac{\partial V}{\partial |\phi|^2} = \mu^2 + 2\lambda|\phi|^2 + \frac{3f_2}{3\Lambda^2}|\phi|^4 \stackrel{!}{=} 0 \quad \Leftrightarrow \quad |\phi|^4 + \frac{2\lambda\Lambda^2}{f_2}|\phi|^2 + \frac{\mu^2\Lambda^2}{f_2} \stackrel{!}{=} 0 \,, \tag{1.78}$$

defining the minimum position $|\phi|^2 = v^2/2$. The two solutions of the quadratic equation for $v^2/2$ are

$$\begin{aligned}
\frac{v^2}{2} &= -\frac{\lambda\Lambda^2}{f_2} \pm \left[\left(\frac{\lambda\Lambda^2}{f_2}\right)^2 - \frac{\mu^2\Lambda^2}{f_2}\right]^{\frac{1}{2}} = \frac{\lambda\Lambda^2}{f_2}\left[-1 \pm \sqrt{1 - \frac{\mu^2 f_2}{\Lambda^2\lambda^2}}\right] \\
&= \frac{\lambda\Lambda^2}{f_2}\left[-1 \pm \left(1 - \frac{f_2\mu^2}{2\lambda^2\Lambda^2} - \frac{f_2^2\mu^4}{8\lambda^4\Lambda^4} + \mathcal{O}\left(\frac{1}{\Lambda^6}\right)\right)\right] \\
&= \begin{cases} -\dfrac{\mu^2}{2\lambda} - \dfrac{f_2\mu^4}{8\lambda^3\Lambda^2} + \mathcal{O}\left(\dfrac{1}{\Lambda^4}\right) = -\dfrac{\mu^2}{2\lambda}\left(1 + \dfrac{f_2\mu^2}{4\lambda^2\Lambda^2} + \mathcal{O}\left(\dfrac{1}{\Lambda^4}\right)\right) \\
\phantom{-\dfrac{\mu^2}{2\lambda} - \dfrac{f_2\mu^4}{8\lambda^3\Lambda^2}} \equiv \dfrac{v_0^2}{2}\left(1 + \dfrac{f_2 v_0^2}{4\lambda\Lambda^2} + \mathcal{O}\left(\dfrac{1}{\Lambda^4}\right)\right) \\ -\dfrac{2\lambda\Lambda^2}{f_2^2} + \mathcal{O}(\Lambda^0) \end{cases}
\end{aligned} \tag{1.79}$$

The first solution we have expanded around the Standard Model minimum, $v_0^2 = -\mu^2/\lambda$. The second, high–scale solution is not the vacuum relevant for our Standard Model. Note that from the W, Z masses we know that $v = 246\,\text{GeV}$ so v is really our first observable in the Higgs sector, sensitive to the higher–dimensional operators.

To compute the Higgs mass as the second observable we could study the second derivative of the potential in the different directions, but we can also simply collect all quadratic terms contributing to the Lagrangian by hand. The regular dimension-4 contributions in terms of the shifted Higgs field $\hat{H}$ are

$$V_{\text{SM}} = \mu^2 \frac{(\hat{H}+v)^2}{2} + \lambda \frac{(\hat{H}+v)^4}{4} = \frac{\mu^2}{2}\left(\hat{H}^2 \cdots\right) + \frac{\lambda}{4}\left(\cdots 6\hat{H}^2 v^2 \cdots\right) . \tag{1.80}$$

Only the terms in the parentheses contribute to the Higgs mass in terms of μ, v and λ. Including the additional potential operator in terms of $\hat{H}$ gives

$$\begin{aligned}
\mathcal{O}_2 &= -\frac{1}{3}(\phi^\dagger\phi)^3 \\
&= -\frac{1}{3}\frac{(\hat{H}+v)^6}{8} \\
&= -\frac{1}{24}\left(\hat{H}^6 + 6\hat{H}^5 v + 15\hat{H}^4 v^2 + 20\hat{H}^3 v^3 + 15\hat{H}^2 v^4 + 6\hat{H} v^5 + v^6\right) .
\end{aligned} \tag{1.81}$$

1.2 The Standard Model

Combining both gives us the complete quadratic mass term to dimension six

$$\begin{aligned}\mathscr{L}_{\text{mass}} &= -\frac{\mu^2}{2}\hat{H}^2 - \frac{3}{2}\lambda v^2 \hat{H}^2 - \frac{f_2}{\Lambda^2}\frac{15}{24}v^4 \hat{H}^2 \\ &= -\frac{1}{2}\left(\mu^2 + 3\lambda v^2 + \frac{5}{4}\frac{f_2 v^4}{\Lambda^2}\right)\hat{H}^2 \\ &= -\frac{1}{2}\left(-\lambda v^2\left(1 + \frac{f_2 v^2}{4\lambda \Lambda^2}\right) + 3\lambda v^2 + \frac{5}{4}\frac{f_2 v^4}{\Lambda^2}\right)\hat{H}^2 \quad \text{replacing } \mu^2 \text{ using Eq. (1.79) twice}\\ &= -\frac{1}{2}\left(2\lambda v^2 - \frac{f_2 v^4}{4\Lambda^2} + \frac{5}{4}\frac{f_2 v^4}{\Lambda^2}\right)\left(1 + \frac{f_1 v^2}{\Lambda^2}\right)^{-1} H^2 \quad \text{replacing } \hat{H} \text{ using Eq. (1.76)}\\ &= -\frac{1}{2}\left(2\lambda v^2 + \frac{f_2 v^4}{\Lambda^2}\right)\left(1 - \frac{f_1 v^2}{\Lambda^2} + \mathcal{O}\left(\frac{1}{\Lambda^4}\right)\right) H^2 \\ &= -\lambda v^2 \left(1 + \frac{f_2 v^2}{2\lambda \Lambda^2}\right)\left(1 - \frac{f_1 v^2}{\Lambda^2} + \mathcal{O}\left(\frac{1}{\Lambda^4}\right)\right) H^2 \\ &= -\lambda v^2 \left(1 - \frac{f_1 v^2}{\Lambda^2} + \frac{f_2 v^2}{2\Lambda^2 \lambda} + \mathcal{O}\left(\frac{1}{\Lambda^4}\right)\right) H^2 \stackrel{!}{=} -\frac{m_H^2}{2} H^2\\ \Leftrightarrow \quad m_H^2 &= 2\lambda v^2 \left(1 - \frac{f_1 v^2}{\Lambda^2} + \frac{f_2 v^2}{2\Lambda^2 \lambda}\right) .\end{aligned} \quad (1.82)$$

Including dimension-6 operators the relation between the vacuum expectation value, the Higgs mass and the factor in front of the $|\phi|^4$ term in the potential changes. Once we measure the Higgs mass at the LHC, we can compute the trilinear and quadrilinear *Higgs self couplings* by collecting the right powers of H in the Higgs potential, in complete analogy to the Higgs mass above. We find

$$\begin{aligned}\mathscr{L}_{\text{self}} = &-\frac{m_H^2}{2v}\left[\left(1 - \frac{f_1 v^2}{2\Lambda^2} + \frac{2 f_2 v^4}{3\Lambda^2 m_H^2}\right) H^3 - \frac{2 f_1 v^2}{\Lambda^2 m_H^2} H\, \partial_\mu H\, \partial^\mu H\right] \\ &-\frac{m_H^2}{8v^2}\left[\left(1 - \frac{f_1 v^2}{\Lambda^2} + \frac{4 f_2 v^4}{\Lambda^2 m_H^2}\right) H^4 - \frac{4 f_1 v^2}{\Lambda^2 m_H^2} H^2\, \partial_\mu H \partial^\mu H\right] .\end{aligned} \quad (1.83)$$

This gives the Feynman rules

$$-i\frac{3 m_H^2}{v}\left(1 - \frac{f_1 v^2}{2\Lambda^2} + \frac{2 f_2 v^4}{3\Lambda^2 m_H^2} + \frac{2 f_1 v^2}{3\Lambda^2 m_H^2}\sum_{j<k}^{3}(p_j p_k)\right) \quad (1.84)$$

and

$$-i\frac{3m_H^2}{v^2}\left(1 - \frac{f_1 v^2}{\Lambda^2} + \frac{4 f_2 v^4}{\Lambda^2 m_H^2} + \frac{2 f_1 v^2}{3\Lambda^2 m_H^2}\sum_{j<k}^{4}(p_j p_k)\right) \quad (1.85)$$

From this discussion we see that in the Higgs sector the Higgs self couplings as well as the Higgs mass can be computed from the Higgs potential and depend on the operators we take into account. As mentioned before, in the Standard Model we use only the dimension-4 operators which appear in the renormalizable Lagrangian and which give us the Higgs mass and self couplings

$$\boxed{m_H^2 = 2\lambda v^2 = -2\mu^2} \quad \text{and} \quad \mathscr{L}_{\text{self}} = -\frac{m_H^2}{2v} H^3 - \frac{m_H^2}{8v^2} H^4 , \quad (1.86)$$

with $v = v_0 = 246\,\text{GeV}$. Given the measured Higgs mass the Higgs self coupling comes out as $\lambda \simeq 1/8$. When the Higgs sector becomes more complicated, not the existence but the form of such relations between masses and couplings will change. With this information we could now start computing Higgs observables at the LHC, but let us first see what else we can say about the Higgs potential from a theoretical point of view.

The Higgs self couplings computed in Eq. (1.85) are structurally different from their dimension-4 counter parts in that their higher dimensional modifications are momentum dependent. They are proportional to f_1, which means they arise from the operator $\mathscr{O}_1$ shown in Eq. (1.75). For large momenta such terms will cause problems, because the momentum in the numerator can exceed the suppression $1/\Lambda$. In our derivation part of this operator is absorbed into a *wave function renormalization*, ensuring the appropriate kinetic term of the Higgs scalar dictated by the definition of outgoing states in an interacting field theory. The question is if we can define a wave function renormalization which also removes the momentum dependent terms for the Higgs self couplings. If $\hat{H}$ is the original Higgs field as part of ϕ we can relate it to the physical Higgs field H using the generally parameterization

$$H = \left(1 + \frac{a_0 v^2}{\Lambda^2}\right)\hat{H} + \frac{a_1 v}{\Lambda^2}\hat{H}^2 + \frac{a_2}{\Lambda^2}\hat{H}^3 . \quad (1.87)$$

The powers of v and Λ are chosen such that all additional terms are related to a dimension-6 operator $(1/\Lambda^2)$ but the a_j have no mass dimension. Unlike for our first attempt we now include powers of the Higgs field in the wave function renormalization. Even higher terms in $\hat{H}$ would be allowed, but it will turn out that we do not need them. The canonically normalized kinetic term for the real scalar field H is

1.2 The Standard Model

$$\begin{aligned}
\mathcal{L}_{\text{kin}} &= \frac{1}{2}\partial_\mu H\, \partial^\mu H \\
&= \frac{1}{2}\partial_\mu\left[\left(1+\frac{a_0 v^2}{\Lambda^2}\right)\hat{H}+\frac{a_1 v}{\Lambda^2}\hat{H}^2+\frac{a_2}{\Lambda^2}\hat{H}^3\right] \\
&\quad \partial^\mu\left[\left(1+\frac{a_0 v^2}{\Lambda^2}\right)\hat{H}+\frac{a_1 v}{\Lambda^2}\hat{H}^2+\frac{a_2}{\Lambda^2}\hat{H}^3\right] \\
&= \left[1+\frac{a_0 v^2}{\Lambda^2}+\frac{2a_1 v}{\Lambda^2}\hat{H}+\frac{3a_2}{\Lambda^2}\hat{H}^2\right]^2\frac{\partial_\mu\hat{H}\partial^\mu\hat{H}}{2} \\
&= \left[1+\frac{2a_0 v^2}{\Lambda^2}+\frac{a_0^2 v^4}{\Lambda^4}+\left(1+\frac{a_0 v^2}{\Lambda^2}\right)\frac{4a_1 v}{\Lambda^2}\hat{H}\right. \\
&\quad \left. +\left(\frac{6a_2}{\Lambda^2}+\frac{2a_0 v^2}{\Lambda^2}\frac{3a_2}{\Lambda^2}+\frac{4a_1^2 v^2}{\Lambda^4}\right)\hat{H}^2+\mathcal{O}(\hat{H}^3)\right]\frac{\partial_\mu\hat{H}\partial^\mu\hat{H}}{2} \\
&= \left[1+\frac{2a_0 v^2}{\Lambda^2}+\frac{4a_1 v}{\Lambda^2}\hat{H}+\frac{6a_2}{\Lambda^2}\hat{H}^2+\mathcal{O}(\hat{H}^3)+\mathcal{O}\left(\frac{1}{\Lambda^4}\right)\right]\frac{\partial_\mu\hat{H}\partial^\mu\hat{H}}{2}.
\end{aligned}$$
(1.88)

Terms of higher mass dimension or including higher powers of $\hat{H}$ will only appear once we go to dimension-8 operators. This general form based on Eq. (1.87) we should use to remove all contributions from the dimension-6 operator $\mathcal{O}_1$ to the kinetic term of the Higgs field $\hat{H}$

$$\begin{aligned}
\mathcal{L}_{\text{kin}} &= \frac{\partial_\mu\hat{H}\partial^\mu\hat{H}}{2}+\frac{f_1}{\Lambda^2}\mathcal{O}_1 \\
&= \left[1+\frac{f_1}{\Lambda^2}\left(v^2+2v\hat{H}+\hat{H}^2\right)\right]\frac{\partial_\mu\hat{H}\partial^\mu\hat{H}}{2}.
\end{aligned}$$
(1.89)

Comparing Eqs. (1.88) and (1.89) we can identify the general pre-factors a_j with the specific f_1 from our dimension-6 ansatz. This gives us $a_0 = f_1/2$, $a_1 = f_1/2$, and $a_2 = f_1/6$. The wave function renormalization equation (1.87) then reads

$$\boxed{H = \left(1+\frac{f_1 v^2}{2\Lambda^2}\right)\hat{H}+\frac{f_1 v}{2\Lambda^2}\hat{H}^2+\frac{f_1}{6\Lambda^2}\hat{H}^3+\mathcal{O}(\hat{H}^4)+\mathcal{O}\left(\frac{1}{\Lambda^4}\right).}$$
(1.90)

In this alternative, *generalized canonical normalization* of the Higgs field we avoid any momentum dependent contributions to the Higgs self couplings. The prize we pay is that we have to apply the shift defined in Eq. (1.90) throughout the entire Standard Model Lagrangian. This means that the higher dimensional operator $\mathcal{O}_1$ leads to multiple Higgs couplings to any pair of massive gauge bosons or massive fermions. This observation is at the heart of the so-called strongly interacting light Higgs (SILH).

Because the wave function renormalization is not a physical observable; the two approaches are physically equivalent and predict the same physical observables to a precision $1/\Lambda^2$. Beyond dimension-6 operators they will become different. However, to properly define the external Higgs states in the interacting theory we need to ensure that the wave function and its commutators are properly defined. Looking at Eq. (1.90) we see that either the field H or the field $\hat{H}$ will induce a wave function normalization and hence a field commutator dependent on the field value.

We will see later that multiple Higgs couplings to other states are a serious challenge to the LHC, as are the momentum dependent terms in the Higgs self couplings. This means that for the interpretation of LHC data these dimension-6 interactions do not pose a serious problem.

1.2.2 Mexican Hat

To understand the well known picture of a Mexican hat which usually illustrates the Higgs mechanism we have to include Goldstone modes again. The Higgs doublets including all degrees of freedom are defined in Eq. (1.50). For our illustration it is sufficient to extend Eq. (1.71) by including all *neutral degrees of freedom* in the Higgs doublet

$$\phi = \frac{1}{\sqrt{2}} \begin{pmatrix} -w_2 - iw_1 \\ v + H + iw_3 \end{pmatrix} \sim \frac{1}{\sqrt{2}} \begin{pmatrix} 0 \\ v + H + iw_3 \end{pmatrix}. \quad (1.91)$$

In this approximation we can again compute the potential defined in Eq. (1.77), but omitting the dimension-6 terms

$$V = \mu^2(\phi^\dagger \phi) + \lambda(\phi^\dagger \phi)^2$$
$$= \frac{\mu^2}{2}\left((v+H)^2 + w_3^2\right) + \frac{\lambda}{4}\left((v+H)^2 + w_3^2\right)^2. \quad (1.92)$$

In the *unbroken phase*, i.e. in the absence of a vacuum expectation value or for $v = 0$ the potential reads

$$V = \frac{\mu^2}{2}\left(H^2 + w_3^2\right) + \frac{\lambda}{4}\left(H^2 + w_3^2\right)^2. \quad (1.93)$$

This form depends only on a radius in the two-dimensional plane formed by the Higgs and Goldstone field values $|\phi| = \sqrt{H^2 + w_3^2}/\sqrt{2}$. Its first derivative with respect to $|\phi|$ is $V' = \mu^2|\phi| + 3\lambda|\phi|^3/4$, so there exists only a minimum at $|\phi| = 0$

1.2 The Standard Model

or equivalently at $H = w_3 = 0$, where the potential becomes $V = 0$. Towards larger field values the potential increases first proportional to $|\phi|^2$ and finally proportional to $|\phi|^4$, always rotationally symmetric in the H-w_3 plane.

In the *broken phase* with $v = 246$ GeV the form of the potential changes. We can follow exactly the derivation following Eq. (1.77) where the minimum condition in the complex plane is $|\phi| = v/\sqrt{2}$. This minimum is rotationally symmetric; on the real H axis it requires $H = \pm v$ while on the imaginary w_3 axis it appears at $w_3 = \pm iv$. For large field values $|\phi| \gg v$ the potential rapidly increases proportional to $|\phi|^4$.

To determine the masses of the particles corresponding to the Higgs and Goldstone fields we choose one point on the degenerate vacuum circle, e.g. $\phi = v$. There we can compute the *mass matrix* from the second derivatives of the potential at the minimum

$$\boxed{\left(\mathcal{M}_H^2\right)_{jk} = \left.\frac{\partial^2 V}{\partial^2 \{H w_3\}}\right|_{\text{minimum}}.} \qquad (1.94)$$

For real fields, where the mass term is proportional to $\mathcal{L} \supset -V \supset -m^2 H^2/2$ there is no factor $1/2$ in this relation. Inserting the full form of the neutral potential given in Eq. (1.92) we first find

$$\frac{\partial V}{\partial H} = \mu^2(v + H) + \frac{\lambda}{2}\left((v + H)^2 + w_3^2\right) 2(v + H)$$

$$= \mu^2(v + H) + \lambda(v + H)\left((v + H)^2 + w_3^2\right)$$

$$\left.\frac{\partial^2 V}{\partial H^2}\right|_{\text{minimum}} = \left.\mu^2 + \lambda\left((v + H)^2 + w_3^2\right) + 2\lambda(v + H)^2\right|_{\text{minimum}}$$

$$= \mu^2 + 3\lambda v^2 = 2\lambda v^2 . \qquad (1.95)$$

In the last step we use the relation $\mu^2 = -\lambda v^2$ at the minimum. This second derivative is identical to the known Higgs mass, but to be save we still compute the complete mass matrix and determine the mass eigenvalues. The second diagonal entry in the neutral Higgs–Goldstone mass matrix is

$$\frac{\partial V}{\partial w_3} = \mu^2 w_3 + \frac{\lambda}{2}\left((v + H)^2 + w_3^2\right) 2w_3$$

$$= \mu^2 w_3 + \lambda w_3\left((v + H)^2 + w_3^2\right)$$

$$\left.\frac{\partial^2 V}{\partial w_3^2}\right|_{\text{minimum}} = \left.\mu^2 + \lambda\left((v + H)^2 + w_3^2\right) + 2\lambda w_3^2\right|_{\text{minimum}}$$

$$= \mu^2 + \lambda v^2 = 0 . \qquad (1.96)$$

The off-diagonal entry of the symmetric mass matrix is

$$\left.\frac{\partial^2 V}{\partial H \partial w_3}\right|_{\text{minimum}} = \left.\frac{\partial}{\partial w_3}\left[\mu^2(v+H) + \lambda(v+H)\left((v+H)^2 + w_3^2\right)\right]\right|_{\text{minimum}}$$

$$= \left.2\lambda(v+H)w_3\right|_{\text{minimum}} = 0 \qquad (1.97)$$

Putting all this together we find that the Higgs and Goldstone basis is identical with the mass eigenstates of the symmetric mass matrix

$$\boxed{\mathcal{M}_H^2 = \begin{pmatrix} 2\lambda v^2 & 0 \\ 0 & 0 \end{pmatrix}} \qquad (1.98)$$

If we fix our vacuum to the positive real axis $\phi = v$ the Higgs mode living on the real axis is massive while the Goldstone mode living orthogonally in the direction of the *flat potential valley* is massless. The fact that we had to fix the vacuum to the real axis is a direct consequence of our linear gauge choice in Eq. (1.49). In the unitary form of Eq. (1.42) the Goldstone mode is aligned with the potential valley over the entire Higgs–Goldstone field plane.

This result confirms Goldstone's theorem, at least in the first step. If we consider the Goldstone modes simply additional scalars they are massless. Historically, this was the big problem with spontaneous symmetry breaking, because such massless scalars with weak charge would have been observed. From Eq. (1.45) we know that after breaking a local gauge symmetry the Goldstones become part of the massive gauge fields, and their mass is not a physical parameter.

1.2.3 Unitarity

If we want to compute transition amplitudes at very high energies the Goldstone modes become very useful. In the V rest frame we can write the three polarization vectors of a massive gauge boson as

$$\epsilon_{T,1}^\mu = \begin{pmatrix} 0 \\ 1 \\ 0 \\ 0 \end{pmatrix} \quad \epsilon_{T,2}^\mu = \begin{pmatrix} 0 \\ 0 \\ 1 \\ 0 \end{pmatrix} \quad \epsilon_L^\mu = \begin{pmatrix} 0 \\ 0 \\ 0 \\ 1 \end{pmatrix}. \qquad (1.99)$$

If we boost V into the z direction, giving it a four-momentum $p^\mu = (E, 0, 0, |\vec{p}|)$, the polarization vectors become

1.2 The Standard Model

$$\epsilon_{T,1}^\mu = \begin{pmatrix} 0 \\ 1 \\ 0 \\ 0 \end{pmatrix} \quad \epsilon_{T,2}^\mu = \begin{pmatrix} 0 \\ 0 \\ 1 \\ 0 \end{pmatrix} \quad \epsilon_L^\mu = \frac{1}{m_V} \begin{pmatrix} |\vec{p}| \\ 0 \\ 0 \\ E \end{pmatrix} \xrightarrow{E \gg m_V} \frac{1}{m_V} \begin{pmatrix} |\vec{p}| \\ 0 \\ 0 \\ |\vec{p}| \end{pmatrix} \equiv \frac{1}{m_V} p^\mu .$$
(1.100)

Very relativistic gauge bosons are dominated by their longitudinal polarization $|\vec{\epsilon}_L| \sim E/m_V \gg 1$. This longitudinal degree of freedom is precisely the Goldstone boson, so at high energies we can approximate the complicated vector bosons $Z, W^\pm$ as scalar Goldstone bosons $w_0, w_\pm$. The problem which gauge–dependent mass value ξm_V to assign to the Goldstone fields does not occur, because in the high energy limit we automatically assume $m_V \to 0$. This simplification comes in handy for example when we talk about unitarity as a constraint on the Higgs sector. This relation between Goldstones and gauge bosons at very high energies is called the *equivalence theorem*.

Based on the equivalence theorem we can compute the amplitude for $W^+W^- \to W^+W^-$ scattering at very high energies ($E \gg m_W$) in terms of scalar Goldstones bosons. Three diagrams contribute to this processes: a four-point vertex, the s-channel Higgs exchange and the t-channel Higgs exchange:

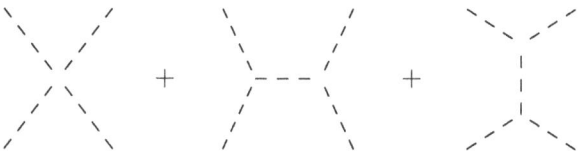

To confirm these Feynman diagrams and to compute the corresponding amplitude we need some basic Feynman rules, for example the Goldstone couplings to the Higgs boson and the four-Goldstone couplings. We start with the Higgs doublet, again including the Goldstone modes in analogy to Sect. 1.2.2

$$\phi = \frac{1}{\sqrt{2}} \begin{pmatrix} -w_2 - iw_1 \\ v + H + iw_3 \end{pmatrix}$$

$$\phi^\dagger \phi = \frac{1}{2} \left(w_1^2 + w_2^2 + w_3^2 + (v+H)^2 \right)$$

$$(\phi^\dagger \phi)^2 = \frac{1}{4} \left(\sum_i w_i^2 \right)^2 + \frac{1}{2}(v+H)^2 \sum_i w_i^2 + \frac{1}{4}(v+H)^4$$

$$= \frac{1}{4} \left(\sum_i w_i^2 \right)^2 + \left(vH + \frac{v^2}{2} + \frac{H^2}{2} \right) \sum_i w_i^2 + \mathcal{O}(w^0) . \quad (1.101)$$

In the last step we neglect all terms without the Goldstone fields. Note that there are no three-Goldstone vertices, only triple dimension-four couplings including the Higgs and a coupling factor v. Only keeping the relevant terms contributing to the four-Goldstone and Higgs–Goldstone–Goldstone couplings at dimension four the potential becomes

$$V = \mu^2 |\phi|^2 + \lambda |\phi|^4 \supset \lambda |\phi|^4 = \frac{m_H^2}{2v^2} |\phi|^4$$

$$= \frac{m_H^2}{2v^2} \left[\frac{1}{4} \left(\sum_i w_i^2 \right)^2 + vH \sum_i w_i^2 + \mathcal{O}(w^0) \right]$$

$$= \frac{m_H^2}{8v^2} \left(\sum_i w_i^2 \right)^2 + \frac{m_H^2}{2v} H \sum_i w_i^2 + \mathcal{O}(w^0) \,. \quad (1.102)$$

Focussing on the scattering of charged Goldstones $w_\pm w_\pm \to w_\pm w_\pm$ we use the corresponding fields $w_\pm = (w_1 \pm iw_2)/\sqrt{2}$ following Eq. (1.11). They appear in the above expression as $w_1^2 + w_2^2 = 2w_+ w_-$, so we find the terms

$$V \supset \frac{m_H^2}{2v^2} w_+ w_- w_+ w_- + \frac{m_H^2}{v} H w_+ w_- \,, \quad (1.103)$$

which fix the two Feynman rules we need. Linking the Lagrangian to the Feynman rule for the quartic coupling involves one complication: for each positively charged Goldstone in the vertex there are two ways we can identify them with the Lagrangian fields. In addition, there are also two choices to identify the two negatively charged Goldstones, which implies an additional combinatorial factor 4 in the Feynman rule. Including a common factor $(-i)$ the two Feynman rules then become $-2im_H^2/v^2$ and $-im_H^2/v$.

The potential in Eq. (1.102) has an interesting feature which has recently lead to some discussions on the computation of Higgs decays to two photons. The question is if the one-loop $H\gamma\gamma$ amplitude mediated by a closed W boson loop should vanish in the limit $m_W \to 0$. This is indeed the case for a closed fermion loop contributing to the same process through the Yukawa coupling. The W loop, in contrast, consists of transverse and longitudinal W modes. The latter we can describe in terms of Goldstone modes which couple to the external Higgs field following Eq. (1.102). Because m_W never appears in this potential there is no reason why the Goldstone modes should decouple, and indeed they do not.

The amplitude for the Goldstone scattering process is given in terms of the Mandelstam variables s and t which describe the momentum flow p^2 through the two Higgs propagators and which we will properly introduce in Sect. 2.1.1

1.2 The Standard Model

$$A = i\,\frac{-2im_H^2}{v^2} + \left(\frac{-im_H^2}{v}\right)^2 \frac{i}{s-m_H^2} + \left(\frac{-im_H^2}{v}\right)^2 \frac{i}{t-m_H^2}$$

$$= \frac{m_H^2}{v^2}\left[2 + \frac{m_H^2}{s-m_H^2} + \frac{m_H^2}{t-m_H^2}\right]. \qquad (1.104)$$

The factor i which ensures that the amplitude is real appears between the transition rate computed from the Feynman rules and the actual transition amplitude.

For this process we want to test the unitarity of the S matrix, which we write in terms of a transition amplitude $S = \mathbb{1} + iA$. The S matrix should be unitary to conserve probability

$$\mathbb{1} \stackrel{!}{=} S^\dagger S = (\mathbb{1} - iA^\dagger)(\mathbb{1} + iA) = \mathbb{1} + i(A - A^\dagger) + A^\dagger A \;\Leftrightarrow\; A^\dagger A = -i(A - A^\dagger). \qquad (1.105)$$

If we sandwich $(A - A^\dagger)$ between *identical asymptotically free fields*, which means that we are looking at forward scattering with a scattering angle $\theta \to 0$, we find in the high energy limit or for massless external particles

$$-i\langle j|A - A^{*T}|j\rangle = -i\langle j|A - A^*|j\rangle = 2\,\text{Im}\,A(\theta = 0)$$

$$\Rightarrow \quad \boxed{\sigma \equiv \frac{1}{2s}\langle j|A^\dagger A|j\rangle = \frac{1}{s}\,\text{Im}\,A(\theta = 0)}. \qquad (1.106)$$

Assuming that our Lagrangian is hermitian this imaginary part corresponds only to absorptive terms in the scattering amplitude. This is the usual formulation of the *optical theorem* reflecting unitarity in terms of the transition amplitude A.

To include the dependence on the scattering angle θ we decompose the transition amplitude into partial waves

$$A = 16\pi \sum_{l=0}^{\infty}(2l+1)\,P_l(\cos\theta)\,a_l \quad \text{with} \quad \int_{-1}^{1} dx\, P_l(x) P_{l'}(x) = \frac{2}{2l+1}\delta_{ll'}, \qquad (1.107)$$

ordered by the orbital angular momentum l. P_l are the Legendre polynomials of the scattering angle θ, which obey an orthogonality condition. The scattering cross section including all prefactors and the phase space integration is then given by

$$\sigma = \int d\Omega\, \frac{|A|^2}{64\pi^2 s}$$

$$= \frac{(16\pi)^2}{64\pi^2 s}\,2\pi \int_{-1}^{1} d\cos\theta \sum_l \sum_{l'} (2l+1)(2l'+1)\,a_l a_{l'}^*\,P_l(\cos\theta) P_{l'}(\cos\theta)$$

$$= \frac{8\pi}{s}\sum_l 2(2l+1)\,|a_l|^2 = \frac{16\pi}{s}\sum_l (2l+1)\,|a_l|^2. \qquad (1.108)$$

The relation between the integral over the scattering angle θ and the Mandelstam variable t we will discuss in more detail in Sect. 2.1.1. Applied to each term in the partial wave expansion the optical theorem requires

$$\frac{16\pi}{s}(2l+1)|a_l|^2 = \frac{1}{s}\text{Im}A(\theta=0)\Big|_t$$
$$= \frac{1}{s}16\pi(2l+1)\text{Im}\,a_l \qquad \Leftrightarrow \qquad |a_l|^2 \stackrel{!}{=} \text{Im}\,a_l\,, \quad (1.109)$$

using $P_l(\cos\theta=1)=1$. This condition we can rewrite as

$$(\text{Re}\,a_l)^2 + \left(\text{Im}\,a_l - \frac{1}{2}\right)^2 = \frac{1}{4} \qquad \Rightarrow \qquad \boxed{|\text{Re}\,a_l| < \frac{1}{2}}\,, \quad (1.110)$$

once we recognize that the condition on $\text{Im}\,a_l$ and on $\text{Re}\,a_l$ is a circle around $a_l = (0,1/2)$ with radius $1/2$.

It is important to remember that in the above argument we have formulated the constraint for each term in the sum over the Legendre polynomials. Mathematically, this is well justified, but of course there might be physics effects which lead to a systematic cancellation between different terms. This is why the constraint we compute is referred to as *perturbative unitary*. For Goldstone scattering we compute the supposedly leading first term in the partial wave expansion from the amplitude

$$a_0 = \frac{1}{16\pi s}\int_{-s}^0 dt\,|A| = \frac{1}{16\pi s}\int_{-s}^0 dt\,\frac{m_H^2}{v^2}\left[2 + \frac{m_H^2}{s-m_H^2} + \frac{m_H^2}{t-m_H^2}\right]$$
$$= \frac{m_H^2}{16\pi v^2}\left[2 + \frac{m_H^2}{s-m_H^2} - \frac{m_H^2}{s}\log\left(1+\frac{s}{m_H^2}\right)\right]$$
$$= \frac{m_H^2}{16\pi v^2}\left[2 + \mathscr{O}\left(\frac{m_H^2}{s}\right)\right]\,. \qquad (1.111)$$

In the high energy limit $s \gg m_H^2$ this translates into an upper limit on the Higgs mass which in Eq. (1.104) enters as the Goldstone coupling in the numerator

$$\frac{m_H^2}{8\pi v^2} < \frac{1}{2} \qquad \Leftrightarrow \qquad \boxed{m_H^2 < 4\pi v^2 = (870\,\text{GeV})^2}\,. \qquad (1.112)$$

This is the maximum value of m_H consistent with perturbative unitarity for $WW \to WW$ scattering. Replacing the Higgs mass by the self coupling we can formulate the same constraint as $\lambda < 2\pi$. The leading term in our analysis of perturbative unitarity is simply the size of the four-Goldstone coupling, the two Higgs diagrams are sub-leading in m_H^2/s. This means that perturbative unitarity seriously probes the

1.2 The Standard Model

limitations of perturbation theory, so we should include higher order effects as well as higher dimensional operators to get a reliable numerical prediction in the range of $m_H \lesssim 1\,\text{TeV}$.

Of course, if we limit s to a finite value this bound changes, and we can compute a maximum scale s_{max} which leaves $WW \to WW$ perturbatively unitary for fixed m_H: for $m_H \lesssim v$ this typically becomes $\sqrt{s_{\text{max}}} \sim 1.2\,\text{TeV}$. This number is one of the motivations to build the LHC as a high energy collider with a partonic center–of–mass energy in the few-TeV range. If something had gone wrong with the Standard–Model–like Higgs sector we could have expected to see something else curing unitarity around the TeV scale. A Higgs boson too heavy to be produced at the LHC would essentially not been able to function as a Higgs boson.

In many discussions of unitarity and the Higgs sector we explain the role of the Higgs boson in the unitarization of WW scattering as a cancellation of the leading divergences through virtual Higgs exchange. Clearly, this is not what we see in our argument. Nevertheless, both answers are correct, because the separation of a gauge–invariant transition amplitude it gauge dependent. Our Higgs–Goldstone gauge assumes the existence of a Higgs boson when we include the coupling strength m_H in the Feynman rules for the Goldstones. With that assumption we are no longer allowed to test the assumption that the Higgs not be there. All we can do is decouple the Higgs by making it heavier, which gives us the limit shown in Eq. (1.112). In the unitary gauge, where the gauge bosons are massive and the Goldstones are eaten, the Higgs is an additional state which we can remove at the expense of ruining renormalizability. At tree level this gauge gives us a cancellation of the leading divergences from gauge boson exchange with the help of the Higgs diagrams. Because the transition amplitude is gauge invariant the limit on the Higgs mass will be identical.

One last but very important comment we need to make: this unitarity argument only works if the WWH coupling is exactly what it should be. While perturbative unitarity only gives us a fairly rough upper limit on m_H, it also uniquely fixes g_{WWH} to its Standard Model value. Any sizeable deviation from this value again means new physics appearing at the latest around the mass scales of Eq. (1.112).

Looking at processes like $WW \to f\bar{f}$ or $WW \to WWH$ or $WW \to HHH$ we can fix *all Higgs couplings* in the Standard Model, including g_{Hff}, g_{HHH}, g_{HHHH}, using exactly the same argument. The most important result of the unitarity test is probably not the upper bound on the Higgs mass, but the underlying assumption that the unitarity test only works in the presence of one Higgs boson if all Higgs couplings look exactly as predicted by the Standard Model.

1.2.4 Renormalization Group Analysis

The unitarity condition derived above is the first of a series of theoretical constraints which we can derive as self consistency conditions on a Higgs boson turning the Standard Model with its particle masses into a renormalizable theory. We can derive two additional theoretical constraints from the renormalization group equation of the Higgs potential, specifically from the *renormalization scale dependence* of the self coupling $\lambda(Q^2)$. Such a scale dependence arises automatically when we encounter ultraviolet divergences and absorb the $1/\epsilon$ poles into a minimal counter term. We will discuss this running of couplings in more detail in Sect. 2.2.1 focussing on the running QCD coupling α_s. In the case of a running quartic Higgs coupling λ the one-loop s, t and u-channel diagrams only depending on λ itself are

Skipping the calculation we quote the complete renormalization group equation including diagrams with the Higgs boson, the top quark and the weak gauge bosons inside the loops

$$\frac{d\lambda}{d\log Q^2} = \frac{1}{16\pi^2}\left[12\lambda^2 + 6\lambda y_t^2 - 3y_t^4 - \frac{3}{2}\lambda\left(3g_2^2 + g_1^2\right) + \frac{3}{16}\left(2g_2^4 + (g_2^2 + g_1^2)^2\right)\right], \tag{1.113}$$

with $y_t = \sqrt{2}m_t/v$. This formula will be the basis of the discussion in this section.

The first regime we study is where the Higgs self coupling λ becomes strong. Fixed order perturbation theory as we use it in the unitarity argument runs into problems in this regime and the renormalization group equation is the appropriate tool to describe it. If we reside in a somewhat strongly interacting regime the leading term in Eq. (1.113) reads

$$\frac{d\lambda}{d\log Q^2} = \frac{1}{2Q}\frac{d\lambda}{dQ} = \frac{1}{16\pi^2}\,12\lambda^2 + \mathcal{O}(\lambda) = \frac{3}{4\pi^2}\lambda^2 + \mathcal{O}(\lambda)\,. \tag{1.114}$$

Because of the positive sign on the right hand side the quartic coupling will become stronger and eventually diverge for large scales Q^2. Obviously, this divergence should not happen in a physical model and will give us a constraint on the maximum value of λ allowed. The approximate renormalization group equation we can solve by replacing $\lambda = g^{-1}$

$$\frac{d\lambda}{d\log Q^2} = \frac{d}{d\log Q^2}\frac{1}{g} = -\frac{1}{g^2}\frac{dg}{d\log Q^2} \stackrel{!}{=} \frac{3}{4\pi^2}\frac{1}{g^2}$$

$$\Leftrightarrow \quad \frac{dg}{d\log Q^2} = -\frac{3}{4\pi^2} \quad \Leftrightarrow \quad g(Q^2) = -\frac{3}{4\pi^2}\log Q^2 + C\,. \tag{1.115}$$

The boundary condition $\lambda(Q^2 = v^2) = \lambda_0$ fixes the integration constant C

1.2 The Standard Model

$$g_0 = \frac{1}{\lambda_0} = -\frac{3}{4\pi^2} \log v^2 + C \quad \Leftrightarrow \quad C = g_0 + \frac{3}{4\pi^2} \log v^2$$

$$\Rightarrow \quad g(Q^2) = -\frac{3}{4\pi^2} \log Q^2 + g_0 + \frac{3}{4\pi^2} \log v^2 = -\frac{3}{4\pi^2} \log \frac{Q^2}{v^2} + g_0$$

$$\Leftrightarrow \quad \boxed{\lambda(Q^2) = \left[-\frac{3}{4\pi^2} \log \frac{Q^2}{v^2} + \frac{1}{\lambda_0}\right]^{-1} = \lambda_0 \left[1 - \frac{3}{4\pi^2}\lambda_0 \log \frac{Q^2}{v^2}\right]^{-1}}.$$
(1.116)

We start from scales $Q \sim v$ where the expression in brackets is close to one. Moving towards larger scales the denominator becomes smaller until λ hits a pole at the critical value Q_{pole}

$$1 - \frac{3}{4\pi^2}\lambda_0 \log \frac{Q_{\text{pole}}^2}{v^2} \overset{!}{=} 0 \quad \Leftrightarrow \quad \frac{3}{4\pi^2}\lambda_0 \log \frac{Q_{\text{pole}}^2}{v^2} = 1$$

$$\Leftrightarrow \quad \log \frac{Q_{\text{pole}}^2}{v^2} = \frac{4\pi^2}{3\lambda_0}$$

$$\Leftrightarrow \quad Q_{\text{pole}} = v \exp \frac{2\pi^2}{3\lambda_0} = v \exp \frac{4\pi^2 v^2}{3m_H^2}$$
(1.117)

Such a pole is called a *Landau pole* and gives us a maximum scale beyond which we cannot rely on our perturbative theory to work. In the upper line of Fig. 1.2 we show Q_{pole} versus the Higgs mass, approximately computed in Eq. (1.117). As a function of the Higgs mass Q_{pole} gives the maximum scale were our theory is valid, so we have to reside below and to the left of the upper line in Fig. 1.2. Turning the argument around, for given Q_{pole} we can read off the maximum allowed Higgs mass which in the limit of large cutoff values around the Planck scale 10^{19} GeV becomes $m_H \lesssim 180$ GeV, in good agreement with the observed Higgs mass around 125 GeV.

This limit is often referred to as the *triviality bound*, which at first glance is precisely not what this theory is—trivial or non–interacting. The name originates from the fact that if we want our Higgs potential to be perturbative at all scales, the coupling λ can only be zero everywhere. Any finite coupling will hit a Landau pole at some scale. Such a theory with zero interaction is called trivial.

After looking at the ultraviolet regime we can go back to the full renormalization group equation of Eq. (1.113) and ask a completely different question: if the Higgs coupling λ runs as a function of the scale, how long will $\lambda > 0$ ensure that our Higgs potential is bounded from below?

This bound is called the *stability bound*. On the right hand side of Eq. (1.113) there are two terms with a negative sign which in principle drive λ through zero. One of them vanishes for small $\lambda \sim 0$, so we can neglect it under the assumption

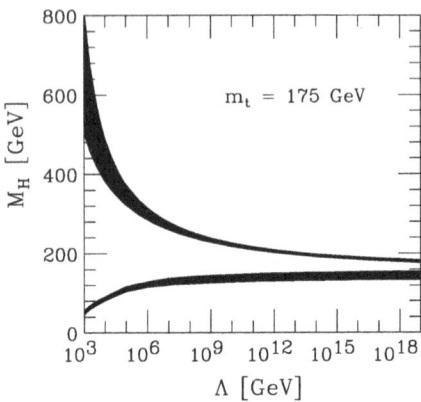

Fig. 1.2 Triviality or Landau pole (upper) and stability bounds (lower) for the Standard Model Higgs boson in the $m_H - Q$ plane. Similar arguments first appeared in Ref. [12], the actual scale dependence can be seen in Refs. [28, 37]

that we only study very weakly interacting Higgs sectors. In the small-λ regime we therefore encounter two finite competing terms

$$\frac{d\lambda}{d\log Q^2} = \frac{1}{16\pi^2}\left[-3\frac{4m_t^4}{v^4} + \frac{3}{16}\left(2g_2^4 + (g_2^2 + g_1^2)^2\right) + \mathcal{O}(\lambda)\right]$$

$$\Leftrightarrow \quad \lambda(Q^2) \simeq \lambda(v^2) + \frac{1}{16\pi^2}\left[-\frac{12m_t^4}{v^4} + \frac{3}{16}\left(2g_2^4 + (g_2^2 + g_1^2)^2\right)\right]\log\frac{Q^2}{v^2}.$$
(1.118)

The usual boundary condition at $\lambda(v^2) = m_H^2/(2v^2)$ is the starting point from which the top Yukawa coupling drives λ through zero. This second critical scale $\lambda(Q^2_{\text{stable}}) = 0$ also depends on the Higgs mass m_H. The second (smaller) contribution from the weak gauge coupling ameliorates this behavior. The condition for a zero Higgs self coupling is

$$\lambda(v^2) = \frac{m_H^2}{2v^2} \stackrel{!}{=} -\frac{1}{16\pi^2}\left[-\frac{12m_t^4}{v^4} + \frac{3}{16}\left(2g_2^4 + (g_2^2 + g_1^2)^2\right)\right]\log\frac{Q^2_{\text{stable}}}{v^2}$$

$$\Leftrightarrow \quad \frac{m_H^2}{v^2} = \frac{1}{8\pi^2}\left[\frac{12m_t^4}{v^4} - \frac{3}{16}\left(2g_2^4 + (g_2^2 + g_1^2)^2\right)\right]\log\frac{Q^2_{\text{stable}}}{v^2}$$

$$\Leftrightarrow \quad m_H = \begin{cases} 70\,\text{GeV} & \text{for} \quad Q_{\text{stable}} = 10^3\,\text{GeV} \\ 130\,\text{GeV} & \text{for} \quad Q_{\text{stable}} = 10^{16}\,\text{GeV} \end{cases}.$$
(1.119)

From Eq. (1.118) we see that only for energy scales below $Q_{\text{stable}}(m_H)$ the Higgs potential is bounded from below and our vacuum stable. For a given maximum validity scale Q_{stable} this stability bound translates into a minimum Higgs mass balancing the negative slope in Eq. (1.118) for which our theory is then well defined. In Fig. 1.2 we show Q_{stable} as the lower curve, above which our consistent theory has to reside.

1.2 The Standard Model

Our discussion of the triviality bound and of the stability of our vacuum has a weak spot: it follows from Eq. (1.113) and assumes that only the renormalizable couplings enter the behavior of the Higgs vacuum at large energy scales. On the other hand, if we start from low energies we should at some point reach energy scales where higher-dimensional operators enter the picture. Even in the Standard Model such operators get induced by loops, and non-perturbative studies indicate that they stabilize the Higgs potential and prevent the sign change in λ. If that should be true it would mean that our perturbative approximation only considering the leading renormalizable operators does not allow us to extrapolate to energy scales beyond 10^{10} GeV or more and that vacuum stability is simply not an issue.

Summarizing what we know about the Higgs mass in the Standard Model we already have indirect experimental as well as theoretical constraints on this otherwise free parameter in the Higgs sector.

Strictly in the Standard Model, electroweak precision data points to the mass range $m_H \lesssim 200$ GeV. This means at the LHC we were either looking for a light Higgs boson or we should have expected a drastic modifications of our Standard Model, altering this picture significantly. If a discovery of a Higgs boson around $m_H = 125$ GeV means good or bad news is in the eye of the beholder. Certainly, at this mass we do not have to expect huge deviations from the Standard Model motivated by the Higgs sector.

From the renormalization group we have two pieces of information on the Higgs mass, again in the renormalizable Standard Model: the Landau pole or triviality bound gives an upper limit on m_H as a function of the cutoff scale. Vacuum stability gives a lower bound on m_H as a function of the cutoff scale. Running both cutoff scales towards the Planck mass $Q_{\text{pole}}, Q_{\text{stable}} \to 10^{19}$ GeV, we see in Fig. 1.2 that only Higgs mass values around $m_H = 130 \cdots 180$ GeV are allowed for a truly fundamental and stable Standard Model. Above this parameter range the Higgs sector interacts too strongly and soon develops a Landau pole, while below this Higgs mass window the Higgs sector is too weakly interacting to give a stable vacuum. The exact numbers including renormalization group running at two loops gives the stability condition in two forms [10]

$$m_H > 129.6\,\text{GeV} + 2\,(m_t - 173.35\,\text{GeV}) - \frac{\alpha_s(m_Z) - 0.1184}{0.0014}\,\text{GeV} \pm 0.3\,\text{GeV}$$

$$m_t < (171.36 \pm 0.46)\,\text{GeV}\,, \tag{1.120}$$

where the error bars have to be taken with a grain of salt, if we follow the strict rules about combining experimental and theory uncertainties described in Sect. 3.4. If vacuum stability should really be a problem, the observed value of the Higgs mass around 125 GeV is at the edge of the vacuum stability bound, again leaving everything open.

1.2.5 Top–Higgs Renormalization Group

The two critical scales in the running of the Higgs self coupling are not the only interesting feature of the renormalization group equation (1.113). If we limit ourselves to only the top and Higgs sector it reads

$$\frac{d\lambda}{d\log Q^2} = \frac{1}{16\pi^2}\left(12\lambda^2 + 6\lambda y_t^2 - 3y_t^4\right). \qquad (1.121)$$

The definition of a fixed point λ_* is that the function $\lambda(Q^2)$ has to stick to this value λ_* once it reaches it. An attractive fixed point is a value λ_* which is automatically approached when the argument Q^2 reaches the corresponding infrared or ultraviolet regime. If we assume that the Higgs self coupling is closely related to the Higgs mass, $m_H = \sqrt{2\lambda}v$, a fixed point really tells us something about the observable Higgs mass.

Including all couplings in Eq. (1.113) we see that there is no obvious fixed point of λ for either large (UV) or small (IR) scales Q. The solution of the RGE for λ alone we compute in Eq. (1.116). In the infrared the scalar four point couplings as well as its derivative vanish,

$$\lim_{\log Q^2 \to -\infty} \lambda(Q^2) = \lambda_* = 0 \qquad \lim_{\log Q^2 \to -\infty} \frac{d\lambda}{d\log Q^2} = \lim_{\log Q^2 \to -\infty} \frac{3\lambda^2}{4\pi^2} = 0. \qquad (1.122)$$

This means that in the infrared the scalar self coupling alone would approach zero. Such a vanishing fixed point $\lambda_* = 0$ is called a *Gaussian fixed point*. Obviously, higher powers of λ in the RGE will not change this infrared pattern. The triviality bound is the first example of an attractive IR fixed point in renormalization group running.

The question is if we can find something interesting when we go beyond the pure Higgs system. The largest electroweak coupling is the top Yukawa, already included in Eq. (1.121). In complete analogy we can compute the Higgs loop corrections to the running of the top Yukawa coupling

$$\frac{d y_t^2}{d\log Q^2} = \frac{9}{32\pi^2} y_t^4. \qquad (1.123)$$

Again, the top Yukawa is closely related to the top mass, $m_t = y_t v/\sqrt{2}$. The top Yukawa also has an attractive Gaussian IR fixed point at $y_{t,*} = 0$, but this is not what we are after. Instead, we define the *ratio of the two couplings* as

$$R = \frac{\lambda}{y_t^2} \qquad (1.124)$$

1.2 The Standard Model

and compute the running of that ratio,

$$\begin{aligned}
\frac{dR}{d\log Q^2} &= \frac{d\lambda}{d\log Q^2}\frac{1}{y_t^2} + \lambda\frac{(-1)}{y_t^4}\frac{dy_t^2}{d\log Q^2} \\
&= \frac{1}{16\pi^2 y_t^2}\left(12\lambda^2 + 6\lambda y_t^2 - 3y_t^4\right) - \frac{9\lambda}{32\pi^2} \\
&= \frac{1}{16\pi^2}\left(12\lambda R + \frac{3}{2}\lambda - 3y_t^2\right) \\
&= \frac{\lambda}{16\pi^2}\left(12R + \frac{3}{2} - 3\frac{1}{R}\right) \\
&= \frac{3\lambda}{32\pi^2 R}\left(8R^2 + R - 2\right) \overset{!}{=} 0 \quad\Leftrightarrow\quad \boxed{R_* = \frac{\sqrt{65}-1}{16} \simeq 0.44}.
\end{aligned}$$
(1.125)

This is not a fixed point in any of the two couplings involved, but a *fixed point in the ratio* of the two. It is broken by the gauge couplings, most notably by the α_s correction to the running top Yukawa or top mass. With the non–Gaussian IR fixed point for the coupling ratio R as well as the Gaussian fixed points for the individual couplings the question is how they are approached. It turns out that the system first approaches the fixed-point region for R and on that line approaches the double zero-coupling limit. In the far infrared this predicts a ratio of the top mass to the Higgs mass of

$$\left.\frac{\lambda}{y_t^2}\right. = \left.\frac{m_H^2}{2v^2}\frac{v^2}{2m_t^2}\right|_{IR} = \left.\frac{m_H^2}{4m_t^2}\right|_{IR} = 0.44 \quad\Leftrightarrow\quad \left.\frac{m_H}{m_t}\right|_{IR} = 1.33$$
(1.126)

At first sight this is not in good agreement with the Standard Model value. On the other hand, the top and Higgs masses we usually quote are not running masses in the far infrared. If the analysis leading to Eq. (1.126) is done properly, including gravitational effects in the ultraviolet, it predicts a pole mass of $m_t = 172\,\text{GeV}$ and a Higgs mass $m_H = 126\,\text{GeV}$. Puzzling.

While the Wetterich fixed point in Eq. (1.125) is the most obvious to discuss in these lecture notes we should also mention that there exists a prototypical fixed point of this kind: the *Pendleton-Ross fixed point* relates the strong coupling and the top mass α_s/y_t^2 in the infrared. It is strictly speaking only valid for non–perturbatively large strong coupling, making it hard to predict a value for the top mass. Together with a more detailed analysis of the actual running the link to the strong couplings predicts a top pole mass in the 100–200 GeV range. What does the obvious quantitative applicability of these fixed points really mean? They suggest that our Standard Model is rooted at high scales and our weak-scale infrared parameters are simply fixed by a renormalization group analysis. Unfortunately, infrared fixed points imply that during the renormalization group evolution we forget all of the high–scale physics. Someone up at high scales wants to know that he/she is in charge, but does not want to reveal any additional information. If we do not find any signature of new physics at the LHC we will have to study such predictions and extract the underlying high–scale structures from the small effect around general fixed point features.

1.2.6 Two Higgs Doublets and Supersymmetry

In Sect. 1.1.5 we indicate how in the Standard Model the $SU(2)_L$ doublet structure of the fermions really involves the Higgs field H and its conjugate $H^\dagger$ to give mass to up–type and down–type fermions. One motivation to use two Higgs doublets instead of one is to avoid the conjugate Higgs field and instead give mass up–type and down–type fermions with one $\Sigma^{(u,d)}$ field each. Such a setup is particularly popular because it appears in supersymmetric theories.

There are (at least) two reasons why supersymmetric models require additional Higgs doublets: first, supersymmetry invariance does not allow us to include superfields and their conjugates in the same Lagrangian. The role which $H^\dagger$ plays in the Standard Model has to be taken on by a second Higgs doublet. Second, the moment we postulate fermionic partners to the weakly charged scalar Higgs bosons we will generate a chiral anomaly at one loop. This means that quantum corrections for example to the effective couplings of two gluons to a pseudoscalar Higgs or Goldstone boson violate the symmetries of the Standard Model. This anomaly should not appear in the limit of massless fermions inside the loop, but it does once we include one supersymmetric Higgsino state. The easiest way to cancel this anomaly is through a second Higgsino with opposite hypercharge.

Two-Higgs–doublet models of type II are not the only way to extend the Standard Model Higgs sector by another doublet. The crucial boundary condition is that the second Higgs doublet should not induce flavor changing neutral currents which have never been observed and are forbidden in the Standard Model. The simplest flavor–compatible approach is to generate all fermion masses with one Higgs doublet and add the second doublet only in the gauge sector. This ansatz is forbidden in supersymmetry and usually referred to as type I. As mentioned above, type II models separate the up–type and down–type masses in the quark and lepton sector and link them to one of the doublets each. Im both cases, type I and type II we can disconnect the lepton sector from the quark sector and for example flip the assignment of the two doublet in the lepton sector. Models of type III use a different way to avoid large flavor changing neutral currents. In an appropriate basis only one of the Higgs doublets develops a vacuum expectation value. It does not induce any flavor changing mass terms. The other doublet couples to two fermions of masses $m_{1,2}$ proportional to $\sqrt{m_1 m_2}$. This way, flavor changing neutral currents are sufficiently suppressed. In the gauge boson sector is it important that none of these three models

1.2 The Standard Model

allow for a $Z^0 W^+ H^-$ coupling, so they obey the custodial symmetry discussed in Sect. 1.1.6 at tree level. In our discussion of fermion masses and couplings we will limit ourselves to a model of type II, though. A supersymmetrized Standard Model Lagrangian together with a type II two-Higgs–doublet model we refer to as the minimal supersymmetric Standard Model (MSSM).

Using *two sigma fields* to generate the gauge boson masses is a straightforward generalization of Eq. (1.28),

$$\mathscr{L}_{D2} = \frac{v_u^2}{2} \operatorname{Tr}\left[V_\mu^{(u)} V^{(u)\mu}\right] + \frac{v_d^2}{2} \operatorname{Tr}\left[V_\mu^{(d)} V^{(d)\mu}\right] . \qquad (1.127)$$

As for one Higgs doublet we define $V_\mu^{(u,d)} = \Sigma^{(u,d)}(D_\mu \Sigma^{(u,d)})^\dagger$ for two Σ fields. In unitary gauge we can compute the corresponding gauge boson masses following Eq. (1.34). The squares of the individual vacuum expectation values add to the observed value of $v = 246$ GeV. This structure can be generalized to any number of Higgs doublets. For two Higgs doublets it allows us to use the known value of v and a new mixing angle θ as a parameterization:

$$v_u^2 + v_d^2 = v^2 \quad \Leftrightarrow \quad v_u = v \sin \beta \quad \text{and} \quad v_d = v \cos \beta . \qquad (1.128)$$

For our type II setup the fermion mass terms in Eq. (1.25) include the two Higgs doublets separately

$$\mathscr{L}_{D3} = -\overline{Q}_L m_{Qu} \Sigma_u \frac{\mathbb{1} + \tau_3}{2} Q_R - \overline{Q}_L m_{Qd} \Sigma_d \frac{\mathbb{1} - \tau_3}{2} Q_R + \dots , \qquad (1.129)$$

with the isospin projectors $(\mathbb{1} \pm \tau_3)/2$.

To study the physical Higgs bosons we express each of the two sigma fields in the usual representation

$$\Sigma^{(u,d)} = \mathbb{1} + \frac{H^{(u,d)}}{v_{u,d}} - \frac{i \vec{w}^{(u,d)}}{v_{u,d}} \qquad \vec{w}^{(u,d)} = w_a^{(u,d)} \tau_a , \qquad (1.130)$$

which means that the longitudinal vector bosons are

$$\vec{w} = \cos \beta \, \vec{w}^{(u)} + \sin \beta \, \vec{w}^{(d)} . \qquad (1.131)$$

Following Eq. (1.50) we can parameterize each of the Higgs doublets in terms of their physical Goldstone and Higgs modes. We first recapitulate the available degrees of freedom. Following the structure Eq. (1.50) we parameterize the two Higgs doublets, now in terms of H and consistently omitting the prefactor $1/\sqrt{2}$.

$$\begin{pmatrix} H_u^+ \\ H_u^0 \end{pmatrix} = \begin{pmatrix} \operatorname{Re} H_u^+ + i \operatorname{Im} H_u^+ \\ v_u + \operatorname{Re} H_u^0 + i \operatorname{Im} H_u^0 \end{pmatrix} \qquad \begin{pmatrix} H_d^0 \\ H_d^- \end{pmatrix} = \begin{pmatrix} v_d + \operatorname{Re} H_d^0 + i \operatorname{Im} H_d^0 \\ \operatorname{Re} H_d^- + i \operatorname{Im} H_d^- \end{pmatrix}$$
$$(1.132)$$

As required by electroweak symmetry breaking we have three Goldstone modes, a linear combination of $\mathrm{Im}H_u^0$ and $\mathrm{Im}H_d^0$ gives the longitudinal Z while a linear combination of H_u^+ and H_d^- gives the longitudinal polarization of $W^\pm$. The remaining five degrees of freedom form physical scalars, one charged Higgs boson $H^\pm$, two neutral CP-even Higgs bosons H_u^0, H_d^0 mixing into the mass eigenstates h^0 and H^0, and a pseudo-scalar Higgs boson A^0 from the remaining imaginary part.

In addition to introducing two Higgs doublets the supersymmetric Standard Model fixes the quartic Higgs coupling λ. From Eq. (1.86) we know that the quartic coupling fixes the Higgs mass to $m_H^2 = 2\lambda v^2$, which means that supersymmetry fixes the Higgs boson mass(es). In broken supersymmetry we have to consider three different sources of scalar self interactions in the Lagrangian:

1. *F terms* from the SUSY-conserving scalar potential $W \supset \mu \cdot H_u H_d$ include four-scalar interactions proportional to Yukawa couplings,

$$\mathscr{L}_W = -\frac{|\mu|^2}{2}\left(|H_u^+|^2 + |H_d^-|^2 + |H_u^0|^2 + |H_d^0|^2\right). \tag{1.133}$$

Note that there is the usual relative sign between the definition of the scalar potential and the Lagrangian.

2. In the Higgs sector the gauge–coupling mediated SUSY-conserving *D terms* involve abelian $U(1)_Y$ terms $D = gH^\dagger H$ as well as non–abelian $SU(2)_L$ terms $D^\alpha = g'H^\dagger \tau^\alpha H$ with the Pauli matrices as $SU(2)_L$ generators,

$$\mathscr{L}_D = -\frac{g^2}{16}\left[\left(|H_u^+|^2 + |H_u^0|^2 - |H_d^-|^2 - |H_d^0|^2\right)^2 + 4\left|H_u^+ H_d^0 + H_u^0 H_d^-\right|^2\right]$$
$$-\frac{g'^2}{16}\left(|H_u^+|^2 + |H_u^0|^2 - |H_d^-|^2 - |H_d^0|^2\right)^2. \tag{1.134}$$

The sign of the D terms in the Lagrangian is indeed predicted to be negative.

3. last, but not least scalar masses and self couplings appear as *soft SUSY breaking* parameters

$$\mathscr{L}_{\text{soft}} = -\frac{m_{H_u}^2}{2}\left(|H_u^+|^2 + |H_u^0|^2\right) - \frac{m_{H_d}^2}{2}\left(|H_d^-|^2 + |H_d^0|^2\right)$$
$$-\frac{b}{2}\left(H_u^+ H_d^- - H_u^0 H_d^0 + \text{h.c.}\right) \tag{1.135}$$

All these terms we can collect into the Higgs potential for a two Higgs doublet model

1.2 The Standard Model

$$V = \frac{|\mu|^2 + m_{H_u}^2}{2} \left(|H_u^+|^2 + |H_u^0|^2 \right) + \frac{|\mu|^2 + m_{H_d}^2}{2} \left(|H_d^0|^2 + |H_d^-|^2 \right)$$
$$+ \frac{b}{2} \left(H_u^+ H_d^- - H_u^0 H_d^0 + \text{h.c.} \right)$$
$$+ \frac{g^2 + g'^2}{16} \left(|H_u^+|^2 + |H_u^0|^2 - |H_d^-|^2 - |H_d^0|^2 \right)^2 + \frac{g^2}{4} |H_u^+ H_d^0 + H_u^0 H_d^-|^2$$
(1.136)

This full form we would like to simplify a little before focussing on the neutral states. Because now we have two Higgs doublets to play with we can first rotate them simultaneously without changing the potential V. We choose $H_u^+ = 0$ at the minimum of V, i.e. at the point given by

$$\frac{\partial V}{\partial H_u^+} = |H_u^+| \left(|\mu|^2 + m_{H_u}^2 \right) + \frac{b}{2} H_d^-$$
$$+ \frac{g^2 + g'^2}{4} |H_u^+| \left(|H_u^+|^2 + |H_u^0|^2 - |H_d^-|^2 - |H_d^0|^2 \right)$$
$$+ \frac{g^2}{2} H_d^0 \left(H_u^+ H_d^0 + H_u^0 H_d^- \right)$$
$$\xrightarrow{H_u^+ = 0} \frac{b}{2} H_d^- + \frac{g^2}{2} H_d^0 H_u^0 H_d^- \overset{!}{=} 0 .$$
(1.137)

This minimization condition can be fulfilled either as $H_d^- = 0$ or as $b + g^2 H_d^0 H_u^0 = 0$. Choosing a field dependent value of the SUSY breaking parameter b is hard to justify—our minimum condition should be a condition on the fields and not on the Lagrangian parameters. The condition $H_d^- = 0$ simplifies the functional form of the potential at the minimum to

$$V\bigg|_{\text{minimum}} = \frac{|\mu|^2 + m_{H_u}^2}{2} |H_u^0|^2 + \frac{|\mu|^2 + m_{H_d}^2}{2} |H_d^0|^2 - b |H_u^0||H_d^0|$$
$$+ \frac{g^2 + g'^2}{16} \left(|H_u^0|^2 - |H_d^0|^2 \right)^2 .$$
(1.138)

At the minimum we absorb the phase of b into a rotation of $H_d^0 H_u^0$, so the entire b term then becomes real.

In this re-rotation we have simply removed the charged Higgs and Goldstone states from the potential. Because there are no charged vacuum expectation values this should not affect the rest of the neutral spectrum. We will use the *simplified supersymmetric Higgs potential* for our study of the neutral Higgs states. Looking for the minimum of the neutral part of the Higgs potential will allow us to relate the two vacuum expectation values $v_{u,d}$ to the parameters in the potential. The minimum conditions are

$$0 \stackrel{!}{=} \left.\frac{\partial V}{\partial |H_u^0|}\right|_{H_j^0 = v_j} = \left(|\mu|^2 + m_{H_u}^2\right)|H_u^0| - b|H_d^0|$$

$$+ \left.\frac{g^2 + g'^2}{4}|H_u^0|\left(|H_u^0|^2 - |H_d^0|^2\right)\right|_{H_i^0 = v_i}$$

$$0 \stackrel{!}{=} \left.\frac{\partial V}{\partial |H_d^0|}\right|_{H_j^0 = v_j} = \left(|\mu|^2 + m_{H_d}^2\right)|H_d^0| - b|H_u^0|$$

$$- \left.\frac{g^2 + g'^2}{4}|H_d^0|\left(|H_u^0|^2 - |H_d^0|^2\right)\right|_{H_i^0 = v_i}$$

$$\Leftrightarrow \quad 0 = |\mu|^2 + m_{H_u}^2 - b\frac{v_d}{v_u} + \frac{g^2 + g'^2}{4}\left(v_u^2 - v_d^2\right)$$

$$0 = |\mu|^2 + m_{H_d}^2 - b\frac{v_u}{v_d} - \frac{g^2 + g'^2}{4}\left(v_u^2 - v_d^2\right) \qquad (1.139)$$

From the Standard Model Higgs sector with custodial symmetry, Eq. (1.38), we know how to replace the gauge couplings squared by the gauge boson masses

$$m_Z^2 = \frac{g^2 + g'^2}{2}\left(v_u^2 + v_d^2\right) \qquad m_W^2 = \frac{g^2}{2}\left(v_u^2 + v_d^2\right). \qquad (1.140)$$

The minimum conditions then read

$$|\mu|^2 + m_{H_u}^2 = b\cot\beta + \frac{m_Z^2}{2}\cos(2\beta) \qquad |\mu|^2 + m_{H_d}^2 = b\tan\beta - \frac{m_Z^2}{2}\cos(2\beta).$$
$$(1.141)$$

These relations can be used to express b and $|\mu|$ in terms of the gauge boson masses and the angle β. This suggests that the extended Higgs sector will be governed by two independent mass scales, $m_Z \sim v_{u,d}$ and $\sqrt{b}$. For now, we will still keep b and $|\mu|$ to shorten our expressions.

The masses of all physical modes as fluctuations around the vacuum state are given by the quadratic approximation to the potential around the vacuum. Because the *interaction eigenstates* $H_{u,d}$ do not have to be *mass eigenstates* for their real or imaginary parts the matrix of second derivatives defines a scalar mass matrix just like in Eq. (1.94)

$$\boxed{\left(\mathcal{M}^2\right)_{jk} = \left.\frac{\partial^2 V}{\partial H_j^0 \partial H_k^0}\right|_{\text{minimum}}} \qquad (1.142)$$

1.2 The Standard Model

We will compute the masses of all three scalar Higgs states, beginning with the compute pseudoscalar mass m_A. If this state is a superposition of $\text{Im}H_u^0$ and $\text{Im}H_d^0$ the relevant terms in Eq. (1.138) are

$$V \supset \frac{|\mu|^2 + m_{H_u}^2}{2}(\text{Im}H_u^0)^2 + \frac{|\mu|^2 + m_{H_d}^2}{2}(\text{Im}H_d^0)^2 + b\,\text{Im}H_u^0\,\text{Im}H_d^0 \quad (1.143)$$

$$+ \frac{g^2 + g'^2}{16}\left[(\text{Re}H_u^0)^2 + (\text{Im}H_u^0)^2 - (\text{Re}H_d^0)^2 - (\text{Im}H_d^0)^2\right]^2$$

$$\frac{\partial V}{\partial(\text{Im}H_u^0)} \supset \left(|\mu|^2 + m_{H_u}^2\right)\text{Im}H_u^0 + b\,\text{Im}H_d^0$$

$$+ \frac{g^2 + g'^2}{4}\text{Im}H_u^0\left[(\text{Re}H_u^0)^2 + (\text{Im}H_u^0)^2 - (\text{Re}H_d^0)^2 - (\text{Im}H_d^0)^2\right]$$

$$\frac{\partial^2 V}{\partial(\text{Im}H_u^0)^2} = \left(|\mu|^2 + m_{H_u}^2\right) + \frac{g^2 + g'^2}{4}\left[(\text{Re}H_u^0)^2 + (\text{Im}H_u^0)^2 - (\text{Re}H_d^0)^2 - (\text{Im}H_d^0)^2\right]$$

$$+ \frac{g^2 + g'^2}{2}(\text{Im}H_u^0)^2.$$

Evaluating this derivative at the minimum of the potential and with both scalar fields replaced by their vacuum expectation values gives us the masses

$$m_{\text{Im}H_u}^2 = \left(|\mu|^2 + m_{H_u}^2\right) + \frac{g^2 + g'^2}{4}\left(v_u^2 - v_d^2\right)$$

$$= b\,\frac{v_d}{v_u} = b\cot\beta \quad \text{and} \quad m_{\text{Im}H_d}^2 = b\tan\beta, \quad (1.144)$$

where we use the minimum condition equation (1.141) and the symmetry under the exchange $H_u \leftrightarrow H_d$. The parameter b indeed has mass dimension two. For the mixed second derivative we find

$$\left.\frac{\partial^2 V}{\partial(\text{Im}H_u^0)\partial(\text{Im}H_d^0)}\right|_{\text{minimum}} = b. \quad (1.145)$$

Without any assumptions the mass matrix for the two CP-odd Higgs and Goldstone mode is symmetric and has the form

$$\mathcal{M}_A^2 = b\begin{pmatrix} \cot\beta & 1 \\ 1 & \tan\beta \end{pmatrix} \quad \text{with the eigenvalues} \quad \begin{cases} m_G^2 = 0 \\ m_A^2 = \dfrac{2b}{\sin(2\beta)} \end{cases}. \quad (1.146)$$

The massive state A^0 is a *massive pseudoscalar Higgs*, while the Goldstone is massless, as expected, and will be absorbed by the massive Z boson. The mixing angle between these two Goldstone/Higgs modes is given by β,

$$\frac{2(\mathcal{M}_A^2)_{12}}{m_A^2 - m_G^2} = \frac{2b}{\frac{2b}{\sin(2\beta)} - 0} = \sin(2\beta) \,. \quad (1.147)$$

Without going into any details we can assume that the Yukawa couplings of the heavy pseudoscalar A^0 will depend on the mixing angle $\tan\beta$. It turns out that its coupling to bottom quarks is enhanced by $\tan\beta$ while the coupling to the top is reduced by the same factor.

Exactly the same calculation as in Eq. (1.143) we can follow for the two CP-even scalar Higgs bosons, starting with $\operatorname{Re} H_u^0$. The relevant quadratic terms in the potential now are

$$V \supset \frac{|\mu|^2 + m_{H_u}^2}{2} (\operatorname{Re} H_u^0)^2$$

$$+ \frac{g^2 + g'^2}{16} \left[(\operatorname{Re} H_u^0)^2 + (\operatorname{Im} H_u^0)^2 - (\operatorname{Re} H_d^0)^2 - (\operatorname{Im} H_d^0)^2 \right]^2 \quad (1.148)$$

$$\frac{\partial V}{\partial (\operatorname{Re} H_u^0)} \supset \left(|\mu|^2 + m_{H_u}^2 \right) \operatorname{Re} H_u^0$$

$$+ \frac{g^2 + g'^2}{4} \operatorname{Re} H_u^0 \left[(\operatorname{Re} H_u^0)^2 + (\operatorname{Im} H_u^0)^2 - (\operatorname{Re} H_d^0)^2 - (\operatorname{Im} H_d^0)^2 \right]$$

$$\frac{\partial^2 V}{\partial (\operatorname{Re} H_u^0)^2} = \left(|\mu|^2 + m_{H_u}^2 \right)$$

$$+ \frac{g^2 + g'^2}{4} \left[(\operatorname{Re} H_u^0)^2 + (\operatorname{Im} H_u^0)^2 - (\operatorname{Re} H_d^0)^2 - (\operatorname{Im} H_d^0)^2 \right]$$

$$+ \frac{g^2 + g'^2}{2} (\operatorname{Re} H_u^0)^2 \,.$$

The mass follows once we evaluate this second derivative at the minimum, which means with both real parts of the scalar Higgs fields replaced by vacuum expectation values:

$$m_{\operatorname{Re} H_u}^2 = \left(|\mu|^2 + m_{H_u}^2 \right) + \frac{g^2 + g'^2}{4} \left(v_u^2 - v_d^2 + 2v_u^2 \right)$$

$$= |\mu|^2 + m_{H_u}^2 + \frac{g^2 + g'^2}{4} \left(3v_u^2 - v_d^2 \right)$$

$$= b \cot\beta - \frac{g^2 + g'^2}{4} \left(v_u^2 - v_d^2 \right) + \frac{g^2 + g'^2}{4} \left(3v_u^2 - v_d^2 \right) \quad \text{using Eq. (1.141)}$$

$$= b \cot\beta + \frac{g^2 + g'^2}{2} v_u^2$$

$$= b \cot\beta + m_Z^2 \sin^2\beta$$

$$m_{\operatorname{Re} H_d}^2 = b \tan\beta + m_Z^2 \cos^2\beta \,.$$

1.2 The Standard Model

As mentioned above, b has the dimension mass squared. Going back to Eq. (1.138) we see that the mixed derivative includes two terms

$$\left.\frac{\partial^2 V}{\partial(\mathrm{Re}\,H_u^0)\partial(\mathrm{Re}\,H_d^0)}\right|_{\text{minimum}} = -b + \frac{g^2 + g'^2}{16}\, 4\,\mathrm{Re}\,H_u^0\,(-2)\,\mathrm{Re}\,H_d^0$$

$$= -b - \frac{g^2 + g'^2}{2} v^2 \sin\beta \cos\beta$$

$$= -b - \frac{m_Z^2}{2}\sin(2\beta)\,. \qquad (1.149)$$

Collecting all double derivatives with respect to the real part of the scalar fields we arrive at the mass matrix for the two CP-even Higgs bosons $\mathrm{Re}\,H_u^0$ and $\mathrm{Re}\,H_d^0$. The Lagrangian parameter b we can replace by the physical Higgs and gauge boson masses and the mixing angle β,

$$\mathcal{M}_{h,H}^2 = \begin{pmatrix} b\cot\beta + m_Z^2 \sin^2\beta & -b - \dfrac{m_Z^2}{2}\sin(2\beta) \\ -b - \dfrac{m_Z^2}{2}\sin(2\beta) & b\tan\beta + m_Z^2 \cos^2\beta \end{pmatrix}$$

$$= \begin{pmatrix} \dfrac{m_A^2}{2}\sin(2\beta)\cot\beta + m_Z^2 \sin^2\beta & -\dfrac{m_A^2 + m_Z^2}{2}\sin(2\beta) \\ -\dfrac{m_A^2 + m_Z^2}{2}\sin(2\beta) & \dfrac{m_A^2}{2}\sin(2\beta)\tan\beta + m_Z^2 \cos^2\beta \end{pmatrix}$$

$$= \begin{pmatrix} m_A^2 \cos^2\beta + m_Z^2 \sin^2\beta & -\dfrac{m_A^2 + m_Z^2}{2}\sin(2\beta) \\ -\dfrac{m_A^2 + m_Z^2}{2}\sin(2\beta) & m_A^2 \sin^2\beta + m_Z^2 \cos^2\beta \end{pmatrix}\,. \qquad (1.150)$$

The mass values for the *mass eigenstates* h^0, H^0, ordered by mass $m_h < m_H$, are

$$2m_{h,H}^2 = m_A^2 + m_Z^2 \mp \sqrt{(m_A^2 + m_Z^2)^2 - 4 m_A^2 m_Z^2 \cos^2(2\beta)}$$

$$\simeq m_A^2 \mp \sqrt{m_A^4 + 2 m_A^2 m_Z^2 (1 - 2\cos^2(2\beta))} \qquad \text{for } m_A \gg m_Z$$

$$\simeq m_A^2 \mp m_A^2 \sqrt{1 - \frac{4 m_Z^2}{m_A^2}\cos(4\beta)}\,. \qquad (1.151)$$

In the limit of a heavy pseudoscalar the supersymmetric Higgs sector with its fixed quartic couplings predicts one light and one heavy scalar mass eigenstate,

$$\boxed{m_{h,H}^2 = \frac{m_A^2}{2} \mp \frac{m_A^2}{2}\left[1 - \frac{2m_Z^2}{m_A^2}\cos(4\beta)\right] = \begin{cases} m_Z^2 \cos(4\beta) \\ m_A^2 \end{cases}}\,. \qquad (1.152)$$

The mass of the lighter of these two states depends on the parameter β, but it is bounded from above to $m_h < m_Z$. As we will see in the following section this upper bound is modified by loop corrections, but it is fair to say that supersymmetry *predicts one light Higgs*. If that is sufficient to claim that the discovery of a 125 GeV Higgs boson is the first discovery predicted by supersymmetry is a little controversial, though.

The mixing angle between the two CP-even scalar states is in general independent of the pseudoscalar mixing angle β. We denote it as α, and it can be computed from the mixing matrix shown in Eq. (1.150). The couplings of the light and heavy scalar Higgs to up–type and down–type quarks are modified both in terms of α and in terms of β, where α appears in the numerator through Higgs mixing and β appears in the denominator of the Yukawas m_q/v, replacing v by v_u and v_d. The correction factors for the light Higgs boson h^0 are $\cos\alpha/\sin\beta$ for up–type quarks and $-\sin\alpha/\cos\beta$ for down–type quarks. The same factors for the heavy Higgs H^0 are $\sin\alpha/\sin\beta$ when coupling to up–type quarks and $\cos\alpha/\cos\beta$ when coupling to down–type quarks.

To keep the equations simple we ignore the *charged Higgs* entirely, even though its existence would be the most striking sign of an extended Higgs sector with (at least) one additional doublet. At tree level a full analysis of the Higgs potential in Eq. (1.136) gives us a massless Goldstone and a massive charged Higgs scalar with

$$m_{H^\pm} = \sqrt{m_W^2 + m_A^2}\,. \tag{1.153}$$

Its Yukawa coupling include up–type and down–type contributions, dependent on the chiralities of the fermions. However, after adding all chiralities the coupling factor typically becomes $(m_d^2 \tan^2\beta + m_u^2 \cot^2\beta)$, very similar to the pseudoscalar case.

In the large-m_A limit $m_A^2 \sim b \gg m_Z^2$ the Higgs mass matrix as shown the first line of Eq. (1.150) is simplified and essentially aligns with its pseudoscalar counter part Eq. (1.146),

$$\mathcal{M}_{h,H}^2 \simeq b \begin{pmatrix} \cot\beta & -1 \\ -1 & \tan\beta \end{pmatrix} \quad \Rightarrow \quad \alpha = \pi - \beta \tag{1.154}$$

This means $\cos\alpha = \sin\beta$ and $\sin\alpha = -\cos\beta$. The correction factors for the h^0 Yukawa couplings become unity while the couplings of the heavy Higgs H^0 are $\tan\beta$ enhanced for down–type quarks and $\tan\beta$ suppressed for up–type quarks. From a phenomenological point of view the light supersymmetric Higgs scalar behaves just like a Standard Model Higgs boson while the heavy scalar and pseudoscalar Higgs bosons are hardly distinguishable. Both of them and the charged Higgs have large masses of order m_A. In Fig. 1.3 we show all masses of the physics Higgs bosons, now including radiative corrections which we will discuss in Sect. 1.2.7. For large pseudoscalar masses we clearly see the decoupling of all three heavy states from the one light Higgs boson.

1.2 The Standard Model

Fig. 1.3 Masses of all supersymmetric Higgs states as a function of the pseudoscalar Higgs mass, computed with FeynHiggs (Figures of this kind can be found for example in Ref. [21])

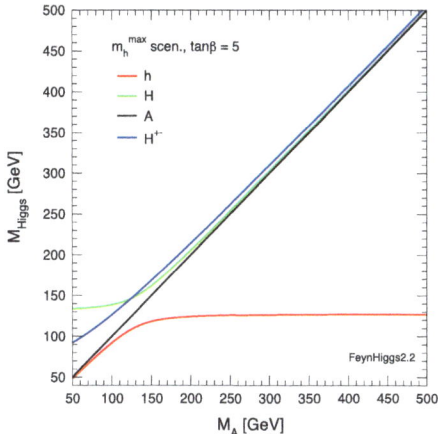

As a matter of fact, this *decoupling regime* where the light supersymmetric Higgs boson is indistinguishable from a Standard Model Higgs of the same mass is exact and includes all couplings and properties. Small deviations from the Standard Model couplings, suppressed by a finite mass ratio m_h/m_A, we need to look for in the Higgs coupling analysis discussed in Sect. 1.8.1.

1.2.7 Coleman–Weinberg Potential

In Sect. 1.2.1 we discuss the form of the Higgs potential, as defined by all allowed renormalizable operators in the Lagrangian. We make it a little more interesting by including dimension-6 operators which are not renormalizable, but we stick to a power series in $\phi^\dagger \phi$. A different kind of contribution to the Higgs potential can arise from loops of states which couple to the Higgs boson.

We start by limiting ourselves to dimension-4 operators and replace the tree level potential Eq. (1.77) by an *effective potential*, including a tree level and a loop contribution. The question is if we can induce spontaneous symmetry breaking with a non–trivial vacuum position of the Higgs field through a loop–induced potential.

Our toy model is a ϕ^4 theory of a single real massive scalar field, a little simpler than the complex Higgs–Goldstone field in the Standard Model

$$\mathscr{L} = \frac{1}{2}(\partial_\mu \phi)^2 - \frac{m_0^2}{2}\phi^2 - \frac{\lambda_0}{4}\phi^4 \qquad (1.155)$$

Using some basic field theory we can elegantly describe this alternative source of spontaneous symmetry breaking. We first review the *generating functional* for a free real scalar field theory (following Mark Srednicki's conventions changed to our metric)

$$Z_0(J) = \int \mathscr{D}\phi \, e^{iS_0(\phi) + i \int d^4x \, J\phi}$$

$$= \int \mathscr{D}\phi \, e^{i \int d^4x \, (\mathscr{L}_0 + J\phi)}$$

$$= e^{\frac{i}{2} \int d^4x_1 d^4x_2 \, J(x_1) \Delta(x_1 - x_2) J(x_2)}$$

$$\text{with} \quad \Delta(x_1 - x_2) = -\int \frac{d^4k}{(2\pi)^4} \frac{e^{-ik(x_1 - x_2)}}{k^2 - m_0^2}$$

$$\tilde{\Delta}(k^2) = -\frac{1}{k^2 - m_0^2} \, . \tag{1.156}$$

In this form we see how we can compute propagators or other time ordered products of field operators using *functional derivatives* on the generating functional

$$i\Delta(x_1 - x_2) \equiv \langle 0 | T\phi(x_1)\phi(x_2) | 0 \rangle = \frac{1}{i} \frac{\delta}{\delta J(x_1)} \frac{1}{i} \frac{\delta}{\delta J(x_2)} Z_0(J) \bigg|_{J=0} . \tag{1.157}$$

The vacuum expectation value of the free field itself is zero, as is the expectation value for any odd number of scalar fields. This is because there will always be one factor J left after the functional derivative which then gets set to zero.

Once we switch on an interaction λ_0 this does not have to be true any longer. Moving from Z to $iW = \log Z$ means we omit the unconnected interaction diagrams. In analogy to the free theory we define an *effective action* Γ in terms of exact propagators and exact vertices as

$$Z_\Gamma(J) = \int \mathscr{D}\phi \, e^{i\Gamma(\phi) + i \int d^4x \, J\phi} \equiv e^{iW_\Gamma(J)} \, . \tag{1.158}$$

This effective action defines a *stationary field configuration* ϕ_J through

$$\frac{\delta}{\delta\phi(x)} \left(\Gamma(\phi) + \int d^4x' \, J(x')\phi(x') \right) = 0 \quad \Leftrightarrow \quad \frac{\delta \Gamma(\phi)}{\delta\phi(x)} \bigg|_{\phi_J} = -J(x) \, . \tag{1.159}$$

Such a stationary point of the exponential allows us to expand the effective action and the corresponding generating functional defined in Eq. (1.158) in terms of the field fluctuations.

At this point it would help if we could make physics sense out of the field configuration $\phi_J(x)$. We only quote that to leading terms in $\hbar$ (i.e. at tree level) we can express the interacting generating functional for connected diagrams in terms of the effective action at this stationary point as a Legendre transform,

$$W(J) = \Gamma(\phi_J) + \int d^4x \, J(x)\phi_J(x) \, . \tag{1.160}$$

1.2 The Standard Model

A proper derivation of this formula can be found in Chapter 21 of Mark Srednicki's field theory book. Using this relation we can speculate about a non–trivial expectation value of an interacting scalar field in the presence of a finite source J. In analogy to Eq. (1.157) we need to compute

$$\langle 0|\phi(x)|0\rangle_J$$

$$= \frac{\delta}{\delta J(x)} W(J)$$

$$= \frac{\delta \Gamma(\phi_J)}{\delta J(x)} + \phi_J(x) + \int d^4x'\, J(x') \frac{\delta \phi_J(x')}{\delta J(x)} \qquad \text{using Eq. (1.160)}$$

$$= \int d^4x'\, \frac{\delta \Gamma(\phi_J)}{\delta \phi_J(x')} \frac{\delta \phi_J(x')}{\delta J(x)} + \phi_J(x) + \int d^4x'\, J(x') \frac{\delta \phi_J(x')}{\delta J(x)}$$

$$= \int d^4x'\, \left(\frac{\delta \Gamma(\phi_J)}{\delta \phi_J(x')} + J(x') \right) \frac{\delta \phi_J(x')}{\delta J(x)} + \phi_J(x) \qquad \text{using Eq. (1.159)}$$

$$= \phi_J(x). \tag{1.161}$$

On the way we apply the definition of the stationary point in Eq. (1.159). The *expectation value* we are looking for is nothing but the stationary point of the effective action Γ. In the limit $J = 0$ this value $\phi_J(x)$ becomes a vacuum expectation value.

Motivated by the general expectation that a classical solution will change much more slowly than the quantum field we assume that ϕ_J is constant,

$$\boxed{\phi(x) = \phi_J + \eta(x)} \qquad \text{and} \qquad e^{iW_\Gamma(J)} = \int \mathcal{D}\eta\, e^{i\Gamma(\phi_J+\eta)+i\int d^4x\, J\,(\phi_J+\eta)}.$$
$$\tag{1.162}$$

The path integral over ϕ_J is trivial and only changes the irrelevant normalization of the generating functional. The expanded exponential around the saddle point with its vanishing first derivative reads

$$\Gamma(\phi) + \int d^4x\, J(x)\phi(x) = \Gamma(\phi_J) + \int d^4x\, J(x)\phi_J(x)$$

$$+ \frac{1}{2} \int d^4x_1 d^4x_2\, \eta(x_1) \left(\frac{\delta^2 \Gamma(\phi)}{\delta\phi(x_1)\delta\phi(x_2)} \right)_{\phi=\phi_J} \eta(x_2)$$

$$\equiv \Gamma(\phi_J) + \int d^4x\, J(x)\phi_J(x)$$

$$+ \frac{1}{2} \int d^4x_1 d^4x_2\, \eta(x_1) \Gamma^{(2)}(\phi_J) \eta(x_2). \tag{1.163}$$

This means the linear term vanishes by definition around the stationary point while the source term does not contribute beyond the linear term. The last step is the definition of $\Gamma^{(2)}(\phi_J)$. Exponentiating this action we can make use of the definition of the functional determinant for real scalar fields,

$$Z(\eta) = \int \mathcal{D}\eta \, e^{-i \int d^n x_1 \, d^n x_2 \, \eta(x_1) M \eta(x_2)} \equiv \frac{2(2\pi)^n}{\det M} \, . \quad (1.164)$$

Inserting this formula into the definition of the generating functional for connected Green functions W_Γ gives us immediately

$$iW_\Gamma(J) = \log\left[e^{i\Gamma(\phi_J) + i \int d^4 x J(x)\phi_J(x)} \int \mathcal{D}\eta \, e^{\frac{i}{2} \int d^4 x_1 d^4 x_2 \, \eta(x_1)\Gamma^{(2)}(\phi_J)\eta(x_2)}\right]$$

$$= i\Gamma(\phi_J) + i\int d^4 x \, J(x)\phi_J(x) + \log\left[\frac{2(2\pi)^n}{\det\left(-\Gamma^{(2)}(\phi_J)\right)}\right]^{1/2} \quad \text{using Eq. (1.164)}$$

$$= i\Gamma(\phi_J) + i\int d^4 x \, J(x)\phi_J(x) + \frac{1}{2}\log\det\left(2^{n+1}\pi^n\right) - \frac{1}{2}\log\det\left(-\Gamma^{(2)}(\phi_J)\right)$$

$$= i\left[\Gamma(\phi_J) + \int d^4 x \, J(x)\phi_J(x) + \frac{i}{2}\operatorname{Tr}\log\left(-\Gamma^{(2)}(\phi_J)\right) + \text{const}\right] \, . \quad (1.165)$$

In the last step we exploit the general operator identity commuting the logarithm and the trace. Finite terms in a potential we can ignore. Comparing this result to Eq. (1.160) we see that the exact generating functional W_Γ includes an additional loop–induced term,

$$W_\Gamma(J) = W(J) + \frac{i}{2}\operatorname{Tr}\log\left(-\Gamma^{(2)}(\phi_J)\right) \, . \quad (1.166)$$

In other words, the Legendre transform of the full effective connected generating functional W_Γ includes an *additional* Tr log *contribution*. We need to translate this loop–induced contribution to the effective action into something we can evaluate for our model. The underlying concept is the effective potential. If ϕ_J does not propagate, its effective action only includes potential terms and no kinetic term. In other words, we can define an effective potential which the propagating field $\eta(x)$ feels as

$$\boxed{V_{\text{eff}} = V_0 + V_{\text{loop}} = -\frac{1}{L^4}\left[\Gamma(\phi_J) + \frac{i}{2}\operatorname{Tr}\log\left(-\Gamma^{(2)}(\phi_J)\right)\right]} \, . \quad (1.167)$$

The relative factor L^4 is the phase space volume which distinguishes the action from the Lagrangian. It will drop out once we compute V_{loop}. In this definition of the effective potential we naively assume that both terms are finite and well defined. It will turn out that this is not the case, so we should add to the definition in Eq. (1.167) something like 'finite terms of' or 'renormalized'.

1.2 The Standard Model

Until now our argument has been very abstract, so let us see if computing the effective potential for our real scalar field equation (1.155) clarifies things. Following the definition in Eq. (1.163) we find

$$-\Gamma^{(2)}(\phi_J) = -\frac{\delta^2}{\delta\eta(x_1)\delta\eta(x_2)} \int d^4x \left[-\frac{1}{2}\eta\partial_\mu^2\eta - \frac{m_0^2}{2}(\phi_J+\eta)^2 - \frac{\lambda_0}{4}(\phi_J+\eta)^4 \right]\bigg|_{\eta=0}$$

$$= -\int d^4x \frac{\delta^2}{\delta\eta(x_1)\delta\eta(x_2)} \left[-\frac{1}{2}\eta\partial_\mu^2\eta - \frac{m_0^2}{2}\left(\eta^2 + \cdots\right) \right.$$

$$\left. - \frac{\lambda_0}{4}\left(\eta^4 + 4\eta^3\phi_J + 6\eta^2\phi_J^2 + \cdots\right) \right]\bigg|_{\eta=0}$$

$$= \partial_1^2 + m_0^2 + 3\lambda_0\phi_J^2 \,. \tag{1.168}$$

The Tr log combination we know how to compute once we assume we know the eigenvalues of the d'Alembert operator ∂^2. Because it will turn out that in four space–time dimensions we need to remove ultraviolet divergences through renormalization we compute it in $n = 4 - 2\epsilon$ dimensions. The formula for this n-dimensional *scalar loop integral* is standard in the literature:

$$\text{Tr} \log \left(\partial^2 + C\right) = \sum_p \log\left(-p^2 + C\right)$$

$$= L^n \int \frac{d^n p}{(2\pi)^n} \log\left(-p^2 + C\right)$$

$$= -iL^n \, \Gamma\left(-\frac{n}{2}\right) \frac{C^n}{(4\pi)^{n/2}} \,. \tag{1.169}$$

The loop-induced contribution to the effective potential, now including the renormalization scale to protect the mass dimension, is then

$$V_{\text{loop}} = -\frac{i}{2L^4} \text{Tr} \log \left(\partial^2 + m_0^2 + 3\lambda_0\phi_J^2\right)$$

$$= -\mu_R^{4-n} L^{4-n} \frac{1}{2(4\pi)^{n/2}} \Gamma\left(-\frac{n}{2}\right) \left(m_0^2 + 3\lambda_0\phi_J^2\right)^{n/2}$$

$$= -\mu_R^{2\epsilon} \frac{1}{2(4\pi)^{2-\epsilon}} \Gamma(-2+\epsilon) \left(m_0^2 + 3\lambda_0\phi_J^2\right)^{2-\epsilon}$$

$$= -\mu_R^{2\epsilon} \frac{1}{2(4\pi)^{2-\epsilon}} \frac{\Gamma(\epsilon)}{(-2+\epsilon)(-1+\epsilon)} \left(m_0^2 + 3\lambda_0\phi_J^2\right)^{2-\epsilon}$$

$$= -\frac{1}{2(4\pi)^2} \frac{1}{2-3\epsilon} \left(\frac{1}{\epsilon} - \gamma_E + \log(4\pi)\right) \left(m_0^2 + 3\lambda_0 \phi_J^2\right)^2$$

$$\left(1 - \epsilon \log \frac{m_0^2 + 3\lambda_0 \phi_J^2}{\mu_R^2} + \mathcal{O}(\epsilon^2)\right)$$

$$= -\frac{1}{64\pi^2} \left(\frac{1}{\epsilon} - \gamma_E + \log(4\pi) + \frac{3}{2}\right) \left(m_0^2 + 3\lambda_0 \phi_J^2\right)^2$$

$$+ \frac{1}{64\pi^2} \left(m_0^2 + 3\lambda_0 \phi_J^2\right)^2 \log \frac{m_0^2 + 3\lambda_0 \phi_J^2}{\mu_R^2} \,. \quad (1.170)$$

In the second to last line we use the simple trick

$$C^\epsilon = e^{\log C^\epsilon} = e^{\epsilon \log C} = 1 + \epsilon \log C + \mathcal{O}(\epsilon^2) \,. \quad (1.171)$$

The expression for V_{loop} is divergent in the limit $\epsilon \to 0$, so we need to renormalize it. In the $\overline{\text{MS}}$ scheme this simply means subtracting the pole $1/\epsilon - \gamma_E + \log(4\pi)$, so the renormalized effective potential or *Coleman–Weinberg potential* becomes

$$\boxed{V_{\text{eff}} = V_0 + V_{\text{loop}}^{(\text{ren})} = V_0 + \frac{1}{64\pi^2} \left(m^2 + 3\lambda \phi_J^2\right)^2 \left(\log \frac{m^2 + 3\lambda \phi_J^2}{\mu_R^2} - \frac{3}{2}\right)} \,.$$
$$(1.172)$$

The bare mass and coupling appearing in Eq. (1.170) implicitly turn into their renormalized counter parts in the $\overline{\text{MS}}$ scheme. Combining the tree level and the loop–induced potentials we see how this additional contribution affects our real scalar ϕ^4 theory defined by Eq. (1.155) in its massless limit

$$V_{\text{eff}} = \frac{\lambda}{4} \phi_J^4 + \frac{9\lambda^2}{64\pi^2} \phi_J^4 \left(\log \frac{3\lambda \phi_J^2}{\mu_R^2} - \frac{3}{2}\right)$$

$$= \frac{\lambda}{4} \phi_J^4 \left[1 + \frac{9\lambda}{16\pi^2} \left(\log \frac{3\lambda \phi_J^2}{\mu_R^2} - \frac{3}{2}\right)\right] \,. \quad (1.173)$$

In the limit where the logarithm including a physical mass scale $\mu_R \equiv M$ becomes large enough to overcome the small coupling λ we can compute where the expression in brackets and hence the whole effective potential passes through zero. Close to this point the potential also develops a *non–trivial minimum*, i.e. a minimum at finite field values,

$$\frac{d}{d\phi_J^2} V_{\text{eff}}(\phi_J) = \frac{\lambda}{2} \phi_J^2 \left[1 + \frac{9\lambda}{16\pi^2} \left(\log \frac{3\lambda \phi_J^2}{M^2} - \frac{3}{2}\right)\right] + \frac{\lambda}{4} \phi_J^4 \frac{9\lambda}{16\pi^2} \frac{1}{\phi_J^2}$$

$$\simeq \frac{\lambda}{2} \phi_J^2 \left[1 + \frac{9\lambda}{16\pi^2} \log \frac{3\lambda \phi_J^2}{M^2} + \frac{9\lambda}{32\pi^2}\right] \qquad \text{with } -\log \frac{\phi_J^2}{M^2} \gg 1$$

1.2 The Standard Model

$$\simeq \frac{\lambda}{2}\phi_J^2 \left[1 + \frac{9\lambda}{16\pi^2} \log \frac{3\lambda \phi_J^2}{M^2}\right] \qquad \text{with } \frac{\lambda}{4\pi^2} \ll 1$$

$$\equiv 0 \quad \Leftrightarrow \quad \phi_{J,\min}^2 = \frac{M^2}{3\lambda} e^{-16\pi^2/(9\lambda)}. \tag{1.174}$$

The finite term $-3/2$ in Eq. (1.173) is numerically sub-leading and hence often omitted. Moreover, compared with Eq. (1.173) the leading contribution only applies the derivative to the over–all factor ϕ_J^4, not to the argument inside the logarithm. This minimum is exclusively driven by the loop contribution to the scalar potential. This means that the loop–induced Coleman–Weinberg potential can break electroweak symmetry, when applied to the Higgs field in the Standard Model. However, the position of this minimum we should take with a grain of salt, because logarithms of the kind $\log \phi_J/M$ will appear in many places of the higher order corrections. The mechanism of generating a physical mass scale through a strong interaction combined with a renormalization group analysis or renormalization scale is called *dimensional transmutation*.

In the Standard Model we can compute the size of the Higgs self coupling, $\lambda = m_H^2/(2v^2) = 0.13$. Forgetting the fact that our toy model is a real scalar theory we can then compute the corresponding field values at the minimum or vacuum expectation value. In the loop–induced minimum it comes out very small,

$$\frac{\phi_{J,\min}^2}{M^2} = 2.6 \times e^{-135} \simeq 10^{-60}. \tag{1.175}$$

To explain the gauge boson masses and the Higgs boson in the Standard Model the Coleman–Weinberg potential is not well suited. However, in the *supersymmetric Higgs sector* discussed in Sect. 1.2.6 it is very useful to compute the mass of the lightest supersymmetric Higgs boson beyond tree level. We left this scenario with the prediction $m_h < m_Z$, which is clearly ruled out by the measured value of $m_H = 125$ GeV. The question is if loop corrections to the supersymmetric Higgs potential can increase this Higgs mass bound, such that it agrees with the measurement.

The toy model we will study is the light supersymmetric Higgs boson combined with a second heavy scalar, the scalar partner of the top quark. *Top squarks* are the supersymmetric partners of the chiral left handed and right handed top quarks. They can mix, which means that we have to define a set of mass eigenstates $\tilde{t}_{1,2}$. From Sect. 1.2.6 we know that there are two independent kinds of four-scalar couplings between the stops and the Higgs bosons: F-term Yukawa interactions proportional to the top Yukawa y_t and D-term gauge interactions. If we limit ourselves to the large top Yukawa corrections and neglect stop mixing we only have to consider one scalar state $\tilde{t}$. To simplify things further we also assume that this scalar be real, neglecting the imaginary part of the electrically and weakly charged supersymmetric top partner.

The Lagrangian we study is the purely real stop–Higgs system with a stop–stop–Higgs–Higgs coupling y_t, extended from the Higgs system given in Eq. (1.86) to

$$\mathscr{L} = \frac{1}{2}\left(\partial_\mu H\right)^2 + \frac{1}{2}\left(\partial_\mu \tilde{t}\right)^2 - \frac{m_{\tilde{t}}^2}{2}\tilde{t}^2 - \frac{y_t^2}{4}\tilde{t}^2 H^2 + \text{Higgs terms}. \quad (1.176)$$

The last term is the renormalizable four-point interaction between the two scalar fields. Its leading coupling strength is the supersymmetric Yukawa coupling, i.e. the F term scalar interaction introduced in Eq. (1.133) with $y_t = \sqrt{2}m_t/v$.

The form of the Higgs potential at tree level is unchanged compared to the Standard Model because the stop $\tilde{t}$ does not have a finite vacuum expectation value. To see what the Coleman–Weinberg effective potential Eq. (1.172) tells us about this case we need to briefly *recapitulate its derivation*. The basis of our derivation is an expansion of the Legendre transformed effective action Eq. (1.163) around a stationary point. This stationary point is the classical or tree level solution, Eq. (1.160), so we can assume that the Coleman–Weinberg potential comes from a loop diagram. Because the only coupling in our scalar ϕ^4 theory is the scalar self coupling λ the relevant diagram must be the scalar one-point diagram. The trace we compute in Eq. (1.169) is linked to a loop integral with one scalar propagator, confirming this interpretation. Finally, the $\overline{\text{MS}}$ renormalization of the loop mass m and the coupling λ in Eq. (1.172) is exactly what we would expect from such a calculation. In the final expression we see that the mass renormalization as well as the coupling renormalization can trigger symmetry breaking. In the example shown in Eq. (1.174) we limit ourselves to the coupling λ alone, to illustrate the loop-induced effect in addition to the tree level ϕ^4 term.

Looking at the Lagrangian equation (1.176) we are instead interested in the effect which an additional massive scalar has on the Higgs potential. This means we have to consider the general Coleman–Weinberg form in Eq. (1.172) in the limit $\lambda = 0$. The mass which appears in the loop-induced potential is the mass which appears in the relevant one-point loop integral. Now, this integral is a closed stop loop coupling to the Higgs propagator through the four-point coupling y_t^2. More generally, the loop-induced or Coleman–Weinberg potential derived in Eq. (1.172) induced only by a massive loop contributing to the Higgs propagator is

$$\boxed{V_{\text{eff}} = \frac{1}{64\pi^2} \sum (-1)^S\, n_{\text{dof}}\, m^4(\phi_J) \left(\log\frac{m^2(\phi_J)}{\mu_R^2} - \frac{3}{2}\right).} \quad (1.177)$$

Spin effects in the closed loop are taken care of by $(-1)^S = +1$ for bosons and $(-1)^S = -1$ for fermions. The number of degrees of freedom is $n_{\text{dof}} = 1$ for a real scalar, 2 for a complex scalar, and 4 for a fermion. The mass m is the $\overline{\text{MS}}$ mass of the particle running inside the loop. One widely used approximation is the generalization of Eq. (1.178) to the Standard Model including the Higgs mode, the Goldstone modes, and the large top Yukawa $m_t = y_t v \sqrt{2} = y_t \phi_J$. In that case the Higgs potential in Eq. (1.77) includes a negative mass term $m^2 \to -2\mu^2$ and a unit prefactor instead of $\lambda/4$,

1.2 The Standard Model

$$V_{\text{eff}} = V_0$$
$$+ \frac{1}{64\pi^2} (-2\mu^2 + 12\lambda\phi_J^2)^2 \left(\log \frac{-\mu^2 + 12\lambda\phi_J^2}{\mu_R^2} - \frac{3}{2} \right)$$
$$+ \frac{3}{64\pi^2} (-2\mu^2 + 4\lambda\phi_J^2)^2 \left(\log \frac{-2\mu^2 + 4\lambda\phi_J^2}{\mu_R^2} - \frac{3}{2} \right)$$
$$- \frac{N_c}{16\pi^2} (y_t \phi_J)^4 \left(\log \frac{y_t^2 \phi_J^2}{\mu_R^2} - \frac{3}{2} \right). \tag{1.178}$$

The MSSM differs from our toy model in two ways. First, the stop is not a single neutral scalar, but a set of two charged scalars. If both of them couple proportional to y_t^2 to the Higgs boson and we omit the sub-leading finite term $-3/2$ we find

$$V_{\text{loop}} = \frac{2N_c}{64\pi^2} (y_t \phi_J)^4 \left(\log \frac{m_{\tilde{t}_1}^2}{\mu_R^2} + \log \frac{m_{\tilde{t}_2}^2}{\mu_R^2} \right) = \frac{N_c}{32\pi^2} (y_t \phi_J)^4 \log \frac{m_{\tilde{t}_1}^2 m_{\tilde{t}_2}^2}{\mu_R^4}. \tag{1.179}$$

The prefactor reflects the complex stop field with its two degrees of freedom. Second, for this model to be complete we need to also take into account the top quark contribution from Eq. (1.178). By definition this includes both chiralities with $n_{\text{dof}} = 4$, so altogether we find

$$V_{\text{loop}}^{(\text{MSSM})} = \frac{N_c}{32\pi^2} (y_t \phi_J)^4 \left(\log \frac{m_{\tilde{t}_1}^2}{\mu_R^2} + \log \frac{m_{\tilde{t}_2}^2}{\mu_R^2} - 2 \log \frac{m_t^2}{\mu_R^2} \right)$$
$$= \frac{N_c}{32\pi^2} (y_t \phi_J)^4 \log \frac{m_{\tilde{t}_1}^2 m_{\tilde{t}_2}^2}{m_t^4}. \tag{1.180}$$

This is the leading loop correction to the *lightest Higgs mass* in the MSSM, lifting the allowed mass range at tree level, $m_h^2 < m_Z^2$, to include the measured value of 125 GeV. Looking in more detail, in Eq. (1.180) we assume the stop mass matrix to be diagonal. If we allow for a non–diagonal stop mass matrix the value increases even further and we find power corrections to m_h proportional to the off-diagonal entries in the stop mass matrix. All these corrections allow the observed Higgs mass around 125 GeV to be consistent with the MSSM prediction—remembering that the light Higgs mass is actually a prediction from the quartic gauge couplings in the MSSM. However, the observed Higgs mass suggests that the additional Higgs bosons is heavy and that the mixing between the stop interaction eigenstates is strong.

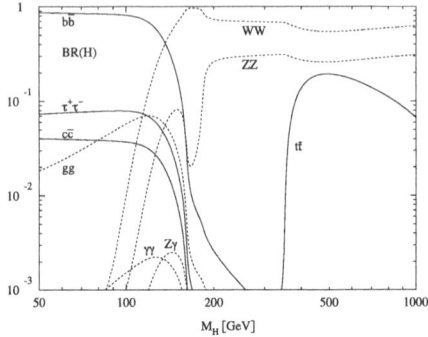

Fig. 1.4 Branching ratios of the Standard-Model Higgs boson as a function of its mass, computed with HDECAY. Off–shell effects in the decays to WW and ZZ are taken into account (Figure found for example in Refs. [15, 35])

1.3 Higgs Decays and Signatures

Signatures for new particles at colliders consist of a production process and a decay pattern. Both, the production and the decay can contribute to unique kinematic features which we can use to extract signal from background events. The actual new particle is then described by a Breit–Wigner propagator for unstable particles which we will discuss in detail in Sect. 2.1.2. Since the Higgs boson is a scalar there are no correlations between production and decay process, which simplifies the calculation and simulation of Higgs signatures. For backgrounds this factorization might of course not hold.

Unlike the production processes the Higgs decay pattern is as simple as it can be. At tree level all decay rates are determined by the Higgs coupling to Standard Model particles, which are *fixed by unitarity*. The rule for different Higgs decays is simple; because by definition the Higgs field couples to all particles (including itself) proportional to their masses it will preferably decay to the heaviest states allowed by phase space. This goes back to the condition $\langle \Sigma \rangle = \mathbb{1}$ translated into the appearance of the combination $(v + H)$ in the Higgs field ϕ and in the Lagrangian.

This behavior we see in Fig. 1.4. Starting at low masses this first applies to decays to $\tau\tau$ and $b\bar{b}$. The relative size of their branching ratios around 10:90 % is given by their Yukawa couplings in the appropriate renormalization scheme ($y_b/y_\tau \sim$ 1.4), times an additional color factor $N_c = 3$ for the bottom quarks. Once the off–shell decays to WW are allowed, they very soon dominate. The dominant decays to bottom pairs and W pairs become equal for Higgs masses around 130 GeV. This is why we can consider ourselves lucky with an observed Higgs mass around 125 GeV: nature has chosen exactly the Higgs mass which allows us to observe the largest number of different Higgs decays and this way extensively study the Higgs sector, as discussed in Sect. 1.8.1.

Because of the small mass difference between the W and Z bosons the decay to ZZ is not as dominant, compared to the WW decay which has two degrees of freedom (W^+W^- and W^-W^+) in the final state. In particular in the region where the W decays first becomes on–shell we see a drop of in the still off–shell Z decays. For large Higgs masses the ratio of $H \to WW$ and $H \to ZZ$ decays is fixed by the relative factor of 2, corresponding to the number of degrees of freedom forming the final state. Above the top threshold the $t\bar{t}$ decay becomes sizeable, but never really dominates.

We can roughly estimate the Higgs width from its decay channels: in general, we know that particles decaying through the weak interaction have a width–to–mass ratio of $\Gamma/m \sim 1/100$. The main Higgs decay is to bottom quarks, mediated by a small bottom Yukawa coupling, $m_b/m_W \lesssim 1/30$. First, this means that in the Standard Model we expect $\Gamma_H/m_H \sim 10^{-5}$, consistent with the exact prediction $\Gamma_H \sim 4$ MeV. Second, loop–induced couplings can compete with such a small tree level decay width. In particular the loop–induced decay to two photons plays an important role in LHC phenomenology. It proceeds via a top and a W triangle which enter with opposite signs in the amplitude and hence interfere destructively. The larger W contribution fixes the sign of the loop–induced coupling.

The structure of the $\gamma\gamma H$ coupling is similar to the production via the loop–induced ggH coupling which we will discuss in Sect. 1.5 and then generalize to the photon case in Sect. 1.5.3. The reason for considering this decay channel are the LHC detectors. To extract a Higgs signal from the backgrounds we usually try to measure the four-momenta of the Higgs decay products and reconstruct their invariant mass. The signal should then peak around m_H while the backgrounds we expect to be more or less flat. The LHC detectors are designed to measure the photon momentum and energy particularly well. The resolution in $m_{\gamma\gamma}$ will at least be a factor of 10 better than for any other decay channel, except for muons. Moreover, photons do not decay, so we can use all photon events in the Higgs search, while for example hadronically decaying $W/Z \to 2$ jets are not particularly useful at the LHC. These enhancement factors make the Higgs decay to two photons a promising signature, in spite of its small branching ratio around $2 \cdot 10^{-3}$. More details of the different decay channels we will give in Sect. 1.5.4.

Because an observed Higgs sector can deviate from the minimal Standard Model assumptions in many ways the LHC or other future colliders will study the different Higgs decays and, as a function of m_H, answer the questions

- Are gauge–boson couplings proportional to $m_{W,Z}$?
- Are these couplings dimension-3 operators?
- Are fermion Yukawa couplings proportional to m_f?
- Is there a Higgs self coupling, i.e. a remnant of the Higgs potential?
- Do λ_{HHH} and λ_{HHHH} show signs of higher–dimensional operators?
- Are there any other unexpected effects, like a Higgs decay to invisible particles?

But before we study the Higgs we need to discover it...

1.4 Higgs Discovery

Of course we cannot discover particles which do not get produced, and for such a discovery we need to understand the production mechanism. On the other hand, once we know the decay signatures of a Higgs boson we should be able to at least roughly understand what the LHC has been looking for. In that sense there is no need to further delay a brief account of the Higgs discovery, as announced on the 4th of July, 2012.

Without knowing any theoretical particle physics we first need to discuss the main feature or problem of hadron collider physics: there is no such thing as a signal without a background. More precisely, there is no kinematic configuration which is unique to signal events and cannot appear as an unlucky combination of uninteresting Standard Model or QCD processes and detector effects. This implies that any LHC measurement will always be a *statistics exercise* based on some kind of event counting combined with a probability estimate for the signal nature of a given event.

Because signals are new things we have not seen before, they are rare compared to backgrounds. Digging out signal events from a large number of background events is the art of LHC physics. To achieve this we need to understand all backgrounds with an incredible precision, at least those background events which populate the signal region of phase space. Such a background description will always be a combination of experimental and theoretical knowledge. The high energy community has agreed that we call a 5σ excess over the known backgrounds a signal discovery

$$\frac{S}{\sqrt{B}} = \#\{\sigma\} > 5 \qquad \text{(Gaussian limit)}$$

$$P_{\text{fluct}} < 5.8 \times 10^{-7} \qquad \text{(fluctuation probability)}. \qquad (1.181)$$

More details on this probability measure we will give later in this section. This statistical definition of a 'discovery' goes back to Enrico Fermi, who asked for 3σ. The number of researchers and analyses in high energy physics has exploded since those days, so nowadays we do not trust anybody who wants to sell you a 3σ evidence as a discovery. Everyone who has been around for a few years has seen a great number of those go away. People usually have reasons to advertize such effects, like a need for help by the community, a promotion, or a wish to get the Stockholm call, but all they are really saying is that their errors do not allow them to make a conclusive statement. On the other hand, in the Gaussian limit the statistical significance improves with the integrated luminosity as $\sqrt{\mathscr{L}}$. So all we need to do is take more data and wait for a 3σ anomaly to hit 5σ, which is what ATLAS and CMS did between the Moriond conference in the Spring of 2012 and the ICHEP conference in the Summer of 2012.

In this section we will go through the results presented by ATLAS (and CMS) after the ICHEP conference 2012. During the press conference following the scientific presentations on the 4th of July 2012 the ATLAS and CMS spokes-people and the CERN general director announced the *discovery of a new particle*, consistent with the Standard Model Higgs boson. To keep it simple, we will limit ourselves to the ATLAS discovery paper [1]—the corresponding CMS publication [14] is very similar.

1.4 Higgs Discovery

To understand the numbers quoted in the Higgs discovery paper we need some basic statistical concepts. This leads us to the general question on how to statistically test hypotheses for example predicting an event rate B_theo (predicted background) or $(S+B)_\text{theo}$ (predicted signal plus background), where the corresponding measured number of events is N. The actual ATLAS and CMS analyses are much more complicated than our argument in terms of event numbers, but our illustration captures most relevant points. For simplicity we assume the usual situation where $(S+B)_\text{theo} > B_\text{theo}$. If we would like to know how badly our background–only prediction is ruled out we need to know if there is any chance a fluctuation around B_theo would be consistent with a measured value N. Note that the index 'theo' does not mean that these predictions are entirely based on the underlying theory. If we largely understand a data set for example in terms of the Standard Model without a Higgs, we can use measurements in regions where we do not expect to see a Higgs effect to improve or even replace the theoretical predictions.

First, an experimental outcome $N < B_\text{theo}$ means that the background prediction is closer than the signal plus background prediction. This means we are done with that signal hypothesis. It gets a little harder when we observe $B_\text{theo} \lesssim N < (S+B)_\text{theo}$. In this situation we need to define a measure which allows us to for example rule out a signal prediction because the measured event rates are close to the background prediction. In this *ruling-out mode* we ask the following question: 'Given that the background prediction and the measurement largely agree, how sure are we that there is no small signal there?'. To answer this question we compute the statistical distribution of event counts N' around the predicted background value B_theo. In the *Gaussian limit* this is symmetric curve centered around B_theo with a standard deviation σ_B,

$$f(N'; B_\text{theo}, \sigma_B) = \frac{1}{\sqrt{2\pi}\sigma_B} e^{-(N'-B_\text{theo})^2/(2\sigma_B^2)}. \quad (1.182)$$

For small event numbers we need to replace this Gaussian with an asymmetric Poisson distribution, which has the advantage that by definition it does not cover negative event numbers. The entire Gaussian integral is normalized to unity, and 68.3 or 95.4 % of it falls within one or two standard deviations σ_B around B_theo. This number of standard deviations is a little misleading because symmetric cuts around the central value B_theo is not what we are interested in. A measure of how well $(S+B)_\text{theo}$ is ruled out by exactly observing $N = B_\text{theo}$ events is the normalized distance from the observed background S_theo/σ_B. To quantify which kinds of small signals would be consistent with the observation of essentially the background rate $N = B_\text{theo}$ we make a choice: if $(S+B)_\text{theo}$ does not fall into the right 5 % tail of the background observation it is fine, if it falls into this tail it is ruled out. Given $N = B_\text{theo}$ this defines a critical number of expected signal events. Any model predicting more signal events is ruled out at the 95 % *confidence level* (CL). These 95 % of the probability distribution are not defined symmetrically, but through integrating from $N' > -\infty$ in Eq. (1.182), so the 95 % confidence level corresponds to something like $S_\text{theo}/\sigma_B < 1.5$. In practice, this makes it relatively easy to translate limits from one signal interpretation to another: all we need to do is compute the number of expected signal events in a given analysis, without any error analysis or other complications. Whenever it comes out above the published critical value the model is ruled out.

A variation of the ruling-out mode is when the observed number of events lies above or below the background prediction $N \neq B_{\text{theo}}$. In this case we still apply the 95 % confidence level condition following Eq. (1.182), but replace the central value B_{theo} by the number of observed events N. This gives us two critical values for S_{theo}, the expected exclusion limit computed around B_{theo} and the observed exclusion limit around N. If the observed background fluctuates below the prediction $N < B_{\text{theo}}$ we rule our more models than expected, when it fluctuates above $N > B_{\text{theo}}$ the exclusion starts to fail. This is the moment when we statistically switch from ruling-out mode to *discovery mode*.

The problem of a discovery is the fundamental insight that it is not possible to prove any scientific hypothesis correct. All we can do is prove all alternatives wrong. In other words, we discover a signal by solidly ruling out the background-only hypothesis. Following the above argument we now observe $B_{\text{theo}} < N \sim (S + B)_{\text{theo}}$. The question becomes: 'How likely is it that the background alone would have fluctuated to the observed number of events?'. To answer this question we need to again compute the statistical probability around the background hypothesis, Eq. (1.182). The difference to the argument above is that in the discovery mode this distribution is entirely hypothetical. A 5σ discovery of a given signal is claimed if at maximum a fraction of 5.8×10^{-7} expected events around the predicted value B_{theo} lie above the measured value $N \sim (S + B)_{\text{theo}}$,

$$p_0 \equiv \int_N^\infty dN' \, f(N'; B_{\text{theo}}, \sigma_B) < 5.8 \times 10^{-7} \,. \quad (1.183)$$

The function f can be close to a Gaussian but does not have to be. For example for small event numbers it should also have a Poisson shape, to avoid negative event numbers contributing to the integral. One interesting aspects in this argument is worth noting: backgrounds at the LHC are usually extracted from data with the help of standard theory tools. An obvious advantage of Eq. (1.183) is that we can immediately generalize it to more than one dimension, with a complicated function f indicating the correlated event numbers for several search channels.

Finally, in Eq. (1.183) the signal does not feature at all. Theorist hardly participate in the actual discovery of a new particle once they have suggested what to look for and delivered an understanding of the background in terms of simulations. On the other hand, experimentalists really only discover for example the 'Higgs boson' because it shows up in the search for a Higgs boson without any obviously weird features. To claim the discovery of a Higgs boson we need the 5σ deviation from the background expectations and a solid agreement of the observed features with the signal predictions. For the Higgs boson this for example means that the observed rates can be mapped on Higgs couplings which agree with the list in Sect. 1.3. Such an analysis we present in Sect. 1.8.1.

1.4 Higgs Discovery

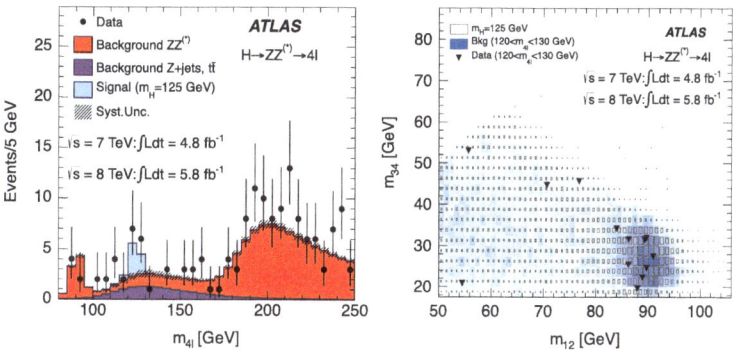

Fig. 1.5 *Left*: Higgs mass peak in the $m_{4\ell}$ spectrum, as shown in the ATLAS Higgs discovery paper [1]. *Right*: the correlation of the reconstructed Z and Z^* masses in the same analysis

Going back to the *ATLAS discovery paper*—the most important information is included in the abstract: ATLAS has discovered something in their search for the Standard Model Higgs boson. The analysis uses the data sets collected in 2011 and 2012, with their respective proton–proton energies of 7 and 8 TeV. The channels which contribute to the statistical discovery are Higgs decays $H \to ZZ \to 4\ell$, $H \to \gamma\gamma$, and $H \to W^+W^- \to 2\ell\,2\nu$. In addition, ATLAS includes the results from the $\tau^+\tau^-$ and $b\bar{b}$ decays, but their impact is negligible. What ATLAS observes is a peak with an invariant $\gamma\gamma$ or 4ℓ mass of 126 GeV and a combined significance of 5.9σ.

What follows after the short introduction are a section on the ATLAS detector (rather useless for us), and a section on the simulation of the signal and background event samples (not all that relevant for the outcome). Next comes the first of the three discovery channels, $H \to ZZ \to 4\ell$, where the four leptons include all possible combinations of two opposite-sign electrons and two opposite-sign muons. The idea of this analysis is to reconstruct the invariant mass of the four leptons $m_{4\ell}$ and observe a signal peak at the Higgs mass value over a relatively flat and well understood background. We discuss more details on this analysis in Sect. 1.5.4. In the left panel of Fig. 1.5 we see that $m_{4\ell}$ distribution. The clearly visible low peak around $m_{4\ell} \sim m_Z \sim 91$ GeV arises from an on-shell Z decay into two leptons plus a radiated photon, which in turn splits into two leptons. Above the threshold for the continuum production process $pp \to ZZ$ the background cross section increases again. In between, we cannot even speak of a flat background distribution, with typically one or two events per mass bin. Nevertheless, the *signal peak* is clearly visible, and we can proceed to compute the signal significance, making sure that we use Poisson statistics instead of Gaussian statistics. The result is quoted in Table 7 of the ATLAS paper—the ZZ decay channel contributes 3.6σ to the Higgs discovery, with a central Higgs mass value around 125 GeV. This number of sigmas really is the p_0 value defined in Eq. (1.183) translated into a Gaussian equivalent number of standard deviations N/σ_B. In the right panel of

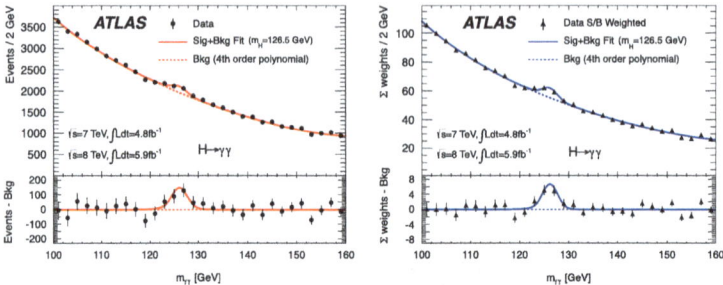

Fig. 1.6 *Left*: Higgs mass peak in the $m_{\gamma\gamma}$ spectrum. *Right*: the same distribution where all events are re-weighted with the value of S/B for the corresponding sub-selection (Both figures from the ATLAS discovery paper [1])

Fig. 1.5 we show an important consistency check of the ZZ sample supposedly coming from a Higgs decay. By definition, all signal events have a combined value around $m_{4\ell} = 125$ GeV. In addition, we know that the four leptons come from, possibly off–shell, Z bosons. In Sect. 2.1.2 we will discuss the functional form of $m_{\ell\ell}$ around the Z-mass pole. Quantum mechanics requires the unstable Z boson to decay exponentially, which corresponds to a Breit–Wigner shape around the resonance. This shape drops rapidly in the vicinity of the pole, but further out develops linear tails. For our case $M_H < 2m_Z$ it is most likely that one of the two Z bosons is on its mass shell while the other one decays into two leptons around $m_{\ell\ell} \sim m_H - m_Z = 35$ GeV. This is precisely what we see in Fig. 1.5. Just as a side remark: in CMS this distribution was originally very different from what we would expect from quantum mechanics.

The second analysis presented in the ATLAS paper is the search for rare $H \to \gamma\gamma$ decays. The basic strategy is to reconstruct the invariant mass of two photons $m_{\gamma\gamma}$ and check if it is flat, as expected from the background, or peaked at the Higgs mass. In the left panel of Fig. 1.6 we show this distribution. The functional shape of the background is flat, so without any derivation from first principles we can approximate the curve outside the peak region by a polynomial. This fit to a flat background we can subtract from the measured data points, to make the peak more accessible to the eye. For the peak shown in the left panel of Fig. 1.6 we could now compute a signal significance, i.e. the probability that the flat background alone fluctuates into the observed data points.

However, this is not how the analysis is done. One piece of information we should include is that we are not equally sure that an experimentally observed photon really is a photon everywhere in the detector. Some of the events entering the distribution shown in the left panel of Fig. 1.6 are more valuable than others. Therefore, ATLAS ranks the photon phase space or the relevant detector regions by their reliability of correctly identifying two photons and measuring their invariant mass. In each of these ten regions, listed in the Table 4 of their paper, they look at the $m_{\gamma\gamma}$ distribution and determine the individual peak significance. If the detector performance were the

1.4 Higgs Discovery

same in all ten regions the combination of these ten significances would be the same as the significance computed from the left panel of Fig. 1.6. Because the individual phase space and detector regimes have different signal and background efficiencies some events shown in the left panel of Fig. 1.6 are more equal than others, i.e. they contribute with a larger weight to the combined significance. This is what ATLAS illustrates in the right panel of Fig. 1.6: here, all events are weighted with the signal–to–background ratio S/B for the respective sub-analysis. It is an interesting development that such a *purely illustrational figure* without any strict scientific value enters a Higgs discovery paper. Obviously, the ATLAS and CMS collaborations feel that their actual results are not beautiful enough even for the scientific community, so they also deliver the public relations version. What is scientifically sound is the measured signal significance of a mass peak centered at 126.5 GeV, quoted as 4.5σ in their Table 7, but it cannot be computed from either of the two curves shown in Fig. 1.6.

The two $\gamma\gamma$ and ZZ analyses are the main ATLAS results shown in the Higgs discovery talk on July 4th. In the discovery paper ATLAS adds a third channel, namely *leptonic* $H \to WW$ decays. Obviously, we cannot reconstruct the $2\ell 2\nu$ mass, which means we need to rely on the transverse momentum balance in the detector to approximately reconstruct the Higgs mass. The details of this transverse mass measurement you can find in Sect. 3.3. Similar to the photon case, the analysis is then split into different regimes, now defined by the number of jets recoiling against the Higgs. The motivation for this observable is first to reject the top pair background with its additional two b jets and second to use the sensitivity of signal and background kinematics to the transverse momentum of the Higgs. However, in Sect. 2.3.4 we will see that perturbative QCD does not allow us to separately study collinear jets, i.e. jets with transverse momentum below the Higgs mass, beyond leading order in α_s. Such an observable violates collinear factorization, induces possibly large logarithms, and this way spoils the application of precision QCD predictions. After fitting the transverse mass distribution instead of simply cutting out the signal region the WW channel contributes 2.8σ to the final significance, but without a good Higgs mass determination.

The next two sections in the ATLAS discovery paper discuss details of the statistical analysis and the correlation of systematic uncertainties. After that, ATLAS combines the three analyses and interprets the result in terms of a *Standard Model Higgs boson*. First, in the *ruling–out mode* described above ATLAS gets rid of models with a Standard Model Higgs boson in the mass ranges 111–122 GeV and 131–559 GeV. This kind of exclusions from LHC (and Tevatron) Higgs analyses are shown in the colors of a great soccer nation. If we know the expected signal and background numbers and the detector performance, we can compute the number of signal events which we expect to exclude with a given amount of data if there were only background and no Higgs events. In the left panel of Fig. 1.7 the dashed line shows the expected exclusion limit as a function of the assumed Higgs mass and in terms of the signal strength normalized to the Standard Model Higgs rate $\mu = (\sigma \times \text{BR})/(\sigma \times \text{BR})_{\text{SM}}$. With the quoted amount of data we would expect to exclude the entire mass range from 110 to 580 GeV, provided this Higgs has Standard Model

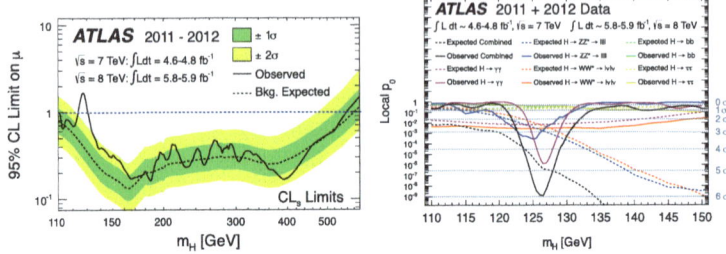

Fig. 1.7 *Left*: exclusion limits on a hypothetical signal strength as a function of the Higgs mass [1]. *Right*: signal significance as computed from the probability of a background fluctuation for the different channels (This figure is from the supplementary documentation to Ref. [1])

production and decay rates. For a hypothetical Higgs boson with only half the number of expected events we only expect to exclude Higgs masses from 120 to 460 GeV. Because this expected exclusion limit is a statistical computation it has error bars which are shown in green (1σ) and yellow (2σ). Replacing our expected event rates with data we find the solid curve in the left panel of Fig. 1.7. Around a hypothetical Higgs mass of 125 GeV the two significantly deviate, so we need to switch to *discovery mode*.

In the right panel of Fig. 1.7 we show the p_0 value computed by ATLAS as a function of the hypothetical Higgs mass. This Higgs mass is not actually needed to predict the background rates, but it enters because we optimize the signal searches for assumed Higgs masses. For example, we attempt to rule out the $\gamma\gamma$ background by determining its shape from a wide $m_{\gamma\gamma}$ range and test it for deviations with a resolution of few GeV. We see that the $H \to \gamma\gamma$ search as well as the $H \to ZZ$ search point towards Higgs masses around 125 GeV. For the $H \to WW$ analysis the assumed Higgs mass is less relevant, so the p_0 value shows a broad excess. Combining all channels gives us the solid black line in the right panel of Fig. 1.7, with a minimum p_0 value around 10^{-9} or 5.9σ.

Again, we can compute the signal significance which we would have predicted for this situation, shown as the dashed black line. The prediction reaches only 5σ, which means that assuming we are observing a Standard Model Higgs boson ATLAS' signal significance is slightly enhanced by upwards fluctuations in the event numbers.

One final technical term in the ATLAS discovery paper is not yet clear: this number of 5.9σ is described as the 'local significance'. The quoted local p_0 value is the probability of the background fluctuating into the observed signal configuration with a Higgs mass around 125 GeV. As mentioned above, ruling out the background should naively not depend on signal properties like the Higgs mass, but it does. Let us assume that we search for a Higgs boson in $m_{\gamma\gamma}$ bins of ± 2 GeV and in the mass range of 110–150 GeV. If all ten analyses in the different mass windows have identical p_0 values of $p_{0,j} = 10^{-9}$ and if they are statistically independent, we can approximately compute the probability of the background faking a signal in at least one of them as

$$p_0^{(\text{global})} = \sum_{j=1}^{10} p_{0,j} = 10 \times 10^{-9} = 10^{-8} \ . \tag{1.184}$$

This means that for a global 5-sigma discovery with $p_0^{\text{global}} < 5.8 \times 10^{-7}$ we need to require a significantly smaller values for the combination of the p_0^{local} for a given Higgs mass. In the above approximation the reduction is simply an effect of independent Poisson processes, its proper treatment is significantly more complicated. It is called *look-elsewhere effect*, where some people correctly point out that the more appropriate name would be look-everywhere effect. Obviously, if we combine the $\gamma\gamma$ and ZZ analyses with the flat p_0 value of the WW search the result is not as easy anymore. The global value $p_0^{(\text{global})}$ which ATLAS quotes corresponds to 5.1σ for an initial Higgs mass range of 110–600 GeV. The remaining discussion of the Higgs excess in the ATLAS paper we postpone to Sect. 1.8.1.

Essentially the same details we can find in the CMS discovery paper [14]. The observed local significance in the three main channels is 5.1σ, with an expected 5.2σ. The additional decay channels $H \to \tau\tau$ and $H \to b\bar{b}$ do not contribute, either by bad luck or expectedly. The main psychological difference between ATLAS and CMS seems to be that by the ordering of the references ATLAS starts with the well established Standard Model, of which the Higgs mechanism is a generic part which one would not mind discovering—while CMS starts with the references to the prediction of the Higgs boson.

1.5 Higgs Production in Gluon Fusion

After discussing the main aspects of the Higgs discovery we now go back to some theoretical physics background. Looking for the Higgs boson at hadron colliders starts with bad news: at tree level the Higgs hardly couples to light-flavor quarks and has no coupling to gluons. This is because the Higgs boson couples to all Standard Model particles proportional to their mass—this is the same operator they get their mass from. Because the $SU(3)_C$ symmetry of QCD is not broken, there is no coupling to gluons at all.

On the other hand, the protons at the LHC contain a lot of gluons, again something we will talk about in more detail in Sect. 2, so the question is if we can find and use a loop–induced coupling of two gluons to the Higgs. In spite of the expected suppression of the corresponding cross section by a one-loop factor $(g^2/(16\pi^2))^2$ we would hope to arrive at an observable production cross section $pp \to H$. Numerically, it will turn out that the production of Higgs bosons in gluon fusion is actually the dominant process at the LHC, as shown in Fig. 1.8.

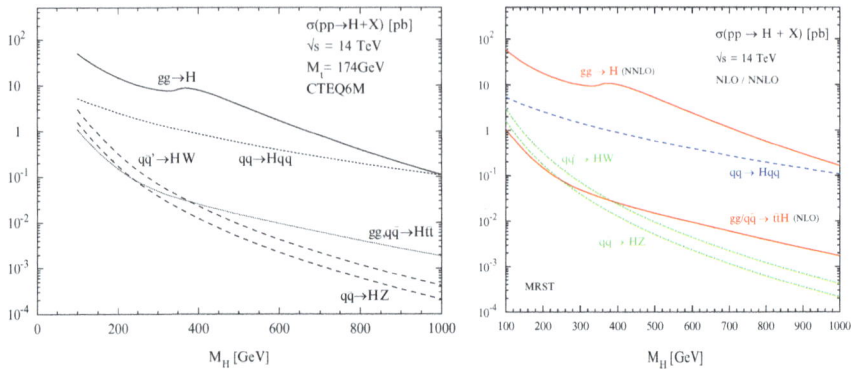

Fig. 1.8 *Left*: production cross section for a Standard-Model Higgs boson at the LHC, as a function of the Higgs mass (Figure from Ref. [35]). *Right*: updated version including higher order corrections

1.5.1 Effective Gluon–Higgs Coupling

If an effective *ggH* coupling should be mediated by a closed Standard Model *particle loop* the top is the perfect candidate: on the one hand it has a strong coupling to gluons, and on the other hand it has the largest of all Standard Model couplings to the Higgs boson, $m_t/v \sim 0.7$. The corresponding Feynman diagram is

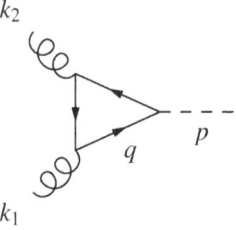

We construct this effective coupling in three steps, starting with the Dirac trace occurring the top loop. All momenta are defined as incoming with $k_1^2 = k_2^2 = 0$ and $p^2 = m_H^2$. The Dirac indices of the two gluons are μ, ν and the loop momentum is q, so in the first step we need to compute

$$T^{\mu\nu} = \text{Tr}\left[(\slashed{q} + m_t)\gamma^\mu (\slashed{q} + \slashed{k}_1 + m_t)\gamma^\nu (\slashed{q} + \slashed{k}_1 + \slashed{k}_2 + m_t)\right] . \qquad (1.185)$$

The calculational problem is the tensor structure of this trace. Because of gauge invariance we can neglect terms proportional to k_1^μ and k_2^ν; they would not survive the multiplication with the transverse gluon polarization ($k \cdot \epsilon = 0$). In a so-called axial gauge we could also get rid of the remaining terms proportional to k_1^ν and k_2^μ.

However, there is a better way to compute this trace. We know that there is no tree level Higgs coupling of the Higgs to two gluons, which would correspond to

1.5 Higgs Production in Gluon Fusion

the fields $HA_\mu A^\mu$ with mass dimension three in the Lagrangian. So we need to find another operator mediating such a coupling, keeping in mind that it is loop induced and can therefore include a mass suppression by powers of the top mass. The Higgs can also couple to the *field strength* in the invariant form $HG_{\mu\nu}G^{\mu\nu}$ with $G_{\mu\nu} \equiv \partial_\mu A_\nu - \partial_\nu A_\mu + \mathcal{O}(A^2)$. This operator has mass dimension five and arises from the dimension-6 gauge–invariant object $\phi^\dagger \phi\, G_{\mu\nu}G^{\mu\nu}$ after breaking $SU(2)_L$.

The factor in front of this term is the effective coupling we are going to compute in this section. Before that it pays to briefly look at the operator itself. Switching from position space and its momentum operator to momentum space $\partial \to ik$ shows that the gauge invariant operator linking exactly two gluon fields to a Higgs field has to be proportional the tensor

$$\begin{aligned}
G^{\mu\nu}G_{\mu\nu} &\xrightarrow{\text{F.T.}} i\left(k_{1\mu}A_{1\nu} - k_{1\nu}A_{1\mu}\right) i\left(k_{2\mu}A_{2\nu} - k_{2\nu}A_{2\mu}\right) + \mathcal{O}(A^3) \\
&= -2\left[(k_1 k_2)(A_1 A_2) - (k_1 A_2)(k_2 A_1)\right] + \mathcal{O}(A^3) \\
&= -2(k_1 k_2)\, A_{1\mu} A_{2\nu} \left[g^{\mu\nu} - \frac{k_1^\nu k_2^\mu}{k_1 k_2}\right] + \mathcal{O}(A^3) \\
&= -\sqrt{2} m_H^2\, A_{1\mu} A_{2\nu}\, P_T^{\mu\nu} + \mathcal{O}(A^3)\,,
\end{aligned} \quad (1.186)$$

where $P_T^{\mu\nu}$ is the transverse tensor

$$P_T^{\mu\nu} = \frac{1}{\sqrt{2}}\left[g^{\mu\nu} - \frac{k_1^\nu k_2^\mu}{(k_1 k_2)}\right]$$

$$P_T^{\mu\nu} P_{T\mu\nu} = 1 \quad \text{and} \quad P_T^{\mu\nu} k_{1\mu} = 0 = P_T^{\mu\nu} k_{2\nu} \qquad (k_1^2 = 0 = k_2^2)\,. \quad (1.187)$$

Based on this known tensor structure of $T^{\mu\nu}$ we can extract the scalar *form factor* F which corresponds to the Dirac trace of Eq. (1.185)

$$T^{\mu\nu} \sim F\, P_T^{\mu\nu} \quad \Leftrightarrow \quad P_{T\mu\nu} T^{\mu\nu} \sim P_{T\mu\nu} P_T^{\mu\nu}\, F = F\,. \quad (1.188)$$

The exact definition of the full form factor F in the Higgs–gluon coupling will obviously include all prefactors and the loop integral. This way we project out the relevant gluon tensor structure or the relevant degrees of freedom of the two gluons contributing to this effective coupling. Terms involving a larger number of gluon fields are related to this ggH coupling by non-abelian $SU(3)$ gauge invariance.

Our projection requires that we first compute $P_{T\mu\nu}T^{\mu\nu}$ based on Eq. (1.185). One thing to mention at this stage is that nobody in the world really computes Dirac traces by hand anymore. There are powerful programs, like FORM, which do this job for us. Using it we find the form factor

$$P_{T\mu\nu}T^{\mu\nu} = \frac{4m_t}{\sqrt{2}}\left(-m_H^2 + 3m_t^2 - \frac{8}{m_H^2}(k_1 q)(k_2 q) - 2(k_1 q) + q^2\right). \quad (1.189)$$

Inside this trace there appears the loop momentum q, which in our second step we have to consider as part of the loop integration. The effective ggH vertex includes the loop integral with the tensor structure from Eq. (1.185) in the numerator,

$$\int \frac{d^4q}{16\pi^4} \frac{P_{T\mu\nu}T^{\mu\nu}}{[q^2 - m_t^2][(q + k_1)^2 - m_t^2][(q + k_1 + k_2)^2 - m_t^2]}$$
$$= \frac{4m_t}{\sqrt{2}} \int \frac{d^4q}{16\pi^4} \frac{q^2 - 2(k_1q) - 8/m_H^2(k_1q)(k_2q) - m_H^2 + 3m_t^2}{[q^2 - m_t^2][(q + k_1)^2 - m_t^2][(q + k_1 + k_2)^2 - m_t^2]} \,. \quad (1.190)$$

The non–trivial q^μ dependence of the numerator observed in Eq. (1.190) we can take care of using a few tricks. For example in the first term we use the relation $q^2/(q^2 - m_t^2) = 1 + m_t^2/(q^2 - m_t^2)$ and then shift q, knowing that the final result for this integral will be finite. This is non–trivial piece of information, because most loop calculations lead to ultraviolet divergences, which need to be removed by first regularizing the integral and then renormalizing the parameters. The reason why we do not see any divergences in this process is that for a renormalization we would need something to renormalize, i.e. a leading order process which receives quantum corrections. However, we only compute this one-loop amplitude because there is no tree level vertex. There is nothing to renormalize, which means there are no ultraviolet divergences.

While these tricks help a little, we still do not know how to remove $(k_1q)(k_2q)$ in the third term of Eq. (1.190). The method of choice is a *Passarino-Veltman reduction* which turns tensor integrals into scalar integrals, where a scalar integral does not have powers of the loop momentum in the numerator. For example, the scalar three-point function is given by

$$C(k_1^2, k_2^2, m_H^2; m_t, m_t, m_t) \equiv \int \frac{d^4q}{i\pi^2} \frac{1}{[q^2 - m_t^2][(q + k_1)^2 - m_t^2][(q + k_1 + k_2)^2 - m_t^2]} \,. \quad (1.191)$$

The integral measure is $d^4q/(i\pi^2)$, or $d^nq/(i\pi^{n/2})$ for an arbitrary number of space–time dimensions. This removes any over–all factors 2 and π from the expression for the scalar integrals, as we will see below. Applied to the tensor integral in Eq. (1.190) the reduction algorithm gives us

$$\int \frac{d^4q}{i\pi^2} \frac{P_{T\mu\nu}T^{\mu\nu}}{[\ldots][\ldots][\ldots]} = \frac{4m_t}{\sqrt{2}} \left[2 + \left(4m_t^2 - m_H^2\right) C(0, 0, m_H^2; m_t, m_t, m_t) \right] \,. \quad (1.192)$$

The first term not proportional to any scalar integral has a curious origin. It comes from a combination of $\mathcal{O}(\epsilon)$ terms from the Dirac trace in $n = 4-2\epsilon$ dimensions and a two-point function which using the integration measure $1/i\pi^{n/2}$ always includes an ultraviolet divergence $1/\epsilon$. Note that these terms appear in the calculation in spite of the fact that the final result for the effective gluon–Higgs coupling is finite.

1.5 Higgs Production in Gluon Fusion

Scalar integrals we can for example calculate using the Feynman parameterization

$$\frac{1}{A_1 A_2 \cdots A_n} = \int_0^1 dx_1 \cdots dx_n \, \delta\left(\sum x_i - 1\right) \frac{(n-1)!}{(x_1 A_1 + x_2 A_2 + \cdots + x_n A_n)^n}, \tag{1.193}$$

but we usually obtain shorter analytical expressions using the *Cutkosky cutting rule* which links the imaginary part of a diagram or a scalar integral to the sum of all cut Feynman graphs.

The cut rule is directly related to the unitarity of the S matrix and the optical theorem discussed in Sect. 1.2.3. Limiting ourselves to scalar amplitudes/integrals, the cut rule tells us that the sum of all cut one-loop or squared scalar diagrams has to vanish, including the two external cuts which correspond to the untouched amplitude A and its complex conjugate A^*. This gives us a very useful expression for the imaginary part of the amplitude

$$-i\left(A - A^*\right) = 2 \operatorname{Im} A \stackrel{!}{=} 16\pi^2 \sum_{\text{cut graphs}} A. \tag{1.194}$$

The factor $16\pi^2$ arises from the generic integral measure $1/(16\pi^4)$ which we replace by $d^4q/(i\pi^2)$ such that the scalar integrals have a typical prefactor of one. Cutting diagrams means replacing all internal propagators by $1/(q^2 - m^2) \to 2\pi\,\theta(q_0)\,\delta(q^2 - m^2)$. Of the four dimensions of the loop integral d^4q the two on-shell conditions cancel two, leaving us with an simple angular integral. This angular integral does not include any kinematic information affecting the pole or cut structure of the scalar diagram.

From this imaginary part we compute the complete amplitude or scalar integral. If we know the pole or cut structure of the amplitude after cutting it, we can make use of the *Cauchy integral* for a complex analytic function $A(z)$

$$A(z) = \frac{1}{2\pi i} \oint_{\text{counter-clockwise}} dz' \, \frac{A(z')}{z' - z}, \tag{1.195}$$

and compute the unknown real part of $A(q^2)$. As an example, let us consider a scalar integral which like a full propagator has a cut on the real axis above $q^2 = m_1^2 + m_2^2$. This cut should not lie inside the integration contour of the Cauchy integral equation (1.195), so we deform the circle to instead follow the cut right above and below the real axis. If no other poles occur in the integral we find

$$\operatorname{Re} A(q^2) = \frac{1}{2\pi i} \oint dq'^2 \, \frac{i \operatorname{Im} A(q'^2)}{q'^2 - q^2}$$

$$= \frac{1}{2\pi} \int_\infty^{(m_1^2 + m_2^2)} dq'^2 \, \frac{\operatorname{Im} A(q'^2 - i\epsilon)}{q'^2 - q^2} + \frac{1}{2\pi} \int_{(m_1^2 + m_2^2)}^\infty dq'^2 \, \frac{\operatorname{Im} A(q'^2 + i\epsilon)}{q'^2 - q^2}$$

$$= \frac{1}{2\pi} \int_{(m_1^2+m_2^2)}^{\infty} dq'^2 \, \frac{\mathrm{Im}A(q'^2+i\epsilon) - \mathrm{Im}A(q'^2-i\epsilon)}{q'^2 - q^2}$$

$$\equiv \frac{1}{2\pi} \int_{(m_1^2+m_2^2)}^{\infty} dq'^2 \, \frac{\mathrm{Im}_+ A(q'^2)}{q'^2 - q^2} \, . \tag{1.196}$$

This step assumes a sufficiently fast convergence on the integration contour for large momenta. This method of computing for example scalar integrals is known to produce the most compact results.

The expression for the finite scalar three point function appearing in our effective coupling equation (1.192) has the form

$$\begin{aligned}
C(0,0,m_H^2;m_t,m_t,m_t) &= \frac{1}{m_H^2} \int_0^1 \frac{dx}{x} \log \frac{m_H^2 x(1-x) - m_t^2}{(-m_t^2)} \\
&= \frac{1}{m_H^2} \int_0^1 \frac{dx}{x} \log \left(1 - x(1-x) \frac{m_H^2}{m_t^2} \right) \\
&= \frac{1}{2m_H^2} \log^2 \left(-\frac{1 + \sqrt{1 - 4m_t^2/m_H^2}}{1 - \sqrt{1 - 4m_t^2/m_H^2}} \right) \qquad \frac{4m_t^2}{m_H^2} \equiv \tau < 1 \, .
\end{aligned} \tag{1.197}$$

For general top and Higgs masses it reads

$$C(0,0,m_H^2;m_t,m_t,m_t) = -\frac{2f(\tau)}{m_H^2} \quad \text{with} \quad f(\tau) = \begin{cases} \left(\arcsin \sqrt{\frac{1}{\tau}} \right)^2 & \tau > 1 \\ -\frac{1}{4} \left(\log \frac{1 + \sqrt{1-\tau}}{1 - \sqrt{1-\tau}} - i\pi \right)^2 & \tau < 1 \end{cases}, \tag{1.198}$$

including imaginary or absorptive terms for $\tau < 1$. The dimensionless variable τ is the appropriate parameter to describe the behavior of this scalar integral. For example the low energy limit of the scalar integral, i.e. the limit in which the top loop becomes heavy and cannot be resolved by the external energy of the order of the Higgs mass, will be given by $\tau \gtrsim 1$ which means $m_H < 2m_t$. In contrast to what many people who use such effective vertices assume, the expression in Eq. (1.198) is valid for *arbitrary Higgs and top masses*, not just in the heavy top limit.

Expressing our Dirac trace and loop integral in terms of this function $f(\tau)$ we find for our effective coupling in Eq. (1.192)

1.5 Higgs Production in Gluon Fusion

$$\int \frac{d^4q}{i\pi^2} \frac{P_{T\mu\nu}T^{\mu\nu}}{[\ldots][\ldots][\ldots]} = \frac{4m_t}{\sqrt{2}} \left(2 - \left(4m_t^2 - m_H^2\right) \frac{2f(\tau)}{m_H^2}\right)$$

$$= \frac{4m_t}{\sqrt{2}} \left(2 - 2(\tau - 1) f(\tau)\right)$$

$$= \frac{8m_t}{\sqrt{2}} \left(1 + (1-\tau) f(\tau)\right) \,. \tag{1.199}$$

Using this result we can as the third and last step of our calculation collect all factors from the propagators and couplings in our Feynman diagram and compute the *effective ggH coupling* now including all pre-factors,

$$F = - i^3 (-ig_s)^2 \frac{im_t}{v} \text{Tr}(T^a T^b) \frac{i\pi^2}{16\pi^4} \int \frac{d^4q}{i\pi^2} \frac{P_{T\mu\nu}T^{\mu\nu}}{[\ldots][\ldots][\ldots]}$$

$$= - i^3 (-ig_s)^2 \frac{im_t}{v} \text{Tr}(T^a T^b) \frac{i\pi^2}{16\pi^4} \frac{8m_t}{\sqrt{2}} (1 + (1-\tau) f(\tau))$$

$$= \frac{g_s^2 m_t}{v} \frac{\delta^{ab}}{2} \frac{i}{16\pi^2} \frac{8m_t}{\sqrt{2}} (1 + (1-\tau) f(\tau))$$

$$= \frac{g_s^2}{v} \frac{\delta^{ab}}{2} \frac{i}{16\pi^2} \frac{8}{\sqrt{2}} \frac{m_H^2 \tau}{4} (1 + (1-\tau) f(\tau))$$

$$= ig_s^2 \, \delta^{ab} \frac{1}{16\sqrt{2}\pi^2} \frac{m_H^2}{v} \tau \, (1 + (1-\tau) f(\tau))$$

$$= i\alpha_s \, \delta^{ab} \frac{1}{4\sqrt{2}\pi} \frac{m_H^2}{v} \tau \, (1 + (1-\tau) f(\tau)) \,. \tag{1.200}$$

The numerical factors originate from the closed fermion loop, the three top propagators, the two top-gluon couplings, the top Yukawa coupling, the color trace, the unmatched loop integration measure, and finally the result computed in Eq. (1.199).

Based on Eq. (1.186) we can write in momentum space as well as in position space

$$F \, P_T^{\mu\nu} A_{1\mu} A_{2\nu} = F \, \frac{-G^{\mu\nu} G_{\mu\nu}}{\sqrt{2} m_H^2} \,. \tag{1.201}$$

In this form we can include the form factor F in an effective Lagrangian and finally define the Feynman rule we are interested in

$$\boxed{\mathscr{L}_{ggH} \supset \frac{1}{v} g_{ggH} \, H \, G^{\mu\nu} G_{\mu\nu}} \quad \text{with} \quad \frac{1}{v} g_{ggH} = -i \frac{\alpha_s}{8\pi} \frac{1}{v} \tau \left[1 + (1-\tau) f(\tau)\right] \,.$$
$$\tag{1.202}$$

after dropping δ^{ab}. It is important to notice that the necessary factor in front of the dimension-5 operator is $1/v$ and not $1/m_t$. This is a particular feature of this coupling, which does not decouple for heavy top quarks because we have included the top Yukawa coupling in the numerator. Without this Yukawa coupling, the heavy top limit $\tau \to \infty$ of the expression would be zero, as we will see in a minute. Unlike one might expect from a general effective theory point of view, the higher dimensional operator inducing the Higgs–gluon coupling is not suppressed by a large energy scale. This means that for example a fourth generation of heavy fermions will contribute to the effective Higgs–gluon coupling as strongly as the top quark, with no additional suppression by the heavy new masses. The breaking of the usual decoupling by a large Yukawa coupling makes it easy to experimentally rule out such an additional generation of fermions, based on the Higgs production rate.

Of course, just like we have three-gluon and four-gluon couplings in QCD we can compute the $gggH$ and the $ggggH$ couplings from the ggH coupling simply using gauge invariance defining the terms we omit in Eq. (1.186). This set of n-gluon couplings to the Higgs boson is again not an approximate result in the top mass. Gauge invariance completely fixes the n-gluon coupling to the Higgs via one exact dimension-5 operator in the Lagrangian. These additional gluon field arise from the commutator of two gluon field in the field strength tensor, so they only exist in non–abelian QCD and cannot be generalized to the photon-photon-Higgs coupling.

1.5.2 Low–Energy Theorem

The general expression for g_{ggH} is not particularly handy, but for light Higgs bosons we can write it in a more compact form. We start with a Taylor series for $f(\tau)$ in the *heavy-top limit* $\tau \gg 1$

$$f(\tau) = \left[\arcsin \frac{1}{\tau^{1/2}}\right]^2 = \left[\frac{1}{\tau^{1/2}} + \frac{1}{6\tau^{3/2}} + \mathcal{O}\left(\frac{1}{\tau^{5/2}}\right)\right]^2$$

$$= \frac{1}{\tau} + \frac{1}{3\tau^2} + \mathcal{O}\left(\frac{1}{\tau^3}\right) \xrightarrow{\tau \to \infty} 0, \qquad (1.203)$$

and combine it with all other τ-dependent terms from Eq. (1.202)

$$\tau\left[1 + (1-\tau)f(\tau)\right] = \tau\left[1 + (1-\tau)\left(\frac{1}{\tau} + \frac{1}{3\tau^2} + \mathcal{O}\left(\frac{1}{\tau^3}\right)\right)\right]$$

$$= \tau\left[1 + \frac{1}{\tau} - 1 - \frac{1}{3\tau} + \mathcal{O}\left(\frac{1}{\tau^2}\right)\right]$$

1.5 Higgs Production in Gluon Fusion

$$= \tau \left[\frac{2}{3\tau} + \mathcal{O}\left(\frac{1}{\tau^2}\right) \right]$$

$$= \frac{2}{3} + \mathcal{O}\left(\frac{1}{\tau}\right), \qquad \text{implying} \qquad \boxed{g_{ggH} = -i \frac{\alpha_s}{12\pi}}. \tag{1.204}$$

In this low energy or heavy top limit we have *decoupled the top quark* from the set of propagating Standard Model particles. The ggH coupling does not depend on m_t anymore and gives a finite result. Computing this finite result in Eq. (1.200) we had to include the top Yukawa coupling from the numerator. We emphasize again that while this low energy approximation is very compact to analytically write down the effective ggH coupling, it is not necessary to numerically compute processes involving the effective ggH coupling.

In this low energy limit we can easily add more Higgs bosons to the loop. Attaching an external Higgs leg to the gluon self energy diagram simply means replacing one of the two top propagators with two top propagators and adding a Yukawa coupling

$$\frac{i}{\slashed{q} - m_t} \rightarrow \frac{i}{\slashed{q} - m_t} \frac{-i\sqrt{2}m_t}{v} \frac{i}{\slashed{q} - m_t} \tag{1.205}$$

We can compare this replacement to a differentiation with respect to m_t

$$\frac{\partial}{\partial m_t} \frac{1}{\slashed{q} - m_t} = \frac{\partial}{\partial m_t} \frac{\slashed{q} + m_t}{q^2 - m_t^2}$$

$$= \frac{(q^2 - m_t^2) - (\slashed{q} + m_t)(-2m_t)}{(q^2 - m_t^2)^2}$$

$$= \frac{q^2 + 2m_t \slashed{q} + m_t^2}{(q^2 - m_t^2)^2} = \frac{\slashed{q}^2 + 2m_t \slashed{q} + m_t^2}{(q^2 - m_t^2)^2} = \frac{(\slashed{q} + m_t)(\slashed{q} + m_t)}{(q^2 - m_t^2)^2}$$

$$\Rightarrow \quad \frac{i}{\slashed{q} - m_t} \rightarrow \frac{-i\sqrt{2}m_t}{v} \frac{1}{\slashed{q} - m_t} \frac{1}{\slashed{q} - m_t} = \frac{-i\sqrt{2}m_t}{v} \frac{\partial}{\partial m_t} \frac{1}{\slashed{q} - m_t}. \tag{1.206}$$

This means that we can replace one propagator by two propagators using a derivative with respect to the heavy propagator mass. The correct treatment including the gamma matrices in $\slashed{q} = \gamma_\mu q^\mu$ involves carefully adding unit matrices in this slightly schematic derivation. However, our shorthand notation gives us an idea how we can in the limit of a heavy top derive the ggH^{n+1} couplings from the ggH^n coupling

$$g_{ggH^{n+1}} = m_t^{n+1} \frac{\partial}{\partial m_t} \left(\frac{1}{m_t^n} g_{ggH^n} \right). \tag{1.207}$$

This relation holds for the scattering amplitude before squaring and assuming that the tensor structure is given by the same transverse tensor in Eq. (1.187). We can for example use this relation to link the two effective couplings with one or two external Higgs legs, i.e. the triangle form factor $g_{ggH} = -i\alpha_s/(12\pi)$ and the box form factor $g_{ggHH} = +i\alpha_s/(12\pi)$. Corrections to this relation appear at the order $1/m_t$.

The question arises if we can even link this form factor to the gluon self energy without any Higgs coupling. The relation in Eq. (1.206) suggests that there should not be any problem as long as we keep the momenta of the incoming and outgoing gluon different. In Sect. 2.2.2 we will see that the gluon self energy loops have a transverse tensor structure, so there is no reason not to use the top loop in the gluon self energy to start the series of ggH^n couplings in the heavy top limit. Note that this does not include the entire gluon self energy diagram, but only the top loop contributing. The so-called beta function does not appear in this effective coupling. If we want to combine the effects of more than just one particle in the gluon self energy we need to integrate out one state after the other. The appropriate effective theory framework then becomes the Coleman–Weinberg potential discussed in Sect. 1.2.7.

To obtain the correct mass dimension each external Higgs field appears as H/v. Using $\log(1 + x) = -\sum_{n=1}(-x)^n/n$ we can eventually resum this series of effective couplings in the form

$$\boxed{\mathscr{L}_{ggH} = G^{\mu\nu}G_{\mu\nu} \frac{\alpha_s}{\pi} \left(\frac{H}{12v} - \frac{H^2}{24v^2} + \cdots\right) = \frac{\alpha_s}{12\pi} G^{\mu\nu}G_{\mu\nu} \log\left(1 + \frac{H}{v}\right).}$$
(1.208)

In Sect. 1.2.3 we note that there should be a relative factor between the Lagrangian and the Feynman rule accounting for more than one way to identify the external legs of the Feynman rule with the fields in the Lagrangian. For n neutral Higgs fields the effective coupling has to include an additional factor of $1/n$ which is precisely the denominator of the logarithm's Taylor series.

Such a closed form of the Lagrangian is very convenient for simple calculations and gives surprisingly exact results for the $gg \to H$ production rate at the LHC, as long as the Higgs mass does not exceed roughly twice the top mass. However, for example for $gg \to H$+jets production its results only hold in the limit that *all* jet momenta are much smaller than m_t. It also becomes problematic for example in the pair production process $gg \to HH$ close to threshold, where the momenta of slow–moving Higgs bosons lead to an additional scale in the process. We will come back to this process later.

1.5.3 Effective Photon-Higgs Coupling

The Higgs coupling to two photons can be computed exactly the same way as the effective coupling to gluons. Specifically, we can compute a form factor in analogy

1.5 Higgs Production in Gluon Fusion

to Eqs. (1.200) and (1.202). The only difference is that there exist two particles in the Standard Model which have a sizeable Higgs coupling (or mass) and electric charge: heavy quarks and the W boson. Both contribute to the partial width of a Higgs decaying to two photons.

To allow for some more generality we give the results for the Higgs decay width for a general set of fermions, gauge bosons, and scalars in the loop:

$$\Gamma(H \to \gamma\gamma) = \frac{G_F \alpha^2 m_H^3}{256\sqrt{2}\pi^3} \left| \sum_f N_c Q_f^2 \, g_{Hff} \, A_{1/2} + g_{HWW} A_1 + \sum_s N_c Q_s^2 \, \frac{g_{Hss}}{m_s^2} A_0 \right|^2 . \tag{1.209}$$

The color factor for fermions without a color charge should be replaced by unity. In the literature, the form factors A are often defined with an additional factor 2, so the prefactor will only include $1/128$. The factor G_F describes the coupling of the loop particles to the Higgs boson, while each factor $\alpha = e^2/(4\pi)$ arises because of the QED coupling of the loop particles to a photon. In the Standard Model the two leading contributions are

$$Q_t^2 g_{Htt} = \left(\frac{2}{3}\right)^2 \times 1 \qquad g_{HWW} = 1 \qquad g_{Hss} = 0 \quad (1.210)$$

This means that the Standard Model Higgs couplings proportional to the particle masses are absorbed into the form factors A. From the discussion in Sect. 1.5.2 we know that the top quark form factor does not decouple for heavy quarks because of the Yukawa coupling in the numerator. The same is true for the W boson where the entire W mass dependence is absorbed into G_F.

The form factors A for scalars, fermions, and gauge bosons we can only quote here. They all include $f(\tau)$ as defined in Eq. (1.198), which is the scalar three-point function and can be computed without any further approximations. As before, we define $\tau = 4m^2/m_H^2$ given the mass m of the particle running in the loop. From Eq. (1.203) we know that in the limit of heavy loop particles $\tau \to \infty$ the scalar integral scales like $f(\tau) \sim 1/\tau + 1/(3\tau^2) \cdots$, which allows us to compute some basic properties of the different loops contributing to the effective Higgs–photon coupling.

$$A_0 = -\frac{\tau}{2}[1 - \tau f(\tau)] \xrightarrow{\tau \to \infty} -\frac{\tau}{2} + \frac{\tau}{2} + \frac{1}{6} = \frac{1}{6}$$

$$A_{1/2} = \tau[1 + (1-\tau)f(\tau)] \longrightarrow \tau + 1 - \tau - \frac{1}{3} = \frac{2}{3}$$

$$A_1 = -\frac{1}{2}[2 + 3\tau + 3(2\tau - \tau^2)f(\tau)] \longrightarrow -1 - \frac{3}{2}\tau - 3 + \frac{3}{2}\tau + \frac{1}{2} = -\frac{7}{2} .$$
$$\tag{1.211}$$

We see that unless a sign appear in the prefactors—and from Eq. (1.209) we know that it does not—the top quark and W boson loops interfere destructively. This is not an effect of the spin alone, because the scalar form factor has the same sign as the fermion expression. In addition, we observe that for equal charges the gauge boson will likely dominate over the fermion, which is correct for the top quark vs W boson in the Standard Model.

1.5.4 Signatures

Different Higgs production and decay processes are not only important for the Higgs discovery, they also allow us to test many properties of the recently discovered new particle. Many aspects of such measurements go beyond our knowledge of hadron collider physics and QCD. For example to discuss the Higgs production in gluon fusion we would normally need to know how to deal with gluons inside the incoming protons, how to parameterize the phase space of the Higgs decay products, and how to kinematically distinguish interesting events from the rest. All of this we will piece by piece introduce in Sect. 2.1. On the other hand, we can try to understand the LHC capabilities in Higgs physics already at this point. In the following sections on Higgs production at the LHC we will therefore limit ourselves to some very basic phenomenological features and postpone any discussion on how to compute these features.

The first quantity we can compute and analyze at colliders is the total number of events expected from a certain production process in a given time interval. For example for our current Higgs studies the event numbers in different Higgs production and decay channels are the crucial input. Such a number of events is the product of the proton–proton LHC luminosity measured in inverse femtobarns, the total production cross section measured in femtobarns, and the detection efficiency measured in per-cent. In other words, a predicted event rate it is split into a collider-specific number describing the initial state, a process-specific number describing the physical process, and a detector-specific efficiency for each particle in the final state.

The latter is the easiest number to deal with: over the sensitive region of the detector, the *fiducial volume*, the detection efficiency is a set of numbers depending on the nature of the detected particle and its energy. This number is very good for muons, somewhere between 90 and 100 %, and less than 1/3 for tau leptons. Other particles typically range somewhere in between.

For theorists luminosity is simply a conversion number between cross sections which we compute for a living and event numbers. People who build colliders use units involving seconds and square meters, but for us inverse femtobarns work better. Typical numbers are: a year of LHC running at design luminosity could deliver up to 10 inverse femtobarns per year in the first few years and three to ten times that later. The key numbers and their orders of magnitude for typical signals are

1.5 Higgs Production in Gluon Fusion

Fig. 1.9 Production rates for signal and background processes at hadron colliders. The discontinuity is due to the Tevatron being a proton–antiproton collider while the LHC is a proton–proton collider. The two colliders correspond to the x–axis values of 2 TeV and something between 7 and 14 TeV (Figure from Ref. [13])

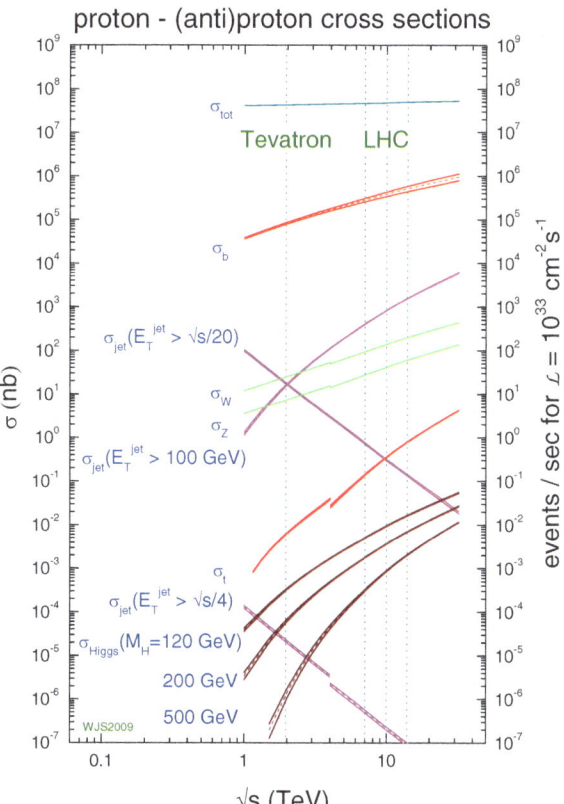

$$N_{events} = \sigma_{tot} \cdot \mathscr{L} \qquad \mathscr{L} = 10 \cdots 300 \,\text{fb}^{-1} \qquad \sigma_{tot} = 1 \cdots 10^4 \,\text{fb} \,. \qquad (1.212)$$

Different cross sections for Tevatron and LHC processes are shown in Fig. 1.9.

Finally, talking about cross sections and how to compute them we need to remember that at the LHC there exist two kinds of processes. The first involves all particles which we know and love, like old-fashioned electrons or slightly more modern W and Z bosons or most recently top quarks. All of these processes we call *backgrounds*. They are described by QCD, which means QCD is the theory of evil. Top quarks have an interesting history, because when I was a graduate student they still belonged to the second class of processes, the *signals*. These either involve particles we have not seen before or particles we want to know something about. By definition, signals are very rare compared to backgrounds. As an example, Fig. 1.9 shows that at the LHC the production cross section for a pair of bottom quarks is larger than 10^5 nb or 10^{11} fb, the typical production rate for W or Z bosons ranges around 200 nb or 2×10^8 fb, the rate for a pair of 500 GeV supersymmetric gluinos would have been 4×10^4 fb, and the Higgs rate can be as big as 2×10^5 fb. This really rare Higgs signal was extracted by ATLAS and CMS with a 5σ significance in the Summer of 2012. If we see such a new particles someone gets a call from Stockholm, while for the rest of the community the corresponding processes instantly turn into backgrounds.

One last aspect we have to at least mention is the *trigger*. Because of the sheer mass of data at the LHC, we will not be able to write every LHC event on tape. As a matter of fact, we could not even write every top pair event on tape. Instead, we have to decide very fast if an event has the potential of being interesting in the light of the physics questions we are asking at the LHC. Only these events we keep. Before a mis-understanding occurs: while experimentalists are reluctant to change triggers these are not carved in stone, so as a functioning high energy physics community we will not miss great new physics just because we forgot to include it in the trigger menu. For now we can safely assume that above an energy threshold we will keep all events with leptons or photons, plus as much as we can events with missing energy, like neutrinos in the Standard Model and dark matter particles in new physics models and jets with high energy coming from resonance decays. This trigger menu reflects the general attitude that the LHC is not built to study QCD, and that very soft final states for example from bottom decays are best studied by the LHCb experiment instead of ATLAS and CMS.

With this minimal background of collider phenomenology we can look at Higgs production in gluon fusion, combined with different Higgs decays. This is the production channels which dominated the Higgs discovery discussed in Sect. 1.4, based on LHC runs with a center–of–mass energy of 7 TeV (2011) and 8 TeV (2012). The total 14 TeV Higgs production cross section through the loop–induced ggH coupling we show in Fig. 1.8. For a reasonably light Higgs boson the cross section ranges around at least 30 pb, which for relevant luminosities starting around 30 fb^{-1} means 10^6 events. The question is: after multiplying with the relevant branching ratio and detection efficiencies, which of the decays can be distinguished from the Standard Model background statistically? Since gluon fusion really only produces the Higgs boson the details of the production process do not help much with the background suppression. The results of experimental simulations, for example by the ATLAS collaboration, are shown in Fig. 1.10. The complete list of possible Higgs decays, ordered by decreasing branching ratio according to Fig. 1.4, is:

- $gg \to H \to b\bar{b}$ is hopeless, because of the sheer size of the QCD continuum background $gg \to b\bar{b}$, which according to Fig. 1.9 exceeds the signal by roughly eight orders of magnitude. In gluon fusion there is little to cut on except for the invariant mass of the $b\bar{b}$ pair with an $\mathcal{O}(10\,\%)$ mass resolution. Such a cut will not reduce the background by more than two or three orders of magnitude, so the signal–to–background ratio will be tiny. Pile–up, i.e. different scattering events between the proton bunches might in addition produce unwanted structures in the m_{bb} distribution for the pure QCD background. The final blow might be that this channel will, as it stands, not be triggered on.
- $gg \to H \to \tau^+\tau^-$ is problematic. If taus decay leptonically we can identify them in the detector, but there will appear one or two neutrinos in their decay. This means that we cannot reconstruct the tau momentum. We will discuss this decay and an approximate mass reconstruction in detail in Sect. 1.6.3.

1.5 Higgs Production in Gluon Fusion

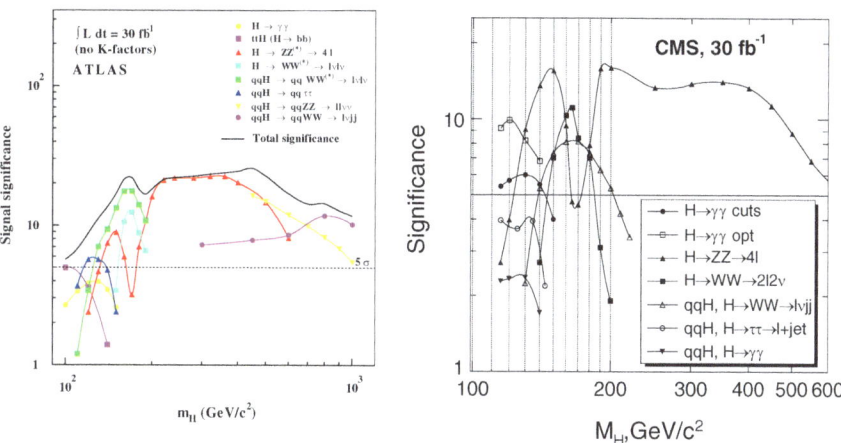

Fig. 1.10 Simulated statistical significance for different Higgs production and decay channels for an integrated luminosity of $30\,\text{fb}^{-1}$ (*left*: ATLAS [2]; *right*: CMS [6]). Five standard deviations over the backgrounds are required for discovery. Given the measured Higgs mass around 125 GeV these figures are of mostly historic interest, except that they illustrate how lucky we are that the Higgs mass lies right in between the fermion and bosonic decays

This approximate reconstruction only works when the neutrinos lead to a measurable two-dimensional missing energy vector. If the Higgs decays at rest its decay production will be mostly back to back, so its low velocity makes the reconstruction of $m_{\tau\tau} \sim m_H$ hard. It is widely assumed that Higgs production in gluon fusion is too close to threshold to see many decays to tau leptons, but in combination with other production channels and including hard jet recoil this channel should have some power.

- $gg \to H \to \gamma\gamma$ is, in spite of the small rate, the main Higgs discovery channel. Because $m_{\gamma\gamma}$ can be reconstructed to $\mathcal{O}(1\,\%)$ this observable has incredibly precise side bins to the left and the to right of the Higgs peak. This is what for example the electromagnetic calorimeter of CMS has been designed for. The main problem is backgrounds for example from pions mistaken for photons, while theory input will play no role in this analysis. A slight problem is that in gluon fusion there is again little to cut on except for $m_{\gamma\gamma}$. The only additional observables which can reduce the physical two-photon background exploit a slightly boosted Higgs kinematics, either through the opening angles between the photons or the transverse momentum of the two photon system. In spite of the small branching ratios to photons the $\gamma\gamma$ channels is therefore limited by physical, irreducible backgrounds and our understanding of their $m_{\gamma\gamma}$ distribution.

The peak in the invariant mass of the photons is great to measure the *Higgs mass*: once we see a Gaussian peak we can determine its central value with a precision of $\Gamma_{\text{detector}}/\sqrt{S}$ (in a signal dominated sample with S signal events), which translates into the per-mille level. The only issue in this measurement are systematic effects from the photon energy calibration. Unlike for electrons and photons there is no Z peak structure in $m_{\gamma\gamma}$, so we would have to calibrate the photon energy from a strongly dropping distribution like $d\sigma/dE_\gamma$. An alternative approach is to use the calibration of another electromagnetic particle, like the electron, and convert the electron energy scale into a photon energy scale using Monte Carlo detector simulations or to look for $\ell^+\ell^-\gamma$ decays of the Z.

- $gg \to H \to W^+W^-$ has a large rate, but once one of the W bosons decays leptonically the Higgs mass is hard to reconstruct. All we can do is reconstruct a transverse mass variable, which we discuss in Sect. 3.3. On the other hand, the backgrounds are electroweak and therefore small. A very dangerous background is top pair production which gives us two W bosons and two relatively hard bottom quarks with typical transverse momenta $p_{T,b} \gtrsim 40$ GeV. We can strongly reduce this background by vetoing jets in additional to the two leptonically decaying W bosons. As we will learn in Sect. 2 such a jet veto is a problem once it covers collinear jet radiation from the incoming hadrons. In that case it breaks collinear factorization, a principle underlying any precision computation of the Higgs production rate at the LHC.

  The $H \to WW$ analysis strongly relies on angular correlations—if the two gauge bosons come from a spin-zero resonance they have to have opposite polarization; because the W coupling to fermions is purely left handed this implies that the two leptons prefer to move into the same direction as opposed to back–to–back. This effect can be exploited either by asking for a small opening angle of the two leptons or asking for a small invariant mass of the two leptons. Note that once we apply this cut we have determined the spin structure of the extracted signal to be a scalar.

  In the original ATLAS and CMS analyses the WW decay looked not very useful for Higgs masses below 150 GeV, i.e. for far off–shell Higgs decays. Because there is not much more to cut on the expected significance dropped sharply with the decreasing branching ratio. However, in the 7 and 8 TeV run this effect was countered by lowering the minimum transverse momentum requirements for the leptons, so the Higgs mass range covered by the WW analysis now extends to the observed mass of 125 GeV.

- $gg \to H \to ZZ$ works great for ZZ to four leptons, in particular muons, because of the fully reconstructed $m_{4\ell} \sim m_{ZZ} \sim m_H$. Of all Higgs channels it requires the least understanding of the LHC detectors. Therefore it is referred to as '*golden channel*'. Experimentally at least the four-muon channel is relatively easy. The electron decays can serve as a useful cross check.

  Its limitation are the leptonic Z branching ratio and the sharp drop in the off–shell Higgs branching ratio towards smaller Higgs masses. Once we include the leptonic Z decays the over–all $H \to 4\ell$ branching ratio for a 125 GeV Higgs is tiny. The good news is that unlike in the two-photon channel there are essentially no irreducible backgrounds. The continuum production $q\bar{q} \to ZZ$ is an electroweak ($2 \to 2$) process and as rare as the Higgs production process. The loop–induced $gg \to ZZ$ process is kinematically very similar to the signal, but even more rare. One useful cut based on the Breit–Wigner propagator shape is the distribution of the two invariant masses of the lepton pairs $m_{\ell\ell}$. For their

1.6 Higgs Production in Weak Boson Fusion   93

Higgs discovery ATLAS and CMS asked for one pair of leptons with $m_{12} \sim m_Z$, all four leptons with $m_{4\ell} = 125\,\text{GeV}$, and the second pair of leptons off–shell, $m_{34} \ll m_Z$.

- $gg \to H \to Z\gamma$ has recently been advertized as theoretically interesting once we link it to the observed loop–induced $H \to \gamma\gamma$ and tree level $H \to ZZ$ decays. It behaves a little like $\gamma\gamma$, but with a smaller rate and a further reduced branching ratio of $Z \to \ell^+\ell^-$. Instead of combining the advantages of $H \to ZZ$ and $H \to \gamma\gamma$ this channel combines more of the disadvantages, so it is not likely to be measured soon. Of course, as for any channel seeing it will give us more information on the Higgs boson, so we should not give up. In addition, for some theoretical ideas it might be useful to determine an upper limit on the $H \to Z\gamma$ branching ratio.
- $gg \to H \to \mu^+\mu^-$ might be the only hope we will ever have to measure a second-generation Yukawa coupling at the LHC. Because of its clear signature and its huge $q\bar{q} \to Z, \gamma \to \mu^+\mu^-$ background this analysis resembles the photons channel, but with a much more rare signal. Eventually, other production processes might help with the Higgs decays to muons, similar to the $H \to \tau^+\tau^-$ case.
- $gg \to H \to$ invisible is not predicted in Standard Model; it is obviously hopeless if only the Higgs is produced, because we would be trying to extract a signal of missing energy and nothing else. 'Absolutely nothing' in addition to some QCD remnant is not a good signature for the trigger.

From the list of above channels we understand that the Higgs discovery is dominated by the 'golden' $H \to 4\ell$ and the 'silver' $H \to \gamma\gamma$ channels. The off-shell and hardly reconstructable $H \to WW$ channel adds only little in terms of a distinctive signal. If we want to learn more about the Higgs boson, we need additional production mechanisms opening more decay signatures. Moreover, at this point it is still not clear why ATLAS and CMS in their Higgs discovery papers separate a Higgs–plus–two–jets signal for example in the photon decay channel.

1.6 Higgs Production in Weak Boson Fusion

Going back to Fig. 1.8 we see that while gluon fusion gives the largest Higgs production rate at the LHC, there are other promising channels to study. In the Standard Model the Higgs has sizeable couplings only to the W and Z bosons and to the top quark, so instead of via the top Yukawa coupling we can produce Higgs bosons via their gauge boson couplings. This induces two channels, the larger of which is weak boson fusion $qq \to qqH$: two incoming quarks each radiate a W or Z boson which merge and form a Higgs. Because the LHC is a pp collider and because the proton mostly contains the valence quarks (uud) and low-x gluons it is important that this process can proceed as $ud \to duH$, where the u radiates a W^+ and the d radiates a W^-. The Feynman diagram for this process is

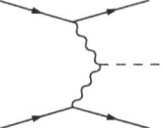

If the Higgs were a Z boson, it could also bremsstrahlung off the incoming or outgoing quarks, but for Higgs production at colliders we safely assume that the first two generation fermions are massless. That is at least unless we discuss a muon collider as a specific way to produce Higgs bosons.

In a way, weak boson fusion looks like double deep inelastic scattering, one from each of the protons. This is one of the key observations which in Sect. 1.6.2 we will use for background suppression via the central jet veto. The double deep inelastic scattering approximation is also a good way to compute corrections to the weak boson fusion production rate, at least provided we neglect kinematic distributions. Just a final comment: the LHC experiments refer to weak boson fusion as vector boson fusion (VBF). However, vector boson fusion includes incoming gluons, which have very different kinematic properties, so in the following we strictly mean weak boson fusion mediated by massive W and Z exchange.

1.6.1 Production Kinematics

In the Feynman diagrams for weak boson fusion Higgs production we encounter intermediate massive gauge boson propagators. They induce a particular shape of the kinematic distributions of the final–state jet. First, we need to quote the exact calculation showing that in the matrix element squared we will usually find one power of p_T in the numerator. With this information we can look for the maximum in the $p_{T,j} = p_{T,W}$ spectrum as a function of the momentum in the beam direction, p_3, and the absolute value of the two-dimensional transverse momentum p_T

$$0 \stackrel{!}{=} \frac{\partial}{\partial p_T} \frac{p_T}{E^2 - p_T^2 - p_L^2 - m_W^2}$$

$$= \frac{1}{E^2 - p_T^2 - p_L^2 - m_W^2} + p_T \frac{(-1)}{(E^2 - p_T^2 - p_L^2 - m_W^2)^2} (-2p_T)$$

$$= \frac{E^2 + p_T^2 - p_L^2 - m_W^2}{(E^2 - p_T^2 - p_L^2 - m_W^2)^2}$$

$$\sim \frac{C m_W^2 + p_T^2 - m_W^2}{(E^2 - p_T^2 - p_L^2 - m_W^2)^2} \qquad \text{with} \quad E^2 \sim p_L^2 \gg m_W^2 \quad \text{but} \quad E^2 - p_L^2 = C m_W^2$$

$$= \frac{p_T^2 - (1-C)m_W^2}{(E^2 - p_T^2 - p_L^2 - m_W^2)^2}$$

$$\Leftrightarrow \quad \boxed{p_T^2 = (1-C)m_W^2} \qquad (1.213)$$

1.6 Higgs Production in Weak Boson Fusion

at the maximum and for some number $C < 1$. This admittedly hand-waving argument shows that in weak boson fusion Higgs production the transverse momenta of the outgoing jets peak at values below the W mass. In reality, the peak occurs around $p_T \sim 30$ GeV. This transverse momentum scale we need to compare to the longitudinal momentum given by the energy scale of valence quarks at the LHC, i.e. several hundreds of GeV.

These two forward jets are referred to as *tagging jets*. They offer a very efficient cut against QCD backgrounds: because of their back–to–back geometry and their very large longitudinal momentum, their invariant mass m_{jj} will for a 14 TeV collider energy easily exceed a TeV. For any kind of QCD background this will not be the case. Compared to Higgs production in gluon fusion the tagging jets are an example how features of the production process which have little or nothing to do with the actual Higgs kinematics can help reduce backgrounds—the largest production rate does not automatically yield the best signatures. The only problem with the weak boson fusion channel is that its distinctive m_{jj} distribution requires a large collider energy, so running the LHC at 7 and 8 TeV for a Higgs discovery was very bad news for this channel.

Moving on to the Higgs kinematics, in contrast to the jets the Higgs and its decay products are expected to hit the detector centrally. We are looking for two forward jets and for example two τ leptons or two W bosons in the central detector. Last but not least, the Higgs is produced with finite transverse momentum which is largely determined by the acceptance cuts on the forward jets and their typical transverse momentum scale $p_{TH} \sim m_W$.

Compared to Higgs production in gluon fusion we buy this distinctive signature and its efficient extraction from the background at the expense of the rate. Let us start with the partonic cross sections: the one-loop amplitude for $gg \to H$ is suppressed by $\alpha_s y_t/(4\pi) \sim (1/10)(2/3)(1/12) = 1/180$. For the production cross section this means a factor of $(1/180)^2 \sim 1/40{,}000$. The cross section for weak boson fusion is proportional to g^6, but with two additional jets in the final state. Including the additional phase space for two jets this roughly translates into $g^6/(16\pi)^2 \sim (2/3)^6 \, 1/(16\pi)^2 = (64/729)(1/2{,}500) \sim 1/25{,}000$. These two numbers governing the main LHC production cross sections roughly balance each other.

The difference in rate which we see in Fig. 1.8 instead arises from the quark and gluon luminosities. In weak boson fusion the two forward jets always combine to a large partonic center–of–mass energy $x_1 x_2 s > (p_{j,1} + p_{j,2})^2 = 2(p_{j,1} p_{j,2})$, with the two parton momentum fractions $x_{1,2}$ and the hadronic center of mass energy $\sqrt{s} = 14$ TeV. Producing a single Higgs in gluon fusion probes the large gluon parton density at typical parton momentum fractions $x \sim m_H/\sqrt{s} \sim 10^{-3}$. This means that each of the two production processes with their specific incoming partons probes its most favorable parton momentum fraction: low-x for gluon fusion and high-x for valence quark scattering. Looking at typical LHC energies, the gluon parton density grows very steeply for $x \lesssim 10^{-2}$. This means that gluon fusion wins: for a 125 GeV Higgs the gluon fusion rate of ~ 50 pb clearly exceeds the weak boson fusion rate of ~ 4.2 pb. On the other hand, these numbers mean little when we battle an 800 pb $t\bar{t}$ background relying on kinematic cuts either on forward jets or on Higgs decay products.

In Fig. 1.10 we see that for large Higgs mass the weak boson fusion rate approaches the gluon fusion rate. The two reasons for this behavior we mentioned already in this section: first of all, for larger x values the rate for $gg \to H$ decreases steeply with the gluon density, while in weak boson fusion the already huge partonic center of mass energy due to the tagging jets ensures that an increase in m_H makes no difference anymore. Even more importantly, there appear *large logarithms* because the low-p_T enhancement of the quark–W splitting. If we neglect m_W in the weak boson fusion process the $p_{T,j}$ distributions will diverge for small $p_{T,j}$ like $1/p_{T,j}$, as we will see in Sect. 2.3.3 After integrating over $p_{T,j}$ this yields a log $p_{T,j}^{\max}/p_{T,j}^{\min}$ dependence of the total rate. With the W mass cutoff and a typical hard scale given by m_H this logarithm becomes

$$\sigma_{\text{WBF}} \propto \left(\log \frac{p_{T,j}^{\max}}{p_{T,j}^{\min}}\right)^2 \sim \left(\log \frac{m_H}{m_W}\right)^2. \tag{1.214}$$

For $m_H = \mathcal{O}(\text{TeV})$ this logarithm gives us an enhancement by factors of up to 10, which makes weak boson fusion the dominant Higgs production process.

Motivated by such logarithms, we will talk about partons inside the proton and their probability distributions for given momenta in Sect. 2.3.3. In the *effective W approximation* we can resum the logarithms appearing in Eq. (1.214) or compute such a probability for W bosons inside the proton. This number is a function of the partonic momentum fraction x and can be evaluated as a function of the transverse momentum p_T. Because the incoming quark inside the proton has negligible transverse momentum, the transverse momenta of the W boson and the forward jet are identical. These transverse momentum distributions in $p_{T,W} = p_{T,j}$ look different for transverse and longitudinal gauge bosons

$$P_T(x, p_T) \sim \frac{g_V^2 + g_A^2}{4\pi^2} \frac{1 + (1-x)^2}{x} \frac{p_T^3}{(p_T^2 + (1-x) m_W^2)^2}$$

$$\to \frac{g_V^2 + g_A^2}{4\pi^2} \frac{1 + (1-x)^2}{x} \frac{1}{p_T}$$

$$P_L(x, p_T) \sim \frac{g_V^2 + g_A^2}{4\pi^2} \frac{1-x}{x} \frac{2 m_W^2 p_T}{(p_T^2 + (1-x) m_W^2)^2}$$

$$\to \frac{g_V^2 + g_A^2}{4\pi^2} \frac{1-x}{x} \frac{2 m_W^2}{p_T^3}. \tag{1.215}$$

The couplings $g_{A,V}$ describe the gauge coupling of the W bosons to the incoming quarks. Looking at large transverse momenta $p_T \gg m_W$ the radiation of longitudinal W bosons falls off sharper than the radiation of transverse W bosons. This different behavior of transverse and longitudinal W bosons is interesting, because it allows us to gain information on the centrally produced particle and which modes it couples to just from the transverse momentum spectrum of the forward jets and without looking at the actual central particle.

However, numerically the effective W approximation does not work well for a 125 GeV Higgs at the LHC. The simple reason is that the Higgs mass is of the order of the W mass, as are the transverse momenta of the W and the final–state jets, and none of them are very small. Neglecting for example the transverse momentum of the W bosons or the final–state jets will not give us useful predictions for the kinematic distributions, neither for the tagging jets nor for the Higgs. For the SSC, the competing design to the LHC in Texas which unfortunately was never built, this might have been a different story, but at the LHC we should not describe W bosons (or for that matter top quarks) as essentially massless partons inside the proton.

1.6.2 Jet Ratios and Central Jet Veto

From the Feynman diagram for weak boson fusion we see that the diagram describing a gluon exchange between the two quark lines multiplied with the Born diagram is proportional to the color factor $\text{Tr}\, T^a \, \text{Tr}\, T^b \delta^{ab} = 0$. The only way to avoid this suppression is the interference of two identical final–state quarks, for example in ZZ fusion. First, this does not involve only valence quarks and second, this assumes a phase space configuration where one of the two supposedly forward jets turns around and goes backwards, so the interfering diagrams contribute in the same phase space region. This means that virtual gluon exchange in weak boson fusion is practically absent.

In Sect. 2 we will see that virtual gluon exchange and real gluon emission are very closely related. Radiating a gluon off any of the quarks in the weak boson fusion process will lead to a double infrared divergence, one because the gluon can be radiated at small angles and one because the gluon can be radiated with vanishing energy. The divergence at small angles is removed by redefining the quark parton densities in the proton. The soft, non–collinear divergence has to cancel between real gluon emission and virtual gluon exchange. However, if virtual gluon exchange does not appear, non–collinear soft gluon radiation cannot appear either. This means that additional QCD jet activity as part of the weak boson fusion process is limited to collinear radiation, i.e. radiation along the beam line or at least in the same direction as the far forward tagging jets. Gluon radiation into the central detector is suppressed by the color structure of the weak boson fusion process.

While it is not immediately clear how to quantify such a statement it is a very useful feature, for example looking at the top pair backgrounds. The $WWb\bar{b}$ final state as a background to qqH, $H \to WW$ searches includes two bottom jets which can mimic the signal's tagging jets. At the end, it turns out that it is much more likely that we will produce another jet through QCD jet radiation, i.e. $pp \to t\bar{t}$+jet, so only one of the two bottom jets from the top decays needs to be forward. In any case, the way to isolate the Higgs signal is to look at additional central jets.

As described above, for the signal additional jet activity is limited to small-angle radiation off the initial–state and final–state quarks. For a background like top pairs this is not the case, which means we can reduce all kinds of background by vetoing jets in the central region above $p_{T,j} \gtrsim 30\,\text{GeV}$. This strategy is referred to as *central jet veto* or *mini-jet veto*. Note that it has nothing to do with rapidity gaps at HERA or pomeron exchange, it is a QCD feature completely accounted for by standard perturbative QCD.

From QCD we then need to compute the probability of not observing additional central jets for different signal and background processes. Postponing the discussion of QCD parton splitting to Sect. 2.3.2 we already know that for small transverse momenta the $p_{T,j}$ spectra for massless states will diverge, as shown in Eq. (1.214). Looking at some kind of n-particle final state and an additional jet radiation we can implicitly define a reference point p_T^{crit} at which the divergent rate for one jet radiation σ_{n+1} starts to exceed the original rate σ_n, whatever the relevant process might be

$$\sigma_{n+1}(p_T^{\text{crit}}) = \int_{p_T^{\text{crit}}}^{\infty} dp_{T,j}\, \frac{d\sigma_{n+1}}{dp_{T,j}} \stackrel{!}{=} \sigma_n \,. \tag{1.216}$$

This condition defines a point in p_T below which our perturbation theory in α_s, i.e. in counting the number of external partons, breaks down. For weak boson fusion Higgs production we find $p_T^{\text{crit}} \sim 10\,\text{GeV}$, while for QCD processes like $t\bar{t}$ production it becomes $p_T^{\text{crit}} = 40\,\text{GeV}$. In other words, jets down to $p_T = 10\,\text{GeV}$ are perturbatively well defined for Higgs signatures, while for the QCD backgrounds jets below $40\,\text{GeV}$ are much more frequent than they should be looking at the perturbative series in α_s. This fixes the p_T range where a central jet veto will be helpful to reject backgrounds

$$p_{T,j} > 30\,\text{GeV} \quad\text{and}\quad \eta_j^{(\text{tag 1})} < \eta_j < \eta_j^{(\text{tag 2})} \,. \tag{1.217}$$

The second condition reminds us of the fact that only central jets will be rare in weak boson fusion. The smaller the p_T threshold the more efficient the central jet veto becomes, but at some point experimental problems as well as non–perturbative QCD effects will force us to stay above 20 or 30 or even $40\,\text{GeV}$.

If we assign a probability pattern to the radiation of jets from the core process we can compute the *survival probability* P_{pass} of such a jet veto. For many years we have been told that higher orders in the perturbative QCD series for the Higgs production cross section is the key to understanding LHC rates. For multi–jet observables like a jet veto this is not necessarily true. As an example we assume NNLO or two-loop precision for the Higgs production rate $\sigma = \sigma_0 + \alpha_s \sigma_1 + \alpha_s^2 \sigma_2$ where we omit the over–all factor α_s^2 in σ_0. Consequently, we define the cross section passing the jet veto $\sigma^{(\text{pass})} = P_{\text{pass}}\, \sigma = \sum_j \alpha_s^j \sigma_j^{(\text{pass})}$. Because the leading order prediction only includes a Higgs in the final state we know that $\sigma_0^{(\text{pass})} = \sigma_0$. Solving this definition for the veto survival probability we can compute

1.6 Higgs Production in Weak Boson Fusion

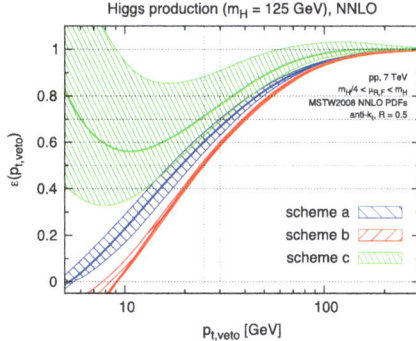

Fig. 1.11 Different predictions for the jet veto survival probability P_{pass} as a function of the maximum allowed $p_{T,j}$. The example process chosen is Higgs production in gluon fusion. The *shaded regions* indicate the independent variation of the factorization and renormalization scales within $[m_H/4, m_H]$ requiring μ_R/μ_F to lie within $[0.5, 2]$ (The figure and the corresponding physics argument are taken from Ref. [3])

$$P_{\text{pass}}^{(a)} = \frac{\sigma^{(\text{pass})}}{\sigma} = \frac{\sigma_0 + \alpha_s \sigma_1^{(\text{pass})} + \alpha_s^2 \sigma_2^{(\text{pass})}}{\sigma_0 + \alpha_s \sigma_1 + \alpha_s^2 \sigma_2}, \quad (1.218)$$

motivated by including the maximum number of terms (NNLO) in the numerator and denominator. The result as a function of the maximum allowed $p_{T,j}$ is shown as 'scheme a' in Fig. 1.11. The shaded region is an estimate of the theoretical uncertainty of this prediction.

Alternatively, we can argue that the proper perturbative observable is the fraction of vetoed events $(1 - P_{\text{pass}})$. Indeed, for small values of α_s the jet radiation probability vanishes and with it $(1 - P_{\text{pass}}) \sim \alpha_s \to 0$. This vetoed event fraction we can compute as $\sigma_j - \sigma_j^{(\text{pass})}$ for $j \geq 0$. However, we need to keep in mind that in the presence of an additional jet the NNLO prediction for the inclusive Higgs production rate reduces to NLO accuracy, so we include the two leading terms in the numerator and denominator,

$$1 - P_{\text{pass}}^{(b)} = \frac{\alpha_s(\sigma_1 - \sigma_1^{(\text{pass})}) + \alpha_s^2(\sigma_2 - \sigma_2^{(\text{pass})})}{\sigma_0 + \alpha_s \sigma_1}$$

$$P_{\text{pass}}^{(b)} = 1 - \frac{\alpha_s(\sigma_1 - \sigma_1^{(\text{pass})}) + \alpha_s^2(\sigma_2 - \sigma_2^{(\text{pass})})}{\sigma_0 + \alpha_s \sigma_1}$$

$$= \frac{\sigma_0 + \alpha_s \sigma_1^{(\text{pass})} + \alpha_s^2 \sigma_2^{(\text{pass})} - \alpha_s^2 \sigma_2}{\sigma_0 + \alpha_s \sigma_1}. \quad (1.219)$$

Obviously, in Eq. (1.219) we can move the term $-\alpha_s^2 \sigma_2$ into the denominator and arrive at Eq. (1.218) within the uncertainty defined by the unknown α_s^3 terms. This defines 'scheme b' in Fig. 1.11.

Finally, we can consistently Taylor expand the definition of P_{pass} as the ratio given in Eq. (1.218). The two leading derivatives of a ratio read

$$\left(\frac{f}{g}\right)' = \frac{f'g - fg'}{g^2} \stackrel{f=g}{=} \frac{f' - g'}{g}$$

$$\left(\frac{f}{g}\right)'' = \left(\frac{f'g}{g^2} - \frac{fg'}{g^2}\right)' = \frac{(f'g)'g^2 - f'g2gg'}{g^4} - \frac{(fg')'g^2 - fg'2gg'}{g^4}$$

$$= \frac{(f'g)' - 2f'g'}{g^2} - \frac{(fg')'g - 2fg'g'}{g^3} = \frac{f''g - f'g'}{g^2} - \frac{fg''g - fg'g'}{g^3}$$

$$\stackrel{f=g}{=} \frac{f'' - g''}{g} - \frac{g'(f' - g')}{g^2} \tag{1.220}$$

In the last steps we assume $f = g$ at the point where we evaluate the Taylor expansion. Applied to the perturbative QCD series for $(1 - P_{\text{pass}})$ around the zero-coupling limit this gives us

$$1 - P_{\text{pass}}^{(c)} = 1 - \frac{\sigma_0 + \alpha_s \sigma_1^{(\text{pass})} + \alpha_s^2 \sigma_2^{(\text{pass})} + \cdots}{\sigma_0 + \alpha_s \sigma_1 + \alpha_s^2 \sigma_2 + \cdots}$$

$$P_{\text{pass}}^{(c)} = 1 + \alpha_s \frac{\sigma_1^{(\text{pass})} - \sigma_1}{\sigma_0} + \alpha_s^2 \frac{\sigma_2^{(\text{pass})} - \sigma_2}{\sigma_0} - \alpha_s^2 \frac{\sigma_1(\sigma_1^{(\text{pass})} - \sigma_1)}{\sigma_0^2}, \tag{1.221}$$

defining 'scheme c' in Fig. 1.11. The numerical results indicate that the three schemes are inconsistent within their theoretical uncertainties, and that the most consistent Taylor expansion around perfect veto survival probabilities is doing particularly poorly. Towards small $p_{T,j}$ veto ranges the fixed order perturbative approach clearly fails. The way to improve the theoretical prediction is a re-organization of the perturbation theory for small jet transverse momenta. We introduce this approach with its leading term, the parton shower, in Sect. 2.5. For now we conclude that our theoretical approach has to go beyond a fixed number of (hard) jets and include the production of any number of jets in some kind of modified perturbative series.

One ansatz for the distribution of any number of radiated jets is motivated by soft photon emission off a hard electron. In Sect. 2.5.2 we derive the *Poisson distribution* in the numbers of jet which follows in the soft limit. If we for now assume a Poisson distribution, the probability of observing exactly n jets given an expected $\bar{n}$ jets is

$$f(n; \bar{n}) = \frac{\bar{n}^n e^{-\bar{n}}}{n!} \quad \Rightarrow \quad \boxed{P_{\text{pass}} \equiv f(0; \bar{n}) = e^{-\bar{n}}}. \tag{1.222}$$

1.6 Higgs Production in Weak Boson Fusion

Note that this probability links rates for exactly n jets, no at least n jets, i.e. it described the exclusive number of jets. The Poisson distribution is normalized to unity, once we sum over all possible jet multiplicities n. It defines the so-called exponentiation model. We consistently fix the expectation value in terms of the inclusive cross sections producing at least zero or at least one jet,

$$\langle n \rangle \equiv \bar{n} = \frac{\sigma_1(p_T^{\min})}{\sigma_0} \,. \tag{1.223}$$

This ensures that the inclusive jet ratio σ_1/σ_0 is reproduced by the ratio of the corresponding Poisson distributions. Including this expectation value $\bar{n}$ into Eq. (1.222) returns a veto survival probability of $\exp(-\sigma_1/\sigma_0)$. This comes out roughly as 88 % for the weak boson fusion signal and as 24 % for the $t\bar{t}$ background. For the signal–to–background ratio this implies a three-fold increase.

An alternative model starts from a constant probability of radiating a jet, which in terms of the inclusive cross sections σ_n, i.e. the production rate for the radiation of at least n jets, reads

$$\frac{\sigma_{n+1}(p_T^{\min})}{\sigma_n(p_T^{\min})} = R^{(\text{incl})}_{(n+1)/n}(p_T^{\min}) \,. \tag{1.224}$$

We derive this pattern in Sect. 2.6. The expected number of jets is then given by

$$\langle n \rangle = \frac{1}{\sigma_0} \sum_{j=1} j(\sigma_j - \sigma_{j+1}) = \frac{1}{\sigma_0} \left(\sum_{j=1} j\sigma_j - \sum_{j=2}(j-1)\sigma_j \right) = \frac{1}{\sigma_0} \sum_{j=1} \sigma_j$$

$$= \frac{\sigma_1}{\sigma_0} \sum_{j=0} (R^{(\text{incl})}_{(n+1)/n})^j = \frac{R^{(\text{incl})}_{(n+1)/n}}{1 - R^{(\text{incl})}_{(n+1)/n}} \,, \tag{1.225}$$

if $R^{(\text{incl})}_{(n+1)/n}$ is a constant. Assuming the series converges this turns into a requirement on $p_T^{\min}$. Radiating jets with such a constant probability has been observed at many experiments, including most recently the LHC, and is in the context of W+jets referred to as *staircase scaling*. We will derive both, the Poisson scaling and the staircase scaling from QCD in Sect. 2.6.1. Even without saying anything on how to calculate exclusive processes with a fixed number of jets we can derive a particular property of the constant probability of staircase scaling: the ratios of the $(n+1)$-jet rate to the n-jet rate for inclusive and exclusive jet rates are identical. We can see this by computing the inclusive $R^{(\text{incl})}_{(n+1)/n}$ in terms of exclusive jet rates

$$R^{(\text{incl})}_{(n+1)/n} = \frac{\sigma_{n+1}}{\sigma_n} = \frac{\sum_{j=n+1}^{\infty} \sigma_j^{(\text{excl})}}{\sigma_n^{(\text{excl})} + \sum_{j=n+1}^{\infty} \sigma_j^{(\text{excl})}}$$

$$= \frac{\sigma_{n+1}^{(\text{excl})} \sum_{j=0}^{\infty} R^j_{(n+1)/n}}{\sigma_n^{(\text{excl})} + \sigma_{n+1}^{(\text{excl})} \sum_{j=0}^{\infty} R^j_{(n+1)/n}} \qquad \text{with} \quad R_{(n+1)/n} = \frac{\sigma_{n+1}^{(\text{excl})}}{\sigma_n^{(\text{excl})}}$$

$$= \frac{\dfrac{R_{(n+1)/n} \sigma_n^{(\text{excl})}}{1 - R_{(n+1)/n}}}{\sigma_n^{(\text{excl})} + \dfrac{R_{(n+1)/n} \sigma_n^{(\text{excl})}}{1 - R_{(n+1)/n}}} = \frac{R_{(n+1)/n}}{1 - R_{(n+1)/n} + R_{(n+1)/n}}$$

$$= R_{(n+1)/n} \,. \tag{1.226}$$

To show that the exponentiation model and staircase scaling are not the only assumptions we can make to compute jet rates we show yet another, but similar ansatz which tries to account for an increasing number of legs to radiate jets off. Based on

$$\frac{\sigma_{j+1}(p_T^{\min})}{\sigma_j(p_T^{\min})} = \frac{j+1}{j} R^{(\text{incl})}_{(n+1)/n}(p_T^{\min}) \,, \tag{1.227}$$

the expectation for the number of jets radiated gives, again following Eq. (1.225)

$$\langle n \rangle = \frac{1}{\sigma_0} \sum_{j=1} j \sigma_j = \frac{1}{\sigma_0} \sigma_0 \sum_{j=1} j (R^{(\text{incl})}_{(n+1)/n})^j$$

$$= R^{(\text{incl})}_{(n+1)/n} \sum_{j=1} j (R^{(\text{incl})}_{(n+1)/n})^{j-1} = \frac{R^{(\text{incl})}_{(n+1)/n}}{(1 - R^{(\text{incl})}_{(n+1)/n})^2} \,. \tag{1.228}$$

All of these models are more or less well motivated statistical approximations. The do not incorporate experimental effects or the non–perturbative underlying event, i.e. additional energy dependent but process independent jet activity in the detectors from many not entirely understood sources. For many reasons none of them is guaranteed to give us a final and universal number. However, by the time we get to Sect. 2.6.2 we will at least be able to more accurately describe the central jet veto in QCD.

For the Poisson distribution and the staircase distribution we can summarize the main properties of the n-jet rates in terms of the upper incomplete gamma function $\Gamma(n, \bar{n})$:

	Staircase scaling	Poisson scaling
$\sigma_n^{(excl)}$	$\sigma_0^{(excl)} e^{-bn}$	$\sigma_0 \dfrac{e^{-\bar{n}} \bar{n}^n}{n!}$
$R_{(n+1)/n} = \dfrac{\sigma_{n+1}^{(excl)}}{\sigma_n^{(excl)}}$	e^{-b}	$\dfrac{\bar{n}}{n+1}$
$R_{(n+1)/n}^{(incl)} = \dfrac{\sigma_{n+1}}{\sigma_n}$	e^{-b}	$\left(\dfrac{(n+1)\, e^{-\bar{n}}\, \bar{n}^{-(n+1)}}{\Gamma(n+1) - n\Gamma(n,\bar{n})} + 1 \right)^{-1}$
$\langle n \rangle$	$\dfrac{1}{2} \dfrac{1}{\cosh b - 1}$	$\bar{n}$
P_{pass}	$1 - e^{-b}$	$e^{-\bar{n}}$

1.6.3 Decay Kinematics and Signatures

For most of the Higgs decays discussed in Sect. 1.3 it does not matter how the Higgs is produced, as long as we only care about the signal events. Many aspects discussed in Sect. 1.5.4 for Higgs production in gluon fusion can be directly applied to weak boson fusion. Serious differences appear only when we also include backgrounds and the kinematic cuts required to separate signal and background events.

The only fundamental difference appears in the reconstruction of Higgs decay into τ pairs. The sizeable transverse momentum of the Higgs in the weak boson fusion process allows us to reconstruct the invariant mass of a $\tau\tau$ system in the *collinear approximation*: if we assume that a τ with momentum $\vec{p}$ decays into a lepton with the momentum $x\vec{p}$ and a neutrino, both moving into the same direction as the tau, we can write the two-dimensional transverse momentum of the two taus from the Higgs decay in two ways

$$\vec{p}_1 + \vec{p}_2 \equiv \frac{\vec{k}_1}{x_1} + \frac{\vec{k}_2}{x_2} \stackrel{!}{=} \vec{k}_1 + \vec{k}_2 + \vec{k} \;. \tag{1.229}$$

The missing transverse momentum $\vec{k}$ is the measured vector sum of the two neutrino momenta. This equation is useful because we can measure the missing energy vector at the LHC in the transverse plane, i.e. as two components, which means Eq. (1.229) is really two equations for the two unknowns x_1 and x_2. Skipping the calculation of solving these two equations for $x_1 x_2$ we quote the result for the invariant $\tau\tau$ mass

$$\boxed{m_{\tau\tau}^2 = 2\,(p_1 p_2) = 2\frac{(k_1 k_2)}{x_1 x_2}} \;. \tag{1.230}$$

For the signal this corresponds to the Higgs mass. From the formula above it is obvious that this approximation does not only require a sizeable $p_i \gg m_\tau$, but also that back–to–back taus will not work—the two vectors contributing to $\vec{k}$ then largely cancel and the computation fails. This is what happens for the inclusive production channel $gg \to H \to \tau\tau$, where the Higgs boson is essentially produced at rest.

Again, we can make a list of *signatures* which work more or less well in connection to weak boson fusion production. These channels are also included in the summary plot by ATLAS, shown in Fig. 1.10.

- $qq \to qqH, H \to b\bar{b}$ is problematic because of large QCD backgrounds and because of the trigger in ATLAS. The signal–to–background ratio is not quite as bad as in the gluon fusion case, but still not encouraging. The most worrisome background is overlapping events, one producing the two tagging jets and the other one two bottom jets. This overlapping scattering gives a non–trivial structure to the background events, so a brute force side-bin analysis will not work.
- $qq \to qqH, H \to \tau^+\tau^-$ had the potential to be a discovery channel for a light Higgs boson with $m_H \lesssim 130\,\text{GeV}$, at least for an LHC energy of 14 TeV. The 2012 run at 8 TeV gives us the opportunity to at least see a small Higgs signal. In early analyses we can limit ourselves to only one jet recoiling against the Higgs, to increase the sensitivity. Eventually, the approximate mass reconstruction might be as good as $\sim 5\,\text{GeV}$, because we can measure the peak position of a Gaussian distribution with a precision of $\Gamma_{\text{detector}}/\sqrt{S}$. This channel is particularly useful in scenarios beyond the Standard Model, like its minimal supersymmetric extension. It guaranteed the discovery of one Higgs boson over the entire supersymmetric parameter space without a dedicated SUSY search.

  Like for almost all weak boson fusion analyses there are two irreducible backgrounds: $Z + n$ jets production at order $G_F \alpha_s^n$ and the same final state at order $G_F^3 \alpha_s^{n-2}$. The latter has a smaller rate, but because it includes weak boson fusion Z production as one subprocess it is much more similar to Higgs production for example from a kinematical or a QCD point of view.
- $qq \to qqH, H \to \gamma\gamma$ should be almost comparable with $gg \to H \to \gamma\gamma$ with its smaller rate but improved background suppression. For 14 TeV the two-jet topology which is already part of the discovery analysis presented in Sect. 1.4 will become more and more important. As a matter of fact, the weak boson fusion channel is usually included in $H \to \gamma\gamma$ analyses, and for neural net analyses zooming in on large Higgs transverse momenta it will soon dominate the inclusive analysis.
- $qq \to qqH, H \to W^+W^-$ contributes to the discovery channel for $m_H \gtrsim 125\,\text{GeV}$, ideally at 14 TeV collider energy. In comparison to $gg \to H \to W^+W^-$ it works significantly better for off–shell W decays, i.e. for Higgs masses below 150 GeV. There, the multitude of background rejection cuts and a resulting better signal–to–background ratio win over the large rate in gluon fusion. Apart from the two tagging jets the analysis proceeds the same way as the gluon fusion analysis. The key background to get rid of are top pairs.

- $qq \to qqH$, $H \to ZZ$ is likely to work in spite of the smaller rate compared to gluon fusion. It might even be possible with one hadronic Z decay, but there are not many detailed studies available. On the other hand, already the gluon–fusion $H \to ZZ$ search, which is one of the backbones of the Higgs discovery, has essentially no backgrounds. This takes away the biggest advantage of weak boson fusion as a Higgs production channel—the improved background reduction.
- $qq \to qqH$, $H \to Z\gamma$ is difficult due to a too small event rate and no apparent experimental advantages compared to the gluon–fusion Higgs production.
- $qq \to qqH$, $H \to \mu^+\mu^-$ sounds very hard, but it might be possible to observe at high luminosities. For gluon fusion the Drell–Yan background $Z \to \mu^+\mu^-$ is very hard to battle only using the reconstructed mass of the muon pair. The two tagging jets and the central jet veto very efficiently remove the leading Z+jets backgrounds and leave us with mostly the G_F^3 process. However, because there is no single highly efficient background rejection cut this analysis will require modern analysis techniques.
- $qq \to qqH$, $H \to$ invisible is the only discovery channel for an invisible Higgs which really works at the LHC. It relies on the pure tagging-jet signature, which means it is seriously hard and requires a very good understanding of the detector and of QCD effects. The irreducible $Z \to \nu\bar{\nu}$ background is not negligible and has to be controlled essentially based on its QCD properties. Jet kinematics as well as jet counting are the key elements of this analysis.

Just a side remark for younger LHC physicists: weak boson fusion was essentially unknown as a production mode for light Higgses until around 1998 and started off as a very successful PhD project. This meant that for example the Higgs chapter in the ATLAS TDR had to be re-written. While it sometimes might appear that way, there is no such thing as a completely understood field of LHC physics. Every aspect of LHC physics continuously moves around and your impact only depends on how good your ideas and their technical realizations are.

1.7 Associated Higgs Production

In Fig. 1.8 there appears a third class of processes at smaller rates: associated Higgs production with heavy Standard Model particles, like W or Z bosons or $t\bar{t}$ pairs. Until the summer of 2008 the Higgs community at the LHC was convinced that (a) we would not be able to make use of WH and ZH production at the LHC and (b) we would not be able to see most of the light Higgs bosons produced at the LHC, because $H \to b\bar{b}$ is no promising signature in gluon fusion or weak boson fusion production.

One key to argument (a) are the two different initial states for signal and background: at the Tevatron the processes $q\bar{q} \to ZH$ and $q'\bar{q} \to WH$ arise from valence quarks. At the LHC with its proton–proton beam this is not possible, so the signal rate will suffer when we go from Tevatron to LHC. The QCD background at leading order is $q\bar{q} \to Zg^* \to Zb\bar{b}$ production, with an off-shell gluon splitting into a low mass bottom quark pair. At next-to-leading order, we also have to consider the t-channel process $q\bar{q} \to Z\bar{b}bg$ and its flipped counter part $qg \to Z\bar{b}bq$. This background becomes more dangerous for larger initial-state gluon densities. Moving from Tevatron to LHC the Higgs signal will decrease while the background increases—not a very promising starting point.

With Ref. [11] the whole picture changed. We will discuss this search strategy in detail in Sect. 3.1.2 in the context of jets and jet algorithms at the LHC. It turns out that searches for *boosted Higgs bosons* are not only promising in the $VH, H \to b\bar{b}$ channel, but might also resurrect the $t\bar{t}H, H \to b\bar{b}$ channel. These new channels are not yet included in the ATLAS list of processes shown in Fig. 1.10 because the simulations are still at an early stage. But we can expect them to play a role in LHC searches for a light Higgs boson.

This is another example of what is possible and what not in LHC phenomenology: it all really depends only on how creative you are; that even applies to a field like Standard Model Higgs searches, which is supposedly studied to death.

1.8 Beyond Higgs Discovery

The prime goal of the LHC was to discover a new light scalar particle which we could then experimentally confirm to be a Higgs boson, either as predicted by the Standard Model or with modifications due to new physics. This has worked great. The discovery of a new particle which was predicted in 1964 purely on the grounds of quantum field theory gives us great confidence in field theories as a description of elementary particles. For the description of this new state we will therefore consistently rely on the Lagrangian as the most basic object of perturbative field theory.

The Standard Model Lagrangian makes many predictions concerning the properties of a Higgs boson; as a matter of fact, all its properties except for its mass are fixed in the minimal one-doublet Higgs sector of the Standard Model. The question is, can we test at least some of these predictions?

In this section we will briefly touch on a few interesting questions relevant to the Higgs Lagrangian. This is where we have seen the most progress in LHC Higgs physics over recent years: not only will we be able to see a light Higgs boson simultaneously in different production and decay channels, as discussed in Sects. 1.5 and 1.6, we can also study many of its properties. In a way this section ties in with the effective theory picture we use to introduce the Higgs mechanism: the obvious requirements to include massive gauge bosons in an effective electroweak gauge theory leads us towards the Standard Model Higgs boson only step by step, and at any of these steps we could have stopped and postulated some alternative ultraviolet completion of our theory.

1.8 Beyond Higgs Discovery

1.8.1 Coupling Measurement

In Sect. 1.4 we present the ATLAS Higgs discovery paper in some detail. While it is clear that the observed $ZZ \to 4\ell$, $\gamma\gamma$, and $WW \to 2\ell\,2\nu$ signals point towards the discovery of a Higgs boson, the nature of the observed excess is not at all clear. For the statistical analysis leading to the discovery the interpretation plays no role. Only some very preliminary information on the observed resonance can be deduced from the fact that it appears in analyses which are designed to look for the Higgs boson.

As a first step in analyzing the Higgs Lagrangian we can for example assume the operator structure of the Standard Model and ask the question how well each of the associated couplings agrees with the Standard Model prediction. If we for a moment forget about the ultraviolet completion of our electroweak theory we observe a scalar particle with a mass around 125 GeV which couples to W and Z bosons, photons, and probably gluons. To describe the Higgs discovery in terms of a Lagrangian we need at least these four terms. Because the Higgs mechanism breaks the weak gauge symmetry the individual operators in terms of the Higgs field do not have to be gauge invariant. Moreover, the coupling measurement in the Standard Model Lagrangian mixed renormalizable couplings to massive gauge bosons and fermions with loop–induced couplings to massless gauge bosons. This mix of renormalizable and dimension-6 operators cannot be expected to descent from a proper effective field theory. If we are interested in the ultraviolet structure of this free couplings model we can for example consider the observed Higgs particle the lightest state in an extended Higgs sector. This way its couplings are allowed to deviate from the Standard Model predictions while renormalizability and unitarity are ensured once we include the additional, heavier Higgs states.

Alternatively, we can define an effective field theory based on all possible Higgs operators to a given mass dimension. In the case of a linear representation this Lagrangian will be based on the Higgs doublet ϕ and by construction $SU(2)_L$ gauge invariant. Some of the dimension-6 potential operators forming this effective theory we study in Sect. 1.2.1. Couplings to W and Z bosons start at mass dimension four, but there exist of course higher–dimensional operators linking the same particles; the same is true for Yukawa couplings. Higgs couplings to gluons and photons, as derived in Sects. 1.5.1 and 1.5.3, start at dimension six and could be supplemented by operators with even higher mass dimension. In an even more general approach we do not even assume that the Higgs boson has spin zero. Instead, we can define effective theories also for spin-one and spin-two Higgs impostors.

Each operator on this extensive list we can equip with a free coupling factor, and this set of couplings we can fit to the LHC measurements. Note that we really mean 'measurements' and not 'event rates', because different operators lead to different kinematic behavior and hence significantly different efficiencies in the LHC analyses. As an example we can quote the angle between the two leptons in the $H \to WW$ analysis or the structure of the tagging jets, which work best for a spin-zero Higgs. Of course, adding all kinds of kinematic distributions will add a huge amount of additional information to the Lagrangian determination, but it is obvious that at least at this stage such an measurement is unrealistic.

Because of this complication we return to the original question, comparing the LHC results to the *Standard–Model–like Higgs Lagrangian*. The different operators essentially fix all Higgs quantum numbers, the production and decay kinematics, and the experimental efficiencies. In our first attempt we will assume a CP-even scalar Higgs boson, where dimension-4 terms in the Lagrangian in general dominate over higher–dimensional terms. Effective couplings to gluons and photons are included at dimension-6 because tree level dimension-4 couplings do not exist. Deviations from this assumptions we discuss in Sect. 1.8.2. Higgs potential terms including the triple and quartic self-couplings we can ignore for now, because LHC analyses will not be sensitive to them for a while. Again, we will discuss possible measurements of the Higgs self-coupling in Sect. 1.8.3. Our basic Lagrangian with Standard model couplings is a combination of Eqs. (1.51), (1.52), and (1.202):

$$\mathcal{L} \supset \frac{1}{2}\left(\partial_\mu H\right)^2 - gm_W\, HW_\mu^+ W^{-\mu} - \frac{gm_Z}{2c_w} HZ_\mu Z^\mu$$

$$+ g_{\gamma\gamma H}\frac{H}{v}A^{\mu\nu}A_{\mu\nu} + g_{ggH}\frac{H}{v}G^{\mu\nu}G_{\mu\nu} - \sum_{\text{fermions}}\frac{y_f}{\sqrt{2}}H\overline{\psi}_f\psi_f$$

$$\equiv \frac{1}{2}\left(\partial_\mu H\right)^2 - g_W\, HW_\mu^+ W^{-\mu} - g_Z\, HZ_\mu Z^\mu + g_\gamma\, HA^{\mu\nu}A_{\mu\nu} + g_g\, HG^{\mu\nu}G_{\mu\nu}$$

$$- \sum_{\text{fermions}} g_f\, H\overline{\psi}_f\psi_f . \tag{1.231}$$

In the second line we have defined each Higgs coupling to a particle x as g_x, which is commonly done in Higgs couplings analyses. Given this ansatz we can measure the *Higgs couplings* or ratios of them, for example in relation to their Standard Model values

$$\boxed{g_x = (1+\Delta_x)\, g_x^{\text{SM}}} \quad\text{and}\quad \frac{g_x}{g_y} = \left(1+\Delta_{x/y}\right)\left(\frac{g_x}{g_y}\right)^{\text{SM}}. \tag{1.232}$$

For the effective couplings we need to separate the parametric dependence on the Standard Model couplings which enter the one-loop expression, so we find for example

$$g_\gamma = \left(1 + \Delta_\gamma^{\text{SM}} + \Delta_\gamma\right) g_\gamma^{\text{SM}} . \tag{1.233}$$

The contribution $\Delta_\gamma^{\text{SM}}$ is the deviation of the effective couplings based on a possible shift in the values of the top and W couplings consistent with their tree level counterparts. This term is crucial once we extract all Higgs couplings from data consistently. Our ansatz in terms of the Δ_x is motivated by the fact that at first sight the observed Higgs signal seems to be in rough agreement with a Standard Model Higgs boson, so the measured values of Δ_x should be small.

1.8 Beyond Higgs Discovery

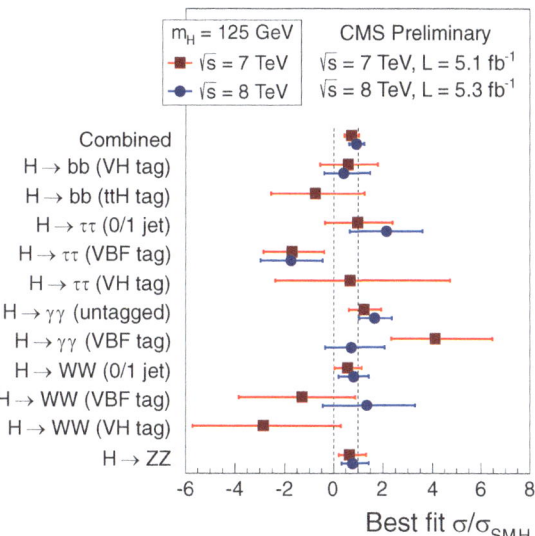

Fig. 1.12 Event rates for the different Higgs signatures relative to the Standard Model expectations as measured by CMS (Figure from the supplementary material to Ref. [14])

A serious complication in the Higgs coupling extraction from event numbers arises through the *total Higgs width*. At the LHC, we will mainly measure event rates in the different Higgs channels as shown in Fig. 1.12. Even though it enters all event rates we will not be able to measure the width of a light Higgs boson as long as it does not exceed $\mathcal{O}(1 \text{ GeV})$. For small deviations from the Standard Model couplings this means that we need to construct the total Higgs width from other observables. The functional dependence of the event count in a production channel p and a decay channel d is

$$N_{\text{events}} = \epsilon \times \sigma_p \times \text{BR}_d \sim \frac{g_p^2 g_d^2}{\Gamma_{\text{tot}}(\{g_j\})} \,. \tag{1.234}$$

The combined efficiencies and the fiducial detector volume we denote as ϵ. The couplings entering the production and decay channels we denote as $g_{p,d}$. The total width, defined as the sum of all partial widths, depends on all relevant Higgs couplings. This functional behavior means that any LHC event number will depend on the Higgs couplings g_j in a highly correlated way, highly sensitive to what we assume for the unobservable Higgs width. An interesting question is if we can scale all Higgs couplings simultaneously without affecting the observables. This means

$$N_{\text{events}} = \lim_{g \to 0} \frac{g^4}{\Gamma_{\text{tot}}} = \lim_{g \to 0} \frac{g^4}{g^2 \frac{\Gamma_{\text{obs}}}{g^2} + \Gamma_{\text{unobs}}} = \lim_{g \to 0} \frac{g^4}{g^2 \frac{\Gamma_{\text{obs}}}{g^2} + \Gamma_{\text{unobs}}} = 0 \,. \tag{1.235}$$

The total width we have generally split into observable and unobservable channels, where the observable channels scale like g_d^2. This means that event numbers are sensitive to more than just the ratio of Higgs couplings. Nevertheless, we need to make some kind of assumption about the Higgs width.

The above argument suggests a theoretically sound assumption we can make on the total Higgs width, once we observe a number of Higgs decay channels and measure the underlying Higgs coupling

$$\Gamma_{\text{tot}} > \sum_{\text{observed}} \Gamma_j \, . \tag{1.236}$$

Each partial Higgs width is independently computed and corresponds to a positive transition amplitude squared. If we assume the Higgs Lagrangian equation (1.231) we can compute each partial width in terms of its unknown coupling. There are no interference effects between Higgs decay channels with different final–state particles, so the total width is strictly bounded from below by the sum of observed channels.

A slightly more tricky assumption gives us an upper limit on at least one Higgs partial width. From our calculation in Sect. 1.2.3 we know that the Standard Model Higgs boson unitarizes the $WW \to WW$ scattering rate. If we overshoot for example in the s-channel Higgs exchange we would need an additional particle which compensates for this effect. However, the amplitude of an such additional particle would be proportional to its Higgs coupling squared, which means it is not clear where the required minus sign would come from. Taken with a grain of salt this tells us

$$g_W < g_W^{\text{SM}} \qquad \text{or} \qquad \Gamma_{H \to WW} < \Gamma_{H \to WW}^{\text{SM}} \, . \tag{1.237}$$

Given that correlations between different Higgs couplings will become a problem in the coupling extraction such a constraint can be very useful.

In the analysis which we present in this section we do not assume an upper limit to any Higgs partial width. Instead, we promote the constraint in Eq. (1.236) to an exact relation:

$$\boxed{\Gamma_{\text{tot}} = \sum_{\text{observed}} \Gamma_x(g_x) + \text{2nd generation} < 2\,\text{GeV}} \, . \tag{1.238}$$

Because at the LHC we will not observe any Higgs couplings to a second generation fermion any time soon we correct for the charm quark contribution using $g_c = m_c/m_t \times g_t^{\text{SM}}(1 + \Delta_t)$. This avoids systematic offsets in the results. The total upper limit on the Higgs width corresponds to very large individual couplings and is an estimate of visible effects in the $H \to \gamma\gamma$ analyses.

In Table 1.1 we list the channels which we can rely on in the extraction of the Higgs couplings. The details of the 2011 results in these channels we discuss

1.8 Beyond Higgs Discovery

Table 1.1 Higgs signatures with significant impact on the Higgs coupling determination in 2011/2012 and beyond. Question marks indicates that beyond 2015 these channels might contribute, but no reliable ATLAS or CMS analysis is currently available. In the top line we indicate the experimental signature while in the second line we show the leading production mode

Experiment theory	H inclusive $gg \to H$	$H + 2$ jets $qq \to qqH$	$H +$ lepton(s) $q\bar{q} \to VH$	$H +$ top(s) $gg/q\bar{q} \to t\bar{t}H$
$H \to ZZ$	2011/2012	$\geq$2015	–	–
$H \to \gamma\gamma$	2011/2012	2011/2012	–	?
$H \to WW$	2011/2012	2011/2012	–	?
$H \to \tau\tau$	?	2012	–	?
$H \to \mu\mu$	–	?	–	–
$H \to b\bar{b}$	–	?	$\geq$2015	$\geq$2015

with the Higgs discovery paper in Sect. 1.4. Note that in Table 1.1 we list the different Higgs production channels which are theoretically well defined but not experimentally observable. For example, the $H \to \gamma\gamma$ analysis is separated into the jet–inclusive and the two-jet analysis, where the inclusive rate is dominated by the gluon–fusion production process while the two-jet rate is dominated by weak boson fusion. Once we can determine the efficiencies for each of the production processes contributing to the different analyses we can rotate the Higgs rates from the experimental basis to the theoretical basis. In the 2011/2012 data set we then have six observable rates from which we want to extract as many Higgs couplings as possible. Before we discuss the physics result of such a coupling extraction let us introduce some of the techniques.

The naive approach to a parameter extraction is a χ^2 *minimization*, experimentally known as 'running MINUIT'. The variable χ^2 measures the quality of a fit or the quality of the theoretical assumptions compared to the measurements for the best possible set of theoretical model parameters. Given an n_{meas}-dimensional vector of measurements $\vec{x}_{\text{meas}}$ and the corresponding model predictions $\vec{x}_{\text{mod}}(\vec{m})$, which in turn depend on an n_{mod}-dimensional vector of model parameters, we define

$$\chi^2(\vec{m}) = \sum_{j=1}^{n_{\text{meas}}} \frac{\left|\vec{x}_{\text{meas}} - \vec{x}_{\text{mod}}(\vec{m})\right|_j^2}{\sigma_j^2}, \qquad (1.239)$$

where σ_j^2 is the variance of the channel $\vec{x}_j$. The best fit point in model parameter space is the vector $\vec{m}$ for which $\chi^2(\vec{m})$ assumes a minimum. In the Gaussian limit, we cannot only compute the minimum value of χ^2 but compare the entire χ^2 distribution for variable model parameters to our expectation. If the theoretical predictions $\vec{x}_{\text{mod}}$ depend on n_{mod} model parameters we can define the normalized or reduced χ^2 distribution

$$\chi^2_{\text{red}}(\vec{m}) = \frac{\chi^2(\vec{m})}{n_{\text{meas}} - n_{\text{mod}} + 1} \equiv \frac{\chi^2(\vec{m})}{\nu}. \qquad (1.240)$$

For this definition of the number of degrees of freedom ν we assume $n_{meas} > n_{mod} + 1$, which means that the system is over–constrained and a measure for the quality of the fit makes sense. If we find min $\chi^2_{red} \sim 1$ the error estimate of σ_j entering Eq. (1.239) is reasonable. The problem with the χ^2 test is that it requires us to know not only the variance σ_j^2, but also the form of the actual $\vec{x}_{meas}$ distributions to apply any test on the final value of χ^2. The natural distribution to assume is a Gaussian, defined in Eq. (1.182) with the covariance σ^2. Technically, a common assumption in the determination of the best fit is that the functional form of χ^2 be quadratic. In that case we can compute the confidence level of a distribution, as defined in Sect. 1.4 from the min χ^2 value.

All this is only true in the Gaussian limit, which means that we cannot use it in our Higgs couplings measurement. Some of the channels involved have a very small event count in the signal or in the background, many uncertainties are heavily correlated, and we have very little control over the form of systematic or theoretical uncertainties.

In Sect. 3.4 we construct the general form of a χ^2 variable including full correlations and without making any assumptions on the form of uncertainties,

$$\chi^2(\vec{m}) \longrightarrow -2\log \mathscr{L}(\vec{m}) = \vec{\chi}_T C^{-1} \vec{\chi} \,, \tag{1.241}$$

with the definition of $\vec{\chi}$ and C given in Eq. (3.33). This variable, evaluated at a model parameter point $\vec{m}$, is the *log-likelihood*. The name likelihood implies that it is some kind of probability, but evaluated over model parameter space and not over experimental outcomes. This means that the normalization of a likelihood is only defined once we agree on an integration measure in model space. Only in the Bayesian approach we ever do that. Here, we construct a completely exclusive log-likelihood map over the n_{mod}-dimensional model parameter space and then reduce the dimensionality one–by–one to show profile likelihoods for individual model parameters or their correlations. We will give details on this procedure in Sect. 3.4. Error bars on the individual couplings can be extracted through toy measurements. These are assumed measurements which we generate to trace the individual uncertainty distribution for each channel and which define a distribution of best fit values. Typically 10^3 toy measurements give us a sufficiently precise distribution around the best–fitting point to construct the model space volume which contains 68 % of the toy measurement distribution below and above the measured central value.

Technically, the log-likelihood can be extracted as a *Markov chain*. This is a set of parameter points which represent their log-likelihood value in their density. The construction of such a Markov chain is simple:

1. Start with any model parameter point $\vec{m}_0$ and compute $\mathscr{L}(\vec{m}_0)$.
2. Random generate a second model parameter point $\vec{m}_1$ according to some suggestion probability. This probability cannot have any memory (detailed balance), should be peaked for $\vec{m}_1 \sim \vec{m}_0$, at the same time give a decent probability to

1.8 Beyond Higgs Discovery

move through the parameter space, and does not have to be symmetric $q(\vec{m}_0 \to \vec{m}_1) \neq q(\vec{m}_1 \to \vec{m}_0)$, We often use a Breit–Wigner or Cauchy distribution defined in Sect. 2.1.2 with a width of 1 % of the entire parameter range.
3. Accept new point as the next point in the Markov chain if $\log \mathscr{L}(\vec{m}_1) > \log \mathscr{L}(\vec{m}_0)$. Otherwise, accept with reduced probability $\log \mathscr{L}(\vec{m}_1)/\log \mathscr{L}(\vec{m}_0)$.
4. Stop once the chain is sufficiently long to cover at least part of the parameter space. Obviously, we can combine several different Markov chains

There are ways to improve such a Markov chain. First, for a Markov chain which carries information only about the log-likelihood itself, i.e. used to estimate the same property we define it through, we can keep the value $\mathscr{L}(\vec{m})$ as a weight to each point. Weighted Markov chains improve the convergence of the final result. Secondly, we can slowly focus on regions of the relevant parameter regions with larger $\mathscr{L}$ values. Following the general idea of simulated annealing we can use cooled Markov chains which include two stages: in the early stage the Markov chain rapidly covers large fractions of the model space while in a second stage it zooms in on the regions with the largest likelihoods. Technically, we change the constant acceptance criterion relative to a linear random number r to a varying

$$\frac{\log \mathscr{L}(\vec{m}_1)}{\log \mathscr{L}(\vec{m}_0)} > r \quad \to \quad \left(\frac{\log \mathscr{L}(\vec{m}_1)}{\log \mathscr{L}(\vec{m}_0)}\right)^{j/10} > r \quad \text{for} \quad j = 1 \ldots 10. \quad (1.242)$$

Because the ratio of log-likelihoods is smaller than unity, for $j = 1$ the weighted ratio becomes larger and the new point is more likely to be accepted. For larger values of j the weighted ratio becomes smaller, which means that the Markov chain will stay more and more localized and mostly move to another point if that one is more likely. Obviously, cooling Markov chains are only reliable if we run several chains to check if they really cover all interesting structures in model space. In the case of the Higgs analysis we use $\mathscr{O}(10)$ chains with 10^6 parameter points each. These 10^7 likelihood values define our completely exclusive likelihood map.

With this technical background we can continue with our Higgs couplings analysis. From Eq. (1.231) we know that in our case the model parameters $\vec{m}$ will be the Higgs couplings g_x or their deviations from the Standard Model values Δ_x as defined in Eq. (1.232). The observables will be the event rates in the Higgs search channels shown in Table 1.1. The errors on signal and background rates include statistical, systematic (correlated), and theoretical (correlated) sources. The *SFitter* version of the Higgs couplings analysis proceeds in a series of independent steps:

1. Construct an exclusive likelihood map over the Higgs couplings model space.
2. Deflate the parameter space to profile likelihoods for one-dimensional and two-dimensional distributions of couplings. This gives us a *global picture* of the Higgs couplings space.

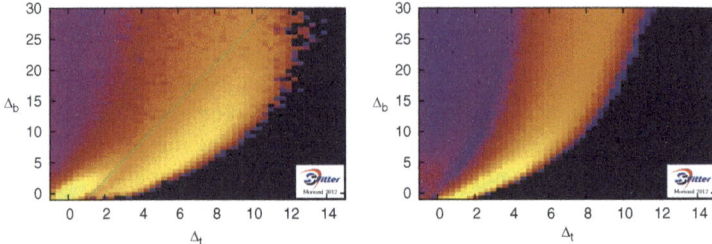

Fig. 1.13 Correlations of the top and bottom Yukawas extracted from the 2011 data set with a center-of-mass energy of 7 TeV. The *left panel* shows the expected likelihood map in the Standard Model, the *right panel* shows the observed one (Figure from Ref. [27])

3. Determine the best–fitting parameter point with high resolution. This means starting from the best point in the Markov chain and using a minimization algorithm on $\log \mathscr{L}(\vec{m})$.
4. Determine the 68 % uncertainty band on each extracted couplings from toy measurements, defining a *local picture* of the Higgs couplings.

The benefits of this approach as well as the impressive progress in the experimental Higgs analyses can be nicely illustrated when we look at the coupling determination from 2011 and early 2012 data.

Based on 2011 data the number of measurements published by ATLAS and CMS and shown in Moriond was relatively small. As listed in Table 1.1 there should have been measurements including a Higgs coupling to gluons and photons through effective operators, plus couplings to W, Z, and tau leptons at the tree level. Dealing with relatively early data we assume that the higher–dimensional couplings are mediated by Standard Model top and W loops, with at best small corrections due to additional states. Before we determine the individual couplings we can look at the global picture, to make sure that everything looks roughly like expected. In the left panel of Fig. 1.13 we show the correlation between the top and bottom Yukawa couplings expected for the 7 TeV run. In the $\Delta_{b,t} = 0 \ldots 1$ range we see the Standard Model coupling range.

For large $\Delta_{b,t} \sim 5 \ldots 10$ another solution appears. Its main features is the simultaneous increase of both quark Yukawas. This behavior can be explained by the indirect handles we have on each of them. The top Yukawa is mostly measured through the effective gluon–Higgs coupling while the bottom Yukawa enters the total width. To keep for example the inclusive $H \to ZZ$ rate constant, both couplings have to increase at roughly the same rate. For a constant inclusive $H \to \gamma\gamma$ rate the discussion in Sect. 1.5.3 we also need to increase Δ_W at the same rate. However, such an increase would be visible in the inclusive and weak boson fusion $H \to WW$ channels. What we see in Fig. 1.13 is that starting from Standard Model values we expect all three couplings to increase. At some point Δ_W hits the limits from other channels, so instead of further increasing it switches back to its Standard Model value, Δ_b follows into the same direction, but Δ_t takes over

1.8 Beyond Higgs Discovery

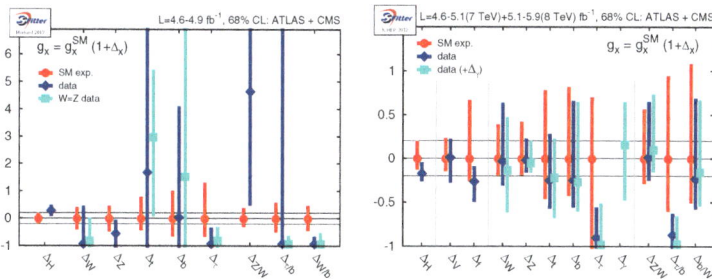

Fig. 1.14 *Left*: Higgs couplings extracted from the 2011 run at 7 TeV. *Right*: couplings extracted from all data published in the Higgs discovery papers by ATLAS and CMS (Figures from Refs. [27, 29])

the increased effective photon–Higgs coupling with a different sign of this effective couplings. From that point on $\Delta_b \sim \Delta_t$ can grow again. If we want to avoid the theoretical problem of a hugely non–perturbative top Yukawa we can limit our couplings extraction to the separable Standard-Model-like solution, as indicated by the green line.

The problem with the same distribution extracted from 2011 data is that this separations does not exist. This can be traced back to essentially missing evidence for a Higgs boson decaying to WW or to $\tau\tau$. If in the argument above we remove the constraints from visible and constraining WW channels the two Yukawa couplings can increase from $\Delta_{b,t} = 0$ to huge values. The Standard Model and the large couplings solutions are blended together. The global picture tells us that we should expect any reasonable Yukawa coupling measurements from 2011 data even if we allow for an indirect determination from the higher–dimensional couplings.

In the left panel of Fig. 1.14 we show the results from the local analysis and see exactly what we expect from the global picture: a universal Higgs coupling modification

$$\Delta_x \equiv \Delta_H \qquad \forall x \qquad (1.243)$$

is determined exactly as we expect from a set of measurements in agreement with the Standard Model. In contrast, $\Delta_W \sim \Delta_\tau \sim -1$, which means that there is no evidence for such a coupling in the 2011 data set. The Higgs gauge coupling to Z bosons is roughly what one would have expected, while the quark Yukawas are very poorly determined.

In the right panel of Fig. 1.14 we repeat the same analysis with all data from the discovery papers published in the Summer of 2012. As mentioned in Sect. 1.4, for ATLAS this includes all results presented in the talks on 4th of July, plus an improved $H \to WW$ analysis.

First, the measurements of a common Higgs form factor for all couplings defined in Eq. (1.243) has improved to a 20 % level, in very good agreement with the Standard Model. If we want to separate this tree level form factor into a universal bosonic and a universal fermionic coupling modification we see that both of them are determined well. All these measurements are within our expectations from the Standard Model for the central values as well as for the uncertainties. The full set of couplings we can determine either limiting the effective Higgs–photon couplings to Standard Model loops or allowing for an additional contribution Δ_γ defined in Eq. (1.233). It turns out that in the absence of any direct quark Yukawa measurements we have no sensitivity to an additional Higgs–gluon contribution.

The most important result of the 2011/2012 coupling analysis is that the Higgs couplings are typically determined at the 20–50 % level. The only coupling measurement still missing entirely is Δ_τ. Allowing for a free Higgs–photon coupling does not affect the other measurements significantly. The central value of Δ_W decreases just sightly, allowing for a barely non-zero central value of Δ_γ. The bottom line that there is no anomaly in the Higgs coupling to photons is in apparent disagreement with other analyses. The difference is that in the SFitter results shown here all Higgs couplings are allowed to vary in parallel. If the event rate for $pp \to H \to \gamma\gamma$ is slightly high, we do not simply translate this effect into a positive value of Δ_γ. Instead, we allow the top and W couplings to absorb some of this effect within their uncertainties from other measurements. Only the part of the $\gamma\gamma$ anomaly which is not compatible with the other Higgs channels then contributes to the Δ_γ measurement shown in Fig. 1.14.

Without showing the numerical outcome we state that the quality of the fit, i.e. the log-likelihood value in the best–fitting parameter point, at this stage includes no useful information. Essentially any model of Higgs couplings, with the exception of a chiral fourth generation, fits the data equally well. As indicated in Table 1.1 we expect significant improvements of the Higgs coupling measurements from the 2015 LHC run and beyond. This will mostly affect the highly sensitive weak boson fusion signatures and a direct measurement of the quark Yukawas based on fat jet techniques described in Sect. 3.1.2. Any kind of measurement at the per-cent level or better makes a very good case for a linear collider, running at enough energy to study the $e^+e^- \to ZH$, $t\bar{t}$ and $t\bar{t}H$ production channels.

1.8.2 Higgs Quantum Numbers

One question for the LHC is: once we see a Higgs boson at the LHC, how do we test its quantum numbers? One such quantum number is its spin, which in the ideal case we want to determine channel by channel in the Higgs observation. Three questions people usually ask really mean the same thing:

- What is the spin of the Higgs boson?
- Is the new resonance responsible for electroweak symmetry breaking?
- What is the form of the Higgs operators in the Lagrangian?

Given that the discovery of the Higgs boson is an impressive confirmation that fundamental interactions are described by quantum field theory, specifically by

1.8 Beyond Higgs Discovery

gauge theories, the last version of the question is the most appropriate. Lagrangians are what describes fields and interactions in such a theory, and what we need to determine experimentally is the Lagrangian of the observed Higgs boson. In that sense, the coupling measurement described in the last section assumes the Higgs Lagrangian of the Standard Model and measures the couplings of the corresponding Higgs interaction terms.

As far as the spin of the new particle is concerned, the spin-one case is easily closed: following the Landau–Yang theorem a spin-one Higgs boson would not couple to two photons, so in the photon decay channel we only need to look at angular distributions for example in weak boson fusion to distinguish spin zero from spin two.

Once we know that we are talking about a scalar field there are a few options, linked to the CP properties of the new particle. The part of the Higgs Lagrangian we are most interested in is the coupling to the massive gauge bosons. In the Standard Model, the fundamental CP-even Higgs boson couples to the W and Z bosons proportional to $g^{\mu\nu}$. For general *CP-even* and *CP-odd* Higgs bosons there are two more ways to couple to W bosons using gauge invariant dimension-6 operators. If inside the dimension-6 operators we replace $(\phi^\dagger \phi)$ by the linear Higgs terms we are interested in, we arrive at the dimension-5 Lagrangian

$$\mathscr{L} \supset -g m_W \, H \, W_\mu W^\mu - \frac{g_{D5}^+ v}{\Lambda^2} \, H \, W_{\mu\nu} W^{\mu\nu} - \frac{g_{D5}^- v}{\Lambda^2} \, A \, W_{\mu\nu} \tilde{W}^{\mu\nu} \, . \qquad (1.244)$$

In this notation H is the scalar Higgs boson, while A is a pseudo-scalar. The coupling to Z bosons is completely analogous to the W case, where $W^{\mu\nu}$ indicates the field strength tensor and $\tilde{V}^{\mu\nu}$ its dual. This set of gauge–invariant terms in the Lagrangian can be translated into Feynman rules for the Higgs coupling to massive gauge bosons. Their tensor structures are

$$\mathcal{O}_{SM}^+ = g^{\mu\nu} \qquad \mathcal{O}_{D5}^+ = g^{\mu\nu} - \frac{p_1^\mu p_2^\nu}{p_1 p_2} \qquad \mathcal{O}_{D5}^- = \epsilon_{\mu\nu\rho\sigma} \, p_1^\rho p_2^\sigma \, . \qquad (1.245)$$

These are the only gauge invariant dimension-5 couplings of two gauge bosons to a (pseudo-) scalar field. The second one appears in the effective one-loop gluon–gluon–Higgs coupling in Eqs. (1.202) and (1.208). This second tensor is not orthogonal to $g^{\mu\nu}$, but we can replace it with The any linear combination with $g^{\mu\nu}$. However, if we trust our description in terms of a Lagrangian we obviously prefer the gauge–invariant basis choice.

The traditional observables reflecting the coupling structure of a massive state decaying to two weak gauge bosons are the *Cabibbo–Maksymowicz–Dell'Aquila–Nelson angles*. They are about the same age as the Higgs boson. We define them in the fully reconstructable decay $X \to ZZ \to e^+ e^- \mu^+ \mu^-$. We already know that one of the two Z bosons will be far off its mass shell, which does not affect the analysis of the decay angles. The four lepton momenta reconstructing the Higgs–like state X are given by

$$p_X = p_{Z_e} + p_{Z_\mu} \qquad p_{Z_e} = p_{e^-} + p_{e^+} \qquad p_{Z_\mu} = p_{\mu^-} + p_{\mu^+} \;.$$
(1.246)

Each of these momenta as well as the beam direction we can boost into the X rest frame and the two $Z_{e,\mu}$ rest frames, defining the corresponding three-momenta $\hat{p}_i$. The spin and CP angles are then defined as

$$\cos\theta_e = \hat{p}_{e^-} \cdot \hat{p}_{Z_\mu}\Big|_{Z_e} \qquad \cos\theta_\mu = \hat{p}_{\mu^-} \cdot \hat{p}_{Z_e}\Big|_{Z_\mu} \qquad \cos\theta^* = \hat{p}_{Z_e} \cdot \hat{p}_{\text{beam}}\Big|_X$$

$$\cos\phi_e = (\hat{p}_{\text{beam}} \times \hat{p}_{Z_\mu}) \cdot (\hat{p}_{Z_\mu} \times \hat{p}_{e^-})\Big|_{Z_e} \qquad \cos\Delta\phi = (\hat{p}_{e^-} \times \hat{p}_{e^+}) \cdot (\hat{p}_{\mu^-} \times \hat{p}_{\mu^+})\Big|_X \;.$$
(1.247)

The index indicates the rest frame where the angles are defined. To distinguish the different spin-zero hypotheses the angular difference $\Delta\phi$ is most useful. Looking at each of the decaying Z bosons defining a plane opened by the two lepton momenta it is the angle between these two planes in the Higgs rest frame is $\Delta\phi$. Its distribution can be written as

$$\frac{d\sigma}{d\Delta\phi} \propto 1 + a\cos\Delta\phi + b\cos(2\Delta\phi) \;.$$
(1.248)

For the CP-odd Higgs coupling to $W^{\mu\nu}\tilde{W}_{\mu\nu}$ we find $a = 0$ and $b = 1/4$, while for the CP-even Higgs coupling $g^{\mu\nu}$ we find $a > 1/4$ depending on m_H. Some example distributions for the decay planes angle we show in Fig. 1.15.

This method only works if we observe the decay $H \to ZZ \to 4\ell$ with a good signal–to–background ratio S/B. We can derive an alternative observable from studying its Feynman diagrams: starting from the $H \to ZZ \to 4\ell$ decay topology we can switch two fermion legs from the final state into the initial state

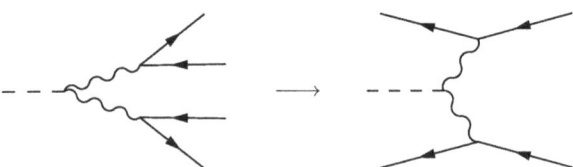

and read the diagram right–to–left. This gives us the Feynman diagram for weak boson fusion Higgs production. The angle between the decay planes gets linked to the angular correlation of two forward the tagging jets. Its advantage is that it gives us a production-side correlation, independent of the Higgs decay signature.

Going back to the transverse momentum spectra for the tagging jets shown in Eq. (1.215) we already know one way of telling apart these couplings: the

1.8 Beyond Higgs Discovery

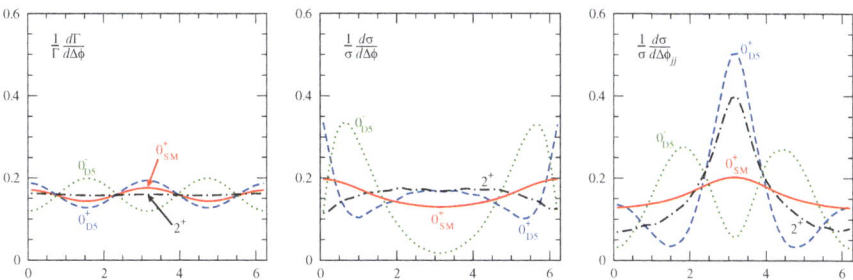

Fig. 1.15 $\Delta\phi$ distributions for $X \to ZZ$ events (*left*), weak boson fusion production in the Breit frame (*center*), and in the laboratory frame (*right*) (Figures from Refs. [18, 30])

dimension-3 *WWH* coupling proportional to $g^{\mu\nu}$ comes out of the Higgs potential as a side product of the *W* mass term, i.e. it couples the longitudinal Goldstone modes in the *W* boson to the Higgs. In contrast, the CP-even dimension-6 operator is proportional to the transverse tensor, which means it couples the transverse *W* polarizations to the Higgs and therefore produces a harder p_T spectrum of the tagging jets. The problem with such an observation is that in the absence of a unitarizing Higgs scalar in *WW* scattering we should expect our theory to generally include momentum dependent form factors instead. Any observable with units of energy will become sensitive to these form factors.

Sticking to angular observables, along the lines of Eq. (1.247), we can first translate all decay angles into the weak boson fusion topology. The problem in this approach are the rest frames: a *W* boson in the *t*-channel is space–like, implying $p_W^2 \equiv t < 0$. This means that we cannot boost into its rest frame. What we can do is boost into the so-called *Breit frame*. It is defined as the frame in which the momentum of the *t*-channel particle only has space–like entries and can be reached through a conventional boost.

Writing the weak boson fusion momenta as $q_1 q_2 \to j_1 j_2 (X \to d\bar{d})$ we can define a modified version of the five angles in Eq. (1.247), namely

$$\cos\theta_1 = \hat{p}_{j_1} \cdot \hat{p}_{V_2}\Big|_{V_1 \text{Breit}} \qquad \cos\theta_2 = \hat{p}_{j_2} \cdot \hat{p}_{V_1}\Big|_{V_2 \text{Breit}} \qquad \cos\theta^* = \hat{p}_{V_1} \cdot \hat{p}_d\Big|_X$$

$$\cos\phi_1 = (\hat{p}_{V_2} \times \hat{p}_d) \cdot (\hat{p}_{V_2} \times \hat{p}_{j_1})\Big|_{V_1 \text{Breit}} \qquad \cos\Delta\phi = (\hat{p}_{q_1} \times \hat{p}_{j_1}) \cdot (\hat{p}_{q_2} \times \hat{p}_{j_2})\Big|_X .$$

(1.249)

In addition, we define the angle $\phi_+ \equiv 2\phi_1 + \Delta\phi$ which typically shows a modulation for spin-two resonances. In Fig. 1.15 we see how $\Delta\phi$ in the Breit frame is closely related to the angle between the *Z* decay planes.

In general, at hadron colliders we describe events using (pseudo-) rapidities and azimuthal angles, suggesting to use the differences $\{\Delta\eta_{mn}, \Delta\phi_{mn}\}$ for $m, n = j_{1,2}, X, d, \bar{d}$ to study properties of a Higgs–like resonance.

A very useful observable is the *azimuthal angle* between the two tagging jets, i.e. the angle separating the two jets in the transverse plane. We can again link it to the angle between the two Z decay planes in $X \to ZZ$ decays: for weak boson fusion it is defined as $\cos \Delta\phi = (\hat{p}_{q_1} \times \hat{p}_{j_1}) \cdot (\hat{p}_{q_2} \times \hat{p}_{j_2})$ in the Higgs candidate's rest frame, as shown in Eq. (1.249). We can links this rest frame to the laboratory frame through a boost with a modest shift in the transverse direction. In the laboratory frame each cross product $(\hat{p}_q \times \hat{p}_j)$ then resides in the azimuthal plane. The difference $\Delta\phi$ is nothing but the azimuthal angle between two vectors which are each orthogonal to one of the tagging jet direction. This is the same as the azimuthal angle between the two tagging jets themselves, $\Delta\phi_{jj}$.

In Fig. 1.15 we finally show this azimuthal angle between the tagging jets, for all three Higgs coupling operators defined in Eq. (1.245) and a sample spin-two coupling. The Standard Model operator predicts an essentially flat behavior, while the other two show a distinctly different modulation. The CP-odd $\epsilon^{\mu\nu\rho\sigma}$ coupling vanishes once two momenta contracted by the Levi–Civita tensor are equal. This explains the zeros at $\phi = 0, \pi$, where the two transverse jet momenta are no more linearly independent. To explain the opposite shape of the CP-even Higgs we use the low transverse momentum limit of the tagging jets. In that case the Higgs production matrix element becomes proportional to the scalar product $(p_T^{\text{tag } 1} p_T^{\text{tag } 2})$ which vanishes for a relative angle $\Delta\phi = \pi/2$.

In addition to the decay plane angle or the azimuthal angle $\Delta\phi$ there is a large number of angular observables we can use to study the quantum numbers. The only thing we have to make sure is that we do not cut on such variables when extracting the Higgs signal from the backgrounds. Examples of such cuts is the angle between the two leptons in the $H \to W^+W^-$ decay or the rapidity difference between the two forward tagging jets in weak boson fusion. In those cases the determination of the Higgs operators in the Lagrangian requires additional information. Also, we would want to test if a mixture of operators is responsible for the observed Higgs–like resonance. In Eq. (1.245) the higher–dimensional CP-even operator $\mathcal{O}_{\text{D5}}^+$ should exist in the Standard Model, but it is too small to be observed. As mentioned above, this question about the structure of the Higgs Lagrangian we should really answer before we measure the Higgs couplings as prefactors to the appropriate operators, following Sect. 1.8.1.

1.8.3 Higgs Self Coupling

If we should ever observe a scalar particle at the LHC, the crucial question will be if this is indeed our Higgs boson arising from electroweak symmetry breaking. In other words, does this observed scalar field have a *potential* with a minimum at a non–vanishing vacuum expectation value?

Of course we could argue that a dimension-3 Higgs coupling to massive W bosons really is a Higgs–Goldstone self coupling, so we see it by finding a Higgs in

1.8 Beyond Higgs Discovery

weak boson fusion. On the other hand, it would be far more convincing to actually measure the self coupling of the actual Higgs field. This trilinear coupling we can probe at the LHC studying Higgs pair production, for example in gluon fusion via the usual top quark loop

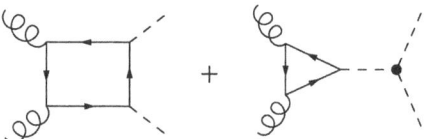

Following exactly the same argument as presented in Sect. 1.5.1 we can derive the two tensor structures contributing to Higgs pair production for transverse gluons

$$\sqrt{2} P_T^{\mu\nu} = g^{\mu\nu} - \frac{k_1^\nu k_2^\mu}{(k_1 k_2)} \tag{1.250}$$

$$\sqrt{2} P_2^{\mu\nu} = g^{\mu\nu} + \frac{k_3^2 k_1^\nu k_2^\mu}{k_T^2 (k_1 k_2)} - \frac{2(k_2 k_3) k_1^\nu k_3^\mu}{k_T^2 (k_1 k_2)} - \frac{2(k_1 k_3) k_2^\mu k_3^\nu}{k_T^2 (k_1 k_2)} + \frac{k_3^\nu k_3^\mu}{k_T^2}$$

$$\text{with} \quad k_T^2 = 2 \frac{(k_1 k_3)(k_2 k_3)}{(k_1 k_2)} - k_3^2.$$

The third momentum is one of the two Higgs momenta, so $k_3^2 = m_H^2$. The two tensors are orthonormal, which means $P_T^2 = 1$, $P_2^2 = 1$ and $P_T \cdot P_2 = 0$. The second tensor structure is missing in single Higgs production, so it only appears in the continuum (box) diagram and turns out to be numerically sub-leading over most of the relevant phase space.

From Sect. 1.5.2 on the effective Higgs–gluon coupling we know that in the low energy limit we can compute the leading form factors associated with the triangle and box diagrams, both multiplying the transverse tensor $g^{\mu\nu} - k_1^\mu k_2^\nu/(k_1 k_2)$ for the incoming gluons

$$g_{ggH} = -g_{ggHH} = -i \frac{\alpha_s}{12\pi} + \mathcal{O}\left(\frac{m_H^2}{4 m_t^2}\right). \tag{1.251}$$

Close to the production threshold $s \sim (2 m_H)^2$ the leading contribution to the loop-induced production cross section for $gg \to HH$ involving the two Feynman diagrams above and the Higgs self coupling derived in Sect. 1.2.1 is then proportional to

$$\boxed{\left[3 m_H^2 \frac{g_{ggH}}{s - m_H^2} + g_{ggHH} \right]^2 = g_{ggH}^2 \left[3 m_H^2 \frac{1}{s - m_H^2} - 1 \right]^2 \sim g_{ggH}^2 \left[3 m_H^2 \frac{1}{3 m_H^2} - 1 \right]^2 \to 0}, \tag{1.252}$$

so the triangle diagram and the box diagram cancel.

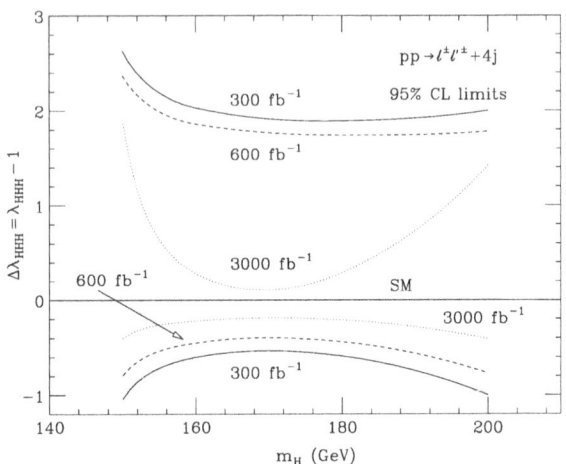

Fig. 1.16 The (parton–level) sensitivity limits for Higgs pair production and a measurement of the Higgs self coupling. The analysis is based on the decay $HH \to WW$ (Figure from Ref. [5])

In this argument we assume that the Higgs self coupling in the Standard Model is proportional to m_H. To see deviations from this self coupling in the first term of Eq. (1.252) we can look at something like the m_{HH} distribution and measure the threshold behavior. In the *absence of any self coupling* this threshold cancellation of the self coupling contribution with the continuum should be absent as well. The threshold contribution to Higgs pair production would be huge. This way a shape analysis of the threshold behavior will allow us to experimentally exclude the case of $\lambda_{HHH} = 0$ which predicts an unobserved large enhancement of the production cross section at threshold. At this point it is still under study if such a measurement will work at the LHC and what kind of LHC luminosities would be required.

As for all Higgs signatures we can go through the different Higgs pair signatures to check which of them might work. All estimates concerning detector performance have to be taken with a grain of salt, because this measurement will only be possible after a significant luminosity upgrade of the LHC, with integrated luminosities of several 1,000 fb^{-1} of data (Fig. 1.16). This might affect identification efficiencies as well as the invariant (Higgs) mass reconstruction:

- $gg \to HH \to b\bar{b}\ b\bar{b}$ is hopeless because of the overwhelming QCD backgrounds.
- $gg \to HH \to b\bar{b}\ W^+W^-$ has a sizeable rate, but the irreducible background is $t\bar{t}$ production. In the very unlikely case that this background can be reduced to an acceptable level, the channel might work.
- $gg \to HH \to b\bar{b}\ \tau^+\tau^-$ might or might not work. The key to this analysis is the reconstruction of the invariant masses of the bottom and tau pairs. Subjet methods along the lines of the Higgs tagger introduced in Sect. 3.1.2 might help, if they survive the pile–up at this luminosity.
- $gg \to HH \to b\bar{b}\ \gamma\gamma$ should benefit from the excellent $m_{\gamma\gamma}$ reconstruction which ATLAS and CMS have already shown in the Higgs discovery. The backgrounds are not huge, either.

- $gg \to HH \to W^+W^- \; W^+W^-$ used to be the best channel for Higgs masses in the 160 GeV range. For lower masses it will be less promising. The most dangerous background is $t\bar{t}$+jets, which at least requires a very careful study.

Other channels should be tested, but at the least they will suffer from small rates once we include the branching ratios. While there exist quite a number of studies for all of the channels listed above, none of them has yet been shown to work for a 125 GeV Higgs boson. While Nature's choice of Higgs mass is excellent when we are interested in measuring as many Higgs couplings to Standard Model particles as possible, it clearly suggests that we not look for the Higgs self coupling.

1.9 Alternatives and Extensions

The Higgs mechanism as discussed up to here is strictly speaking missing two aspects: a reason why the Higgs field exists and formal consistency of such a scalar field. In the following two sections we will present two example models: first, we will show how technicolor avoids introducing a new fundamental scalar field and instead relies on a dynamic breaking of the electroweak gauge symmetry. The problem is that technicolor does not predict the light scalar which ATLAS and CMS discovered in the Summer of 2012. In addition, we will briefly introduce the hierarchy problem, or the question why the Higgs is light after including loop corrections to the Higgs mass. Little Higgs models are one example for models stabilizing the Higgs mass, supersymmetry is another.

1.9.1 Technicolor

Technicolor is an alternative way to break the electroweak symmetry and create masses for gauge bosons essentially using a non–linear sigma model, as introduced in Sect. 1.1.4. There, we give the scalar field ϕ a vacuum expectation value v through a potential, which is basically the Higgs mechanism. However, we know another way to break (chiral) symmetries through *condensates*—realized in QCD. So let us review very few aspects of QCD which we will need later.

First, we illustrate why an asymptotically free theory like QCD is a good model to explain electroweak symmetry breaking. This is what will guide us to technicolor as a mechanism to break the electroweak gauge symmetry. As we will see in Sect. 2.2.2 the inherent mass scale of QCD is $\Lambda_{QCD} \sim 200$ MeV. It describes the scale below which the running QCD coupling constant $\alpha_s = g_s^2/(4\pi)$ becomes large, which means that perturbation theory in α_s breaks down and quarks and gluons stop being QCD's physical degrees of freedom. This reflects the fact that QCD is not scale invariant. We introduce a renormalization scale in our perturbative expansion. The running of a dimensionless coupling constant can be translated into an inherent mass scale. This mass scale characterizes the theory in the sense that $\alpha_s(p^2 = \Lambda_{\text{QCD}}^2) \sim 1$; for scales below Λ_{QCD} the theory will become strongly interacting and hit its Landau pole. Note that first of all this scale could not appear if for some reason $\beta \simeq 0$ and that it secondly does not depend on any mass scale in the theory. This phenomenon of a logarithmically running coupling introducing a mass scale in the theory is called *dimensional transmutation*. If at a high scale we start from a strong coupling in the $10^{-2} \cdots 10^{-1}$ range the QCD scale will arrive at its known value without any tuning.

The symmetry QCD breaks is the chiral symmetry. Just including the quark doublets and the covariant derivative describing the qqg interaction the QCD Lagrangian reads

$$\mathscr{L}_{\text{QCD}} \supset \overline{\Psi}_L \, i \slashed{D} \Psi_L + \overline{\Psi}_R \, i \slashed{D} \Psi_R \tag{1.253}$$

From Eqs. (1.15) and (1.16) we know that the Lagrangian in Eq. (1.253) it is symmetric under a chiral–type $SU(2)_L \times SU(2)_R$ transformation. Quark masses are not allowed, the chiral symmetry acts as a custodial symmetry for the tiny quark masses we measure for example for the valence quarks u, d. In Sect. 2.2.2 we will elaborate more on the beta function of QCD. It is defined as

$$\frac{d\alpha}{d \log p^2} = \beta_{\text{QCD}} = -\frac{\alpha_s^2}{4\pi}\left(\frac{11}{3} N_c - \frac{2}{3} n_f\right) \propto N_c \,, \tag{1.254}$$

where the N_c scaling is only true in the pure Yang–Mills theory, but gives asymmetric freedom due to $\beta_{\text{QCD}} < 0$. Towards small energies the running strong coupling develops a Landau poles at Λ_{QCD}. Because QCD is asymptotically free, at energies below Λ_{QCD} the essentially massless quarks form condensates, which means two-quark operators will develop a vacuum expectation value $\langle \overline{\Psi}\Psi \rangle$. This operator spontaneously breaks the $SU(2)_L \times SU(2)_R$ symmetry into the (diagonal) $SU(2)$ of isospin. This allows us to write down massive constituents which are the different composite color–singlet mesons and baryons become the relevant physical degrees of freedom. Their masses are of the order of the nucleon masses $m_{\text{nucleon}} \sim 1$ GeV. The only remaining massless particles are the Goldstone bosons from the breaking of $SU(2)_L \times SU(2)_R$, the pions. Their masses are not strictly zero, because the valence quarks do have a small mass of a few MeV. Their coupling strength (or decay rate) is governed by f_π. It is defined via

$$\langle 0 | j_\mu^5 | \pi \rangle = i f_\pi p_\mu \,, \tag{1.255}$$

and parameterizes the breaking of the chiral symmetry via breaking the axialvector–like $U(1)_A$. The axial current can be computed as $j_\mu^5 = \delta \mathscr{L}/\delta(\partial_\mu \pi)$ and in the $SU(2)$ basis reads $j_\mu^5 = \overline{\psi}\gamma_\mu \tau \psi/2$. From the measured decays of the light color–singlet QCD pion into two leptons we know that $f_\pi \sim 100$ MeV.

1.9 Alternatives and Extensions

To generalize them to technicolor we write the QCD observables in terms of two QCD model parameters: the size of the gauge group, N_c, and the scale at which the asymptotically free theory becomes strongly interacting, Λ_{QCD}. It is hard to compute observables like m_{nucleon} or f_π as a function of N_c and Λ_{QCD}. Instead, we derive simple scaling rules.

The Λ_{QCD} dependence simply follows from the mass dimension which for the vacuum expectation value is given by the mass dimension 3/2 of each fermion field,

$$f_\pi \sim \Lambda_{\text{QCD}} \qquad \langle \overline{Q} Q \rangle \sim \Lambda_{\text{QCD}}^3 \qquad m_{\text{fermion}} \sim \Lambda_{\text{QCD}} \,. \qquad (1.256)$$

The N_c dependence of f_π can be guessed from color factors: the pion decay rate is by definition proportional to f_π^2. Leaving aside the strongly interacting complications parameterized by the appearance of f_π, the Feynman diagrams for this decay are the same as for the Drell–Yan process $q\bar{q} \to \gamma, Z$. The color structure of this process leads to an explicit factor of N_c, to be combined with an averaging factor of $1/N_c$ for each of the incoming quarks. Together, the pion decay rate is proportional to f_π^2/N_c. We therefore postulate the pion decay constant to scale like $f_\pi \sim \sqrt{N_c}$. The vev–operator represents two quarks exchanging a gluon at energy scales small enough for α_s to become large. The color factor without any averaging over initial states) simply sums over all colors states for the color–singlet condensate, so it is proportional to N_c. The fermion masses have nothing to do with color states and hence should not depend on the number of colors. For details you should ask a lattice gauge theorist, but we already get the idea how we can construct our *high–scale version of QCD* through

$$f_\pi \sim \sqrt{N_c}\, \Lambda_{\text{QCD}} \qquad \langle \overline{Q} Q \rangle \sim N_c\, \Lambda_{\text{QCD}}^3 \qquad m_{\text{fermion}} \sim \Lambda_{\text{QCD}} \,. \qquad (1.257)$$

These scaling laws will allow us to make predictions for technicolor, in spite of the fact that we cannot compute any of these observables perturbatively.

Let us work out the idea that a mechanism just like QCD condensates could be the underlying theory of the non–linear σ model. In contrast to QCD we now have a gauged custodial symmetry forbidding weak gauge boson masses. The longitudinal modes of the massive W and Z bosons come from the Goldstone modes of the condensate's symmetry breaking called technipions. The corresponding mass scale would have to be

$$f_T \sim v = 246\,\text{GeV} \,. \qquad (1.258)$$

Fermion masses we postpone to the next section—in the 1970s, when technicolor was developed, the top quark was not yet known. All known fermions had masses of the order of GeV or much less, so they were to a good approximation massless compared to the gauge bosons.

To induce W and Z masses we write down the non–linear sigma model in its $SU(2)$ version. In Sect. 1.1.5 we re-write the linear sigma model using the Higgs field. Omitting the Goldstone modes Eqs. (1.46) and (1.50) read

$$\Sigma = \left(1 + \frac{H}{v}\right)\mathbb{1} \qquad \phi = \frac{1}{\sqrt{2}}\begin{pmatrix} 0 \\ v+H \end{pmatrix} = \frac{1}{\sqrt{2}}\begin{pmatrix} 0 \\ v\Sigma \end{pmatrix}. \quad (1.259)$$

In the non–linear sigma model, defined in Eq. (1.42), we replace v by f_T and find

$$\phi = \frac{1}{\sqrt{2}}\begin{pmatrix} 0 \\ f_T\Sigma \end{pmatrix} = \frac{1}{\sqrt{2}} e^{-i(\pi\cdot\tau)/f_T}\begin{pmatrix} 0 \\ f_T \end{pmatrix} = \frac{1}{\sqrt{2}}\begin{pmatrix} 0 \\ f_T - i(\pi\cdot\tau) + \mathscr{O}\left(\frac{1}{f_T}\right) \end{pmatrix} \quad (1.260)$$

As basis vectors we use the three Pauli matrices $\{\tau_a, \tau_b\} = 2\delta_{ab}$ which according to Eq. (1.10) fulfill $(\tau\cdot\pi_1)(\tau\cdot\pi_2) = (\pi_1\cdot\pi_2)$. The $SU(2)$-covariant derivative in the charge basis of the Pauli matrices defined in Eq. (1.8) gives, when to simplify the formulas we for a moment forget about the $U(1)_Y$ contribution and only keep the non-zero upper entry in Eq. (1.260):

$$iD^\mu\phi\bigg|_{\text{lower}} = \left[i\partial^\mu - \frac{g}{2}(\tau\cdot W^\mu)\right]\frac{1}{\sqrt{2}}\left[f_T - i(\tau\cdot\pi) + \mathscr{O}\left(\frac{1}{f_T}\right)\right]$$

$$= \frac{1}{\sqrt{2}}\left[\partial^\mu(\tau\cdot\pi) - \frac{gf_T}{2}(\tau\cdot W^\mu) + \mathscr{O}(f_T^0)\right]$$

$$(D_\mu\phi)^\dagger(D^\mu\phi) = -\frac{1}{2}\left[\partial_\mu(\tau\pi) - \frac{gf_T}{2}(\tau\cdot W_\mu) + \mathscr{O}(f_T^0)\right]$$

$$\times\left[\partial^\mu(\tau\pi) - \frac{gf_T}{2}(\tau\cdot W^\mu) + \mathscr{O}(f_T^0)\right]$$

$$\supset -\frac{1}{2}(\partial\pi)^2 + \frac{gf_T}{2}(W_\mu\cdot(\partial^\mu\pi)) + \mathscr{O}(f_T^0). \quad (1.261)$$

If we also include the generator of the hypercharge $U(1)$ we find a mixing term between the technipions and the $SU(2)$ gauge bosons

$$\boxed{\mathscr{L} \supset \frac{gf_T}{2} W^+_\mu \partial^\mu \pi^- + \frac{gf_T}{2} W^-_\mu \partial^\mu \pi^+ + f_T\left(\frac{g}{2} W^3_\mu + \frac{g'}{2} B_\mu\right)\partial^\mu \pi^0}. \quad (1.262)$$

This is precisely the mixing term from the massive–photon example of Eq. (1.3) which we need to absorb the Goldstone modes into the massive vector bosons.

We have strictly speaking not shown that the f_T appearing in the scalar field ϕ is really the correctly normalized decay constant of the technipions and there is a lot of confusion about factors $\sqrt{2}$ in the literature which we will ignore in this sketchy argument. Nevertheless, if we assume the correct normalization the massive

1.9 Alternatives and Extensions

Lagrangian equation (1.262) with a mixing term proportional to $gf_T/2$ generates $m_W = gf_T/2$. We know from Eq. (1.38) that $m_W = gv/2$, so electroweak symmetry breaking might well be a scaled-up version of f_π.

Once we are convinced that we can scale up QCD and break electroweak symmetry we want to check what kind of predictions for the electroweak observables come out of technicolor. We can study this scaling in the general case, where technicolor involves a gauge group $SU(N_T)$ instead of $SU(N_c)$ and we have N_D left handed fermion doublets in the fundamental representation of $SU(N_T)$. To be able to write down Dirac masses for the fermions at the end of the day we also need $(2N_D)$ right handed fermion singlets. From the case of more than one Higgs field contributing to v we know that if we have N_D separate condensates their squares have to add up to g^2v^2; for equal vacuum expectation values they scale like $v \sim \sqrt{N_D} f_T$. The scaling rules of Eq. (1.257) then give

$$\frac{v}{\sqrt{N_D}} = f_T \sim \sqrt{\frac{N_T}{N_c}} \frac{\Lambda_T}{\Lambda_{\text{QCD}}} f_\pi \quad \Leftrightarrow \quad \frac{\Lambda_T}{\Lambda_{\text{QCD}}} \sim \frac{v}{f_\pi} \sqrt{\frac{N_c}{N_D N_T}} \sim \frac{246\,\text{GeV}}{130\,\text{MeV}} \sqrt{\frac{N_c}{N_D N_T}}. \tag{1.263}$$

One simple example for this technicolor setup is the *Susskind–Weinberg model*. Its gauge group is $SU(N_T) \times SU(3)_c \times SU(2)_L \times U(1)_Y$. As matter fields forming the condensate which in turn breaks the electroweak symmetry we include one doublet ($N_D = 1$) of charged color–singlet technifermions $(u^T, d^T)_{L,R}$. In some ways this doublet and the two singlets look like a fourth generation of chiral fermions, but with different charges under all Standard Model gauge groups: for example, their hypercharges Y need to be chosen such that gauge anomalies do not occur and we do not have to worry about non–perturbatively breaking any symmetries, namely $Y = 0$ for the left handed doublet and $Y = \pm 1$ for u_R^T and d_R^T. The usual formula $q = \tau_3/2 + Y/2$ then defines the electric charges $\pm 1/2$ for the heavy fermions u^T and d^T.

The additional $SU(N_T)$ gauge group gives us a running gauge coupling which becomes large at the scale Λ_T. Its beta function is modelled after the QCD case

$$\beta_{\text{QCD}} = -\frac{\alpha_s^2}{4\pi}\left(\frac{11}{3}N_c - \frac{2}{3}n_f\right) \qquad \beta_T = -\frac{\alpha_T^2}{4\pi}\left(\frac{11}{3}N_T - \frac{4}{3}N_D\right), \tag{1.264}$$

keeping in mind that N_D counts the doublets, while n_f counts the number of flavors at the GUT scale. This relation holds for a simple model, where quarks are only charged under $SU(3)_c$ and techniquarks are only charged under $SU(N_T)$. Of course, both of them can carry weak charges. As a high–scale boundary condition we can for example choose $\alpha_s(M_{\text{GUT}}) = \alpha_T(M_{\text{GUT}})$. Using Eq. (2.73) for Λ_{QCD} we find

$$\frac{\Lambda_T^2}{\Lambda_{QCD}^2} = \exp\left[+\frac{\alpha_T(m_{GUT})}{\beta_T}\right]\exp\left[-\frac{\alpha_s(m_{GUT})}{\beta_{QCD}}\right]$$

$$= \exp\left[\alpha_s(m_{GUT})\left(\frac{1}{\beta_T} - \frac{1}{\beta_{QCD}}\right)\right]$$

$$= \exp\left[-\frac{4\pi}{\alpha_s(m_{GUT})}\left(\frac{1}{\frac{11}{3}N_T - \frac{4}{3}N_D} - \frac{1}{11-4}\right)\right]. \quad (1.265)$$

Such a GUT-inspired model based on an $SU(4)$ gauge group with $\alpha_s(M_{GUT}) \sim 1/30$ and $N_D = 1$ does not reproduce the scale ratio required by Eq. (1.263). However, for example choosing $N_T = N_D = 4$ predicts $\Lambda_T/\Lambda_{QCD} \sim 830$ and with it exactly the measured ratio of v/f_π.

At this stage, our fermion construction has two global chiral symmetries $SU(2) \times SU(2)$ and $U(1) \times U(1)$ protecting the technifermions from getting massive, which we will of course break together with the local weak $SU(2)_L \times U(1)_Y$ symmetry. Details about fermion masses we postpone to the next sections. Let us instead briefly look at the *particle spectrum* of our minimal model:

- Techniquarks: from the scaling rules we know that the techniquark masses will be of the order Λ_T as give above. Numerically, the factor $\Lambda_T/\Lambda_{QCD} \sim 800$ pushes the usual quark constituent masses to around 700 GeV for the minimal model with $N_T = 4$ and $N_D = 1$. Because of the $SU(N_T)$ gauge symmetry there should exist four–techniquark bound states (technibaryons) which are stable due to the asymptotic freedom of the $SU(N_T)$ symmetry. Those are not preferred by standard cosmology, so we should find ways to let them decay.
- Goldstone modes: from the breaking of the global chiral $SU(2) \times SU(2)$ and the $U(1) \times U(1)$ we will have four Goldstone modes. The three $SU(2)$ Goldstones are massless technipions, following our QCD analogy. Because we gauge the remaining Standard Model subgroup $SU(2)_L$, they become the longitudinal polarizations of the W and Z boson, after all this is the entire idea behind this construction. The remaining $U(1)$ Goldstone mode also has an equivalent in QCD (η'), and its technicolor counter part acquires a mass though non–perturbative instanton breaking. Its mass can be estimates to ~ 2 TeV, so we are out of trouble.
- More exotic states: just like in QCD we will have a whole zoo of additional technicolor vector mesons and heavy resonances, but all we need to know about them is that they are heavy (and therefore not a problem for example for cosmology).

Before we move on, let us put ourselves into the shoes of the technicolor proponents in the 1970s. They knew how QCD gives masses to protons, and the Higgs mechanism has nothing to do with it. Just copying the appealing idea of dimensional transmutation without any hierarchy problem they explained the measured W and Z masses. And just like in QCD, the masses of the four light quarks and the leptons

1.9 Alternatives and Extensions

are well below a GeV and could be anything, but not linked to weak–scale physics. Then, people found the massive bottom quark and the even more massive top quark and it became clear that at least the top mass was very relevant to the weak scale. In this section we will very briefly discuss how this challenge to technicolor basically removed it from the list of models people take seriously—until extra dimensions came and brought it back to the mainstream.

Extended technicolor is a version of the original idea of technicolor which attempts to solve two problems: create fermion masses for three generations of quarks and leptons and let the heavy techniquarks decay, to avoid stable technibaryons. From the introduction we in principle know how to obtain a fermion mass from Yukawa couplings, but to write down the Yukawa coupling to the sigma field or to the TC condensate we need to write down some Standard Model and technifermion operators. This is what ETC offers a framework for.

First, we need to introduce some kind of multiplets of matter fermions. Just as before, the techniquarks, like all matter particles have $SU(2)_L$ and $U(1)_Y$ or even $SU(2)_R$ quantum numbers. However, there is no reason for them all to have a $SU(3)_c$ charge, because we would prefer not to change β_{QCD} too much. Similarly, the Standard Model particles do not have a $SU(N_T)$ charge. This means we can write matter multiplets with explicitly assigned color and technicolor charges as

$$\left(Q^T_{a=1..N_T},\ Q^{(1)}_{j=1,...,N_c},\ Q^{(2)}_{j=1,...,N_c},\ Q^{(3)}_{j=1,...,N_c},\ L^{(1)},\ L^{(2)},\ L^{(3)} \right). \quad (1.266)$$

These multiplets replace the usual $SU(2)_L$ and $SU(2)_R$ singlets and doublets in the Standard Model. The upper indices denote the generation, the lower indices count the N_T and N_c fundamental representations. In the minimal model with $N_T = 4$ this multiplet has $4 + 3 + 3 + 3 + 1 + 1 + 1 = 16$ entries. In other words, we have embedded $SU(N_T)$ and $SU(N_c)$ in a local gauge group $SU(16)$. If without further discussion we also extend the Standard Model group by a $SU(2)_R$ gauge group, the complete ETC symmetry group is $SU(16) \times SU(2)_L \times SU(2)_R$, where we omit the additional $U(1)_{B-L}$ throughout the discussion.

A technicolor condensate will now break $SU(2)_L \times SU(2)_R$, while leaving $SU(3)_c$ untouched. If we think of the generators of the ETC gauge group as (16×16) matrices we can put a (4×4) block of $SU(N_T)$ in the upper left corner and then three (3×3) copies of $SU(N_c)$ on the diagonal. The last three rows/columns can be the unit matrix. Once we break $SU(16)_{ETC}$ to $SU(N_T)$ and the Standard Model gauge groups, the Goldstone modes corresponding to the broken generators obtain masses of the order of Λ_{ETC}. This breaking should on the way produce the correct fermion masses. The remaining $SU(N_T) \times SU(2)_L \times U(1)_Y$ will then break the electroweak symmetry through a $SU(N_T)$ condensate and create the measured W and Z masses as described in the last section.

In this construction we will have ETC gauge bosons which for example in the quark sector form currents of the kind $(\overline{Q^T} \gamma_\mu T_{ETC} Q^T)$, $(\overline{Q^T} \gamma_\mu T_{ETC} Q)$, or $(\overline{Q} \gamma_\mu T_{ETC} Q)$. Here, T_{ETC} stands for the $SU(16)_{ETC}$ generators. The multiplets Q^T and Q replace the $SU(2)_{L,R}$ singlet and doublets. Below the ETC breaking scale

Λ_{ETC} these currents become four-fermion interactions, just like a Fermi interaction in the electroweak theory,

$$\frac{(\overline{Q}^T\gamma_\mu T^a_{\text{ETC}} Q^T)(\overline{Q}^T\gamma^\mu T^b_{\text{ETC}} Q^T)}{\Lambda^2_{\text{ETC}}} \quad \frac{(\overline{Q}^T\gamma_\mu T^a_{\text{ETC}} Q)(\overline{Q}\gamma^\mu T^b_{\text{ETC}} Q^T)}{\Lambda^2_{\text{ETC}}}$$

$$\frac{(\overline{Q}\gamma_\mu T^a_{\text{ETC}} Q)(\overline{Q}\gamma^\mu T^b_{\text{ETC}} Q)}{\Lambda^2_{\text{ETC}}} \; . \quad (1.267)$$

The mass scale in this effective theory can be linked to the mass of the ETC gauge bosons and their gauge coupling and should be of the order $1/\Lambda_{\text{ETC}} \sim g_{\text{ETC}}/M_{\text{ETC}}$. Let us see what this kind of interaction predicts at energy scales below Λ_{ETC} or around the weak scale. Because currents are hard to interpret, we Fierz–rearrange these operators and then pick out three relevant classes of scalar operators.

Let us briefly recall this *Fierz transformation*. The complete set of four-fermion interactions is given by the structures

$$\mathscr{L} \supset (\overline{\psi} A_j \psi)(\overline{\psi} A^j \psi) \qquad \text{with} \qquad A_j = \mathbb{1},\; \gamma_5,\; \gamma_\mu,\; \gamma_5\gamma_\mu,\; \sigma_{\mu\nu} \; . \quad (1.268)$$

The multi–index j implies summing over all open indices in the diagonal combination $A_j A^j$. These five types of (4×4) matrices form a basis of all real (4×4) matrices which can occur in the Lagrangian. If we now specify the spinors and cross them in this interaction we should be able to write the new crossed (1,4,3,2) scalar combination (or any new term, for that matter) as a linear combination of the basis elements ordered as (1,2,3,4):

$$(\overline{\psi}_1 A_i \psi_4)(\overline{\psi}_3 A_i \psi_2) = \sum_j C_{ij} (\overline{\psi}_1 A_j \psi_2)(\overline{\psi}_3 A_j \psi_4) \; . \quad (1.269)$$

In this notation we ignore the normal–ordering of the spinors in the Lagrangian. It is easy to show that $C \cdot C = \mathbb{1}$. The coefficients C_{ij} we list for completeness reasons:

	$\mathbb{1}$	γ_5	γ_μ	$\gamma_5\gamma_\mu$	$\sigma_{\mu\nu}$
$\mathbb{1}$	$-1/4$	$-1/4$	$-1/4$	$1/4$	$-1/8$
γ_5	$-1/4$	$-1/4$	$1/4$	$-1/4$	$-1/8$
γ_μ	-1	1	$1/2$	$1/2$	0
$\gamma_5\gamma_\mu$	1	-1	$1/2$	$1/2$	0
$\sigma_{\mu\nu}$	-3	-3	$1/2$	0	$1/2$

(1.270)

Applying a Fierz transformation to the three quark–techniquark *four-fermion operators* given in Eq. (1.271) we obtain scalar ($A = \mathbb{1}$) operators,

1.9 Alternatives and Extensions

$$\frac{(\overline{Q^T}\, T^a_{\text{ETC}}\, Q^T)\, (\overline{Q^T}\, T^b_{\text{ETC}}\, Q^T)}{\Lambda^2_{\text{ETC}}} \qquad \frac{(\overline{Q^T_L}\, T^a_{\text{ETC}}\, Q^T_R)\, (\overline{Q_R}\, T^b_{\text{ETC}}\, Q_L)}{\Lambda^2_{\text{ETC}}}$$

$$\frac{(\overline{Q_L}\, T^a_{\text{ETC}}\, Q_R)\, (\overline{Q_R}\, T^b_{\text{ETC}}\, Q_L)}{\Lambda^2_{\text{ETC}}}\,. \quad (1.271)$$

In these examples we pick certain chiralities of the Standard Model fields and the technifermions. Let us go through these operators one by one. While not all of these operators will be our friends we need them to give masses to the Standard Model fermions, so we have to live with the constraints.

1. Once technicolor forms condensates of the kind $\langle \overline{Q^T} Q^T \rangle$ the first operator in Eq. (1.271) will give masses to technicolor generators. This happens through loops involving technicolor states. We only quote the result for example for technipions which receive masses of the order $m \sim N_T \Lambda^2_T / \Lambda_{\text{ETC}}$. For $\Lambda_T \sim v$ and $\Lambda_{\text{ETC}} \sim 3$ TeV this not a large mass value, but nevertheless this first scalar operator clearly is our friend.

2. The scaling rules in Eq. (1.257) require the condensate to be proportional to $N_T \Lambda^3_T$, The scalar dimension-6 operator adds a factor to $1/\Lambda^2_{\text{ETC}}$, so dimensional analysis tells us that the resulting Standard Model fermion masses will be of the order

$$\mathcal{L} \supset \frac{N_T \Lambda^3_T}{\Lambda^2_{\text{ETC}}} \overline{Q_L} q_R \equiv m_Q \overline{Q_L} q_R \; \Leftrightarrow \; \Lambda_{\text{ETC}} \sim \sqrt{\frac{N_T \Lambda^3_T}{m_Q}} \sim \begin{cases} 3.7\,\text{TeV} & m_Q = 1\,\text{GeV} \\ 300\,\text{GeV} & m_Q = 150\,\text{GeV} \end{cases}$$
(1.272)

for $N_T = 4$ and $\Lambda_T = 150$ GeV. This operator appears to be our friend for light quarks, but it becomes problematic for the top quark, where $\Lambda_{\text{ETC}} \sim v$ comes out too small.

The top mass operator can be fierzed into a left handed fermion–technifermion current $(\overline{Q^T_L}\gamma_\mu Q_L)(\overline{Q_L}\gamma^\mu Q^T_L)$. Because of the custodial $SU(2)_L \times SU(2)_R$ symmetry, which will turn out crucial to avoid electroweak precision constraints, we can rotate the top quarks into bottom quarks,

$$\frac{g^2_{\text{ETC}}}{M^2_{\text{ETC}}} \left(\overline{Q^T_L}\gamma_\mu b_L\right) \left(\overline{Q_L}\gamma^\mu Q^T_L\right)\,. \quad (1.273)$$

This operator induces a coupling of a charged ETC gauge boson to $T_L b_L$ which induces a one-loop contribution to the *decay* $Z \to b\bar{b}$. It contributes to the effective bbZ coupling at the order $v^2/\Lambda^2_{\text{ETC}} \sim \mathcal{O}(1)$, considerably too big for the LEP measurement of $R_b = \Gamma_Z(b\bar{b})/\Gamma_Z(\text{hadrons})$. Note that such a constraint will affect any theory which induces a top mass through a partner of the top quark and allows for a general set of fierzed operators corresponding to this mass term, not just extended technicolor.

3. The third operator in Eq. (1.271) does not include any techniquarks, but all combinations of four-quark couplings. In the Standard Model such operators are very strongly constrained, in particular when they involve different quark flavors. *Flavor–changing neutral currents* essentially force operators like

$$\frac{1}{\Lambda_{\text{ETC}}^2} (\bar{s}\gamma^\mu d)(\bar{s}\gamma_\mu d) \qquad \frac{1}{\Lambda_{\text{ETC}}^2} (\bar{\mu}\gamma^\mu e)(\bar{e}\gamma_\mu \mu) \qquad (1.274)$$

to vanish. The currently strongest constraints come from kaon physics, for example the mass splitting between the K^0 and the $\overline{K}^0$. Its limit $\Delta M_K \lesssim 3.5 \cdot 10^{-12}$ MeV implies $M_{\text{ETC}}/(g_{\text{ETC}}\theta_{sd}) \gtrsim 600$ TeV in terms of the Cabibbo angle θ_{sd}. We can translate this bound on Λ_{ETC} into an upper bound on fermion masses we can construct in our minimal model. $\Lambda_{\text{ETC}} > 10^3$ TeV simply translates in a maximum fermion mass which we can explain in this model: $m \lesssim 4$ MeV for $\Lambda_{\text{T}} \lesssim 1$ TeV. This is obviously not good news, unless we find a flavor symmetry to protect us from unwanted dimension-6 operators.

Let us collect all the evidence we have against technicolor, in spite of its very appealing numerical analogy to QCD. First and most importantly, it predicts no light Higgs resonance, but a zoo of heavy techni-particles. Both of these predictions are in disagreement with current LHC data. In the next section we will use Goldstone's theorem to break this degeneracy and generate a single light Higgs scalar. This Goldstone protection of a single light state can be applied to many models, including technicolor and other strongly interacting Higgs sectors.

In addition, technicolor is strongly constrained by electroweak precision constraints described in Sect. 1.1.6. If we introduce new particles with $SU(2)_L \times U(1)_Y$ quantum numbers, all of these particles will contribute to gauge boson self energies. Contributions from different states largely add, as we can see for example for the S and T parameters in Eq. (1.67). In technicolor the singlet techniquarks will each contribute as $\Delta S \sim N_T/(6\pi) \sim 4/20$, assuming $N_D = 1$. More realistic models easily get to $\Delta S \sim \mathcal{O}(1)$, which is firmly ruled out, no matter what kind of ΔT we manage to generate. The way out of some technicolor problems is so-called walking technicolor, which still does not predict a light narrow Higgs resonance. Nevertheless, it is instructive to understand dynamic electroweak symmetry breaking because we know that this mechanism is realized elsewhere in Nature and might eventually enter our Higgs sector in some non–trivial way. After all, we did not yet manage to answer the question where the Higgs field and the Higgs potential come from.

1.9.2 Hierarchy Problem and the Little Higgs

Before we introduce the little Higgs mechanism of breaking electroweak symmetry we first need to formulate a major problem with the Higgs boson as a fundamental scalar. Let us start by assuming that the Standard Model is a renormalizable theory.

1.9 Alternatives and Extensions

At next–to–leading order, the bare leading order Higgs mass gets corrected by loops involving all heavy Standard Model particles. Even within the Higgs sector alone we can for example compute the four-point Higgs loop proportional to the coupling given in Eq. (1.85), namely $-3im_H^2/v^2$. Introducing a cutoff scale Λ and sending it to infinity is certainly a valid physical regularization scheme. We can implement the cutoff using the *Pauli–Villars regularization*,

$$\frac{1}{q^2 - m_H^2} \to \frac{1}{q^2 - m_H^2} - \frac{1}{q^2 - \Lambda^2} = \begin{cases} \dfrac{1}{q^2 - m^2} & q^2 \ll \Lambda^2 \\ \dfrac{1}{q^2} - \dfrac{1}{q^2} = 0 & q^2 \gg \Lambda^2. \end{cases} \quad (1.275)$$

The one-loop integral mediated by the Higgs self coupling then reads, modulo prefactors,

$$\frac{3m_H^2}{v^2} \int^\Lambda \frac{d^4q}{(2\pi)^4} \frac{1}{q^2 - m_H^2} \to \frac{3m_H^2}{v^2} \int \frac{d^4q}{(2\pi)^4} \left(\frac{1}{q^2 - m_H^2} - \frac{1}{q^2 - \Lambda^2} \right)$$

$$= \frac{3m_H^2}{v^2} \left(m_H^2 - \Lambda^2 \right) \int \frac{d^4q}{(2\pi)^4} \frac{1}{(q^2 - m_H^2)(q^2 - \Lambda^2)}$$

$$= \frac{3m_H^2}{v^2} \left(m_H^2 - \Lambda^2 \right) \frac{C}{16\pi^2}, \quad (1.276)$$

where C is a numerical constant coming from the computation of the four-dimensional integral. When we separate a remaining factor $1/(16\pi^2)$ from the integral measure $1/(2\pi)^4$ it comes out of order unity. The problem is that Eq. (1.276) contributes to the Higgs mass, as we will show in some detail in Sect. 2.1.2. This means that we observe a divergent one-loop contribution to the Higgs mass $\Delta m_H^2 \propto \Lambda^2$. Including all heavy Standard Model loops we find a full set of quadratically divergent corrections to the bare Higgs mass $m_{H,0}$

$$\boxed{m_H^2 = m_{H,0}^2 + \frac{3g^2}{32\pi^2} \frac{\Lambda^2}{m_W^2} \left[m_H^2 + 2m_W^2 + m_Z^2 - \frac{4}{3}m_t^2 \right].} \quad (1.277)$$

This form of the Higgs mass corrections has one interesting new aspect, when we compare it to the known behavior of the fermion masses: the loop corrections to the mass and hence the quadratic divergence are not proportional to the Higgs mass. Unlike fermion masses which are linked to an approximate chiral symmetry, a finite Higgs mass can appear entirely at loop level. We have already exploited this feature writing down the Coleman–Weinberg mechanism in Sect. 1.2.7.

The naive solution $m_H^2 + 2m_W^2 + m_Z^2 - 4n_f m_t^2/3 = 0$, called Veltman's condition, assumes that fermionic and bosonic loop corrections are regularized the same way, which is not realistic.

Why is the quadratic divergence in Eq. (1.277) a problem? Dimensional regularization using $n = 4 - 2\epsilon$ space–time dimensions does not distinguish between logarithmic and quadratic divergences. And we know that all masses develop poles $1/\epsilon$ which reflects the fact that the bare masses in the Lagrangian have to be renormalized. In that sense dimensional regularization is not a solution to our problem, but an approach which does not even see the issue.

In an effective theory approach there exist many physical scales at which we need to add new effects to the Standard Model. This could for example be a see-saw scale to generate right handed neutrinos or some scale where the quark flavor parameters are generated. In such an effective theory we should be able to use a cutoff and matching scheme around the high mass scale Λ. Varying this matching scale slightly should not make a big difference. However, the quadratic divergence of the Higgs mass implies that we have to compensate for a large matching scale dependence of an observable mass on the ultraviolet side of the matching. In other words, we need to seriously fine-tune the ultraviolet completion of the effective Standard Model.

Alternatively, we can argue that in the presence of any large energy scale the Higgs mass wants to run to this high scale. This is only true for a fundamental scalar particle, fermion masses only run logarithmically. This means that while the Higgs mechanism only works for a light Higgs mass around the electroweak scale, the Higgs naturally escapes, if we let it. Keeping the Higgs mass stable in the presence of a larger physical energy scale is called the *hierarchy problem*.

We can quantify the level of fine tuning, which would be required to remove the huge next–to–leading order contributions using a counter term,

$$m_{H,0}^2 + \frac{3g^2}{32\pi^2} \frac{\Lambda^2}{m_W^2} \left[m_H^2 + 2m_W^2 + m_Z^2 - \frac{4}{3}m_t^2 \right] - \delta m_H^2 \stackrel{!}{=} m_{H,0}^2 . \quad (1.278)$$

Assuming $\Lambda = 10\,\text{TeV}$ the different Standard Model contributions require

$$\delta m_H^2 = \begin{cases} -\dfrac{3}{8\pi^2} \lambda_t^2 \Lambda^2 \sim -(2\,\text{TeV})^2 & t \text{ loop} \\ \dfrac{1}{16\pi^2} g^2 \Lambda^2 \sim (100\,\text{TeV})^2 & W \text{ loop} \\ \dfrac{1}{16\pi^2} \lambda^2 \Lambda^2 \sim (500\,\text{TeV})^2 & H \text{ loop}. \end{cases} \quad (1.279)$$

Varying the cutoff scale Λ we need to ensure

$$m_H = m_{H,0}^2 - \delta m_H^2 + \begin{cases} (-250 + 50 + 25)\,(125\,\text{GeV})^2 & \text{for } \Lambda = 10\,\text{TeV} \\ (-25{,}000 + 2{,}500 + 1{,}250)\,(125\,\text{GeV})^2 & \text{for } \Lambda = 100\,\text{TeV} \\ \cdots \end{cases} \quad (1.280)$$

While we need to emphasize that the hierarchy problem is a mathematical or even esthetic problem of the Higgs sector, it might guide us to a better understanding of the Higgs sector. The best–known solution to the hierarchy problem

1.9 Alternatives and Extensions

is supersymmetry, with the modified Higgs sector discussed in Sect. 1.2.6. Alternatively, extra space–time dimensions, flat or warped, offer a solution. Finally, we will show how little Higgs models protect the Higgs mass based on concepts related to Goldstone's theorem.

Trying to solve this hierarchy problem using broken symmetries will lead us to *little Higgs models*. This mechanism of stabilizing a small Higgs mass is based on Goldstone's theorem: we make the Higgs a Goldstone mode of a broken symmetry at some higher scale. This way the Higgs mass is forbidden by a symmetry and cannot diverge quadratically at large energy scales. More precisely, the Higgs has to be a pseudo–Goldstone, so that we can write down a Higgs mass and potential. This idea has been around for a long time, but for decades people did not know how to construct such a symmetry.

Before we solve this problem via the little Higgs mechanism, let us start by constructing an example symmetry which protects the Higgs mass from quadratic divergences at one loop. We break a for now global $SU(3)$ symmetry to $SU(2)_L$. The number of generators which are set free when we break $SU(N) \to SU(N-1)$ is

$$(N^2 - 1)^2 - ((N-1)^2 - 1) = 2N - 1 \,. \tag{1.281}$$

The $SU(2)$ generators are the Pauli matrices given in Eq. (1.8). For $SU(3)$ the basis is given by the traceless hermitian and unitary *Gell–Mann matrices*,

$$\lambda^1 = \begin{pmatrix} \tau^1 & 0 \\ & 0 \\ 0\ 0 & 0 \end{pmatrix} \quad \lambda^2 = \begin{pmatrix} \tau^2 & 0 \\ & 0 \\ 0\ 0 & 0 \end{pmatrix} \quad \lambda^3 = \begin{pmatrix} \tau^3 & 0 \\ & 0 \\ 0\ 0 & 0 \end{pmatrix} \quad \lambda^8 = \frac{1}{\sqrt{3}} \begin{pmatrix} \mathbb{1} & 0 \\ & 0 \\ 0\ 0 & -2 \end{pmatrix}$$

$$\lambda^4 = \begin{pmatrix} 0 & 1 \\ 0 & \\ 1\ 0 & 0 \end{pmatrix} \quad \lambda^5 = \begin{pmatrix} 0 & -i \\ 0 & \\ i\ 0 & 0 \end{pmatrix} \quad \lambda^6 = \begin{pmatrix} 0 & 0 \\ & 1 \\ 0\ 1 & 0 \end{pmatrix} \quad \lambda^7 = \begin{pmatrix} 0 & 0 \\ & -i \\ 0\ i & 0 \end{pmatrix} \,. \tag{1.282}$$

We can arrange all generators of $SU(3)$ which are not generators of $SU(2)$, and hence turn into Goldstones, in the outside column and row of the 3×3 matrix

$$\begin{pmatrix} SU(2) & w_1 \\ & w_2 \\ w_1^* & w_2^* & w_0 \end{pmatrix} \equiv \begin{pmatrix} SU(2) & \phi \\ \phi^\dagger & w_0 \end{pmatrix} \,. \tag{1.283}$$

The entry w_0 is fixed by the requirement that the matrix has to be traceless when we include $\mathbb{1}$ as the fourth $SU(2)$ matrix in the top–left corner. The corresponding field is an $SU(2)$ singlet and can be ignored for now.

We now assume that the $SU(2)_L$ doublet ϕ formed by the broken $SU(3)$ generators is the Standard Model Higgs doublet. Normalization factors $1/\sqrt{2}$ we omit in this section. The Higgs can then only acquire a mass at the electroweak scale, where $SU(2)_L$ is broken. Based on Eq. (1.283) we define a sigma field as in Eq. (1.42). The only difference to Eq. (1.42) is that Σ is now a triplet and includes a symmetry breaking scale $f > v$

$$\Sigma = \exp\left[-\frac{i}{f}\begin{pmatrix} 0_{2\times 2} & \phi \\ \phi^\dagger & 0 \end{pmatrix}\right]\begin{pmatrix} 0_2 \\ f \end{pmatrix}$$

$$= \left[\mathbb{1} - \frac{i}{f}\begin{pmatrix} 0 & \phi \\ \phi^\dagger & 0 \end{pmatrix} - \frac{1}{2}\left(\frac{-1}{f}\right)^2 \begin{pmatrix} 0 & \phi \\ \phi^\dagger & 0 \end{pmatrix}\begin{pmatrix} 0 & \phi \\ \phi^\dagger & 0 \end{pmatrix} + \mathscr{O}\left(\frac{1}{f^3}\right)\right]\begin{pmatrix} 0 \\ f \end{pmatrix}$$

$$= \begin{pmatrix} 0 \\ f \end{pmatrix} - \begin{pmatrix} i\phi \\ 0 \end{pmatrix} - \frac{1}{2f^2}\begin{pmatrix} 0 \\ \phi^\dagger\phi f \end{pmatrix} + \mathscr{O}\left(\frac{1}{f^3}\right)$$

$$= \begin{pmatrix} 0 \\ f \end{pmatrix} - \begin{pmatrix} i\phi \\ \phi^\dagger\phi/(2f) \end{pmatrix} + \mathscr{O}\left(\frac{1}{f^3}\right). \tag{1.284}$$

Only in the first line we indicate which of the zeros in the 3×3 matrix is a 2×2 sub-matrix. This is easy to keep track of if we remember that the Higgs field ϕ is a doublet, while $\phi^\dagger\phi$ is a scalar number. The kinetic term for the triplet field Σ becomes

$$|\partial_\mu \Sigma|^2 = (i\partial_\mu \phi^*)_i(-i\partial^\mu \phi)_i + \frac{1}{4f^2}(\partial_\mu \phi^\dagger \phi)(\partial^\mu \phi^\dagger \phi) = |\partial_\mu \phi|^2\left(1 + \frac{\phi^\dagger \phi}{f^2}\right), \tag{1.285}$$

where we skip the non–trivial intermediate steps. The second term in the brackets includes two Higgs fields which we can link to a propagator, generating a one-loop correction to the Higgs propagator. We know that our theory is an effective non–renormalizable field theory, so we can apply a cutoff to the divergent loop diagram. The result we already know from Eq. (1.276). Ignoring the constant C but keeping the factor $1/(4\pi)^2$ from the integral measure we can write down the condition under which the second term in Eq. (1.285) does not dominate the tree level propagator,

$$\boxed{\frac{\Lambda^2}{(4\pi)^2 f^2} \lesssim 1} \quad \Leftrightarrow \quad \Lambda \lesssim 10 \times f. \tag{1.286}$$

Once the loop–induced effect exceeds the tree level propagator at high energies we consider the theory strongly interacting. Our perturbative picture of the little Higgs theory breaks down. Without even writing out a model for a Higgs mass protected by Goldstone's theorem we already know that its ultraviolet completion will not be perturbative and hence not predictive, and that it's range of validity will be rather limited. Accepting these limitations we now introduce a coupling to the $SU(2)$ gauge bosons and see what happens to the Higgs mass. Of course, from the discussion of Goldstone's theorem in Sect. 1.1 we already know that we will not be able to generate the Higgs mass or potential in a straightforward way, but it is constructive to see the problems which will arise.

As a *first attempt* we simply add $g\,(W^\mu \cdot \tau)$ as part of the covariant derivative to the kinetic term. In other words, we gauge the $SU(2)$ subgroup of the global $SU(3)$ group. This automatically creates a four-point coupling of the kind $g^2|\vec{W}_\mu \phi|^2$. As we

1.9 Alternatives and Extensions

did for Eq. (1.285) we combine the two W bosons to a propagator and generate a one–loop Higgs mass term of the kind

$$\mathscr{L} \supset \frac{g^2 \Lambda^2}{(4\pi)^2} \phi^\dagger \phi \,. \qquad (1.287)$$

This term gives the quadratically divergent Higgs mass we know from Eq. (1.277). Our ansatz does not solve or even alleviate the hierarchy problem, so we discard it. What we learn from it is that we cannot just write down the Standard Model $SU(2)_L$ gauge sector and expect the hierarchy problem to vanish.

In a *second attempt* we therefore write the same interaction in terms of the triplet field Σ, just leaving the third entry in the gauge–boson matrix empty,

$$g^2 \left| \begin{pmatrix} (W_\mu \cdot \tau) & 0 \\ 0 & 0 \end{pmatrix} \Sigma \right|^2 \,. \qquad (1.288)$$

We can again square this interaction term contributing to the Higgs mass and find schematically

$$g^2 \Sigma^\dagger \begin{pmatrix} \mathbb{1} & 0 \\ 0 & 0 \end{pmatrix} \begin{pmatrix} \mathbb{1} & 0 \\ 0 & 0 \end{pmatrix} \Sigma \sim g^2 \Sigma^\dagger \begin{pmatrix} \mathbb{1} & 0 \\ 0 & 0 \end{pmatrix} \Sigma \sim g^2 \phi^\dagger \phi \,, \qquad (1.289)$$

so the self energy contribution with the two W fields linked now reads

$$\mathscr{L} \supset \frac{g^2 \Lambda^2}{(4\pi)^2} \Sigma^\dagger \begin{pmatrix} \mathbb{1} & 0 \\ 0 & 0 \end{pmatrix} \Sigma = \frac{g^2 \Lambda^2}{(4\pi)^2} \phi^\dagger \phi \,. \qquad (1.290)$$

This is precisely Eq. (1.287) and leads us to also discard this second attempt. This outcome is not surprising because we really only write the same thing in two different notations, either using $\Sigma^\dagger \Sigma$ or $\phi^\dagger \phi$. Embedding the $SU(2)_L$ gauge sector into a $SU(3)$ structure can only improve our situation when the $SU(3)$ gauge group actually extends beyond $SU(2)_L$.

Learning from these two failed attempts we can go for a *third attempt*, where we add a proper covariant derivative including all $SU(3)$ degrees of freedom. Closing all of them into loops we obtain in a proper basis

$$\mathscr{L} \supset \frac{g^2 \Lambda^2}{(4\pi)^2} \Sigma^\dagger \mathbb{1} \Sigma = \frac{g^2 \Lambda^2}{(4\pi)^2} f^2 \,. \qquad (1.291)$$

This is no contribution to the Higgs mass because the now massive $SU(3)$ gauge bosons ate the Goldstones altogether. According to Eq. (1.281) the numbers match for the breaking of $SU(N)$ to $SU(N-1)$. However, this attempt brings us closer to solving Higgs–Goldstone problem. We are stuck between either including only the $SU(2)_L$ covariant derivative and finding quadratic divergences or including the

Fig. 1.17 Feynman diagrams contributing to the Higgs mass in little Higgs models (Beautiful figure from Ref. [33])

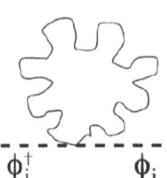
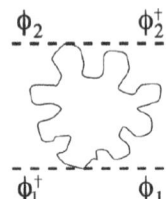

$SU(3)$ covariant derivative and turning the Higgs into a Goldstone mode which gives a mass of scale f to the heavy gauge bosons. What we need is a mix of an extended $SU(3)$ gauge sector and a global symmetry where the Goldstone modes are not eaten.

In our *fourth and correct attempt* we find a way our of this dilemma by using two independent sets of $SU(3)$ generators. We break them to our $SU(2)_L$ gauge group through a combination of spontaneous and explicit breaking. This way will get eaten Goldstones which make the $SU(3)$ gauge bosons heavy and at the same time uneaten Goldstones which can form our Higgs, provided we only gauge one $SU(3)$ gauge group. Naively, we will be able to distribute $8 + 8 - 3 = 13$ Goldstones this way. However, we have to be careful not to double count three of them in the case where we identify both $SU(2)$ subgroups of the two original $SU(3)$ groups; in this case we are down to ten Goldstone modes. The art will be to arrange the spontaneous and hard symmetry breakings into a workable model.

First, we write each of the set of $SU(3)$ generators the same way as shown in Eq. (1.284) and identify those degrees of freedom which we hope will include the Higgs field

$$\Sigma_j = \exp\left[-\frac{i}{f}\begin{pmatrix} 0_{2\times 2} & \phi_j \\ \phi_j^\dagger & 0 \end{pmatrix}\right]\begin{pmatrix} 0_2 \\ f \end{pmatrix} \qquad j = 1, 2. \qquad (1.292)$$

For simplicity we set $f_1 \equiv f_2 \equiv f$. Each of the two Σ fields couples to the one set of $SU(3)$ gauge bosons through the usual covariant derivative

$$\mathscr{L} \supset |D_\mu \Sigma_1|^2 + |D_\mu \Sigma_2|^2 \supset g_1^2 |W_\mu \Sigma_1|^2 + g_2^2 |W_\mu \Sigma_2|^2. \qquad (1.293)$$

The gauge boson fields we can linked to form propagators in loop diagrams of the kind shown in the left panel of Fig. 1.17. From our attempt number three we know that for a universal coupling $g_1 = g_2 = g$ these diagrams give us

$$\mathscr{L} \supset \frac{\Lambda^2}{(4\pi)^2}\left(g_1^2\, \Sigma_1^\dagger \Sigma_1 + g_2^2\, \Sigma_2^\dagger \Sigma_2\right) = \frac{2g^2\Lambda^2}{(4\pi)^2}\, f^2. \qquad (1.294)$$

However, these are not the only loop diagrams we can generate with two sets of Goldstones. For example, we can write diagrams like the one in the right panel

1.9 Alternatives and Extensions

of Fig. 1.17, where we couple Σ_1 to Σ_2 directly through a gauge–boson loop. Counting powers of momentum we can guess that it will only be logarithmically divergent, so its contribution to the Lagrangian should be of the kind

$$\mathscr{L} \supset \frac{g_1^2 g_2^2}{(4\pi)^2} \log \frac{\Lambda^2}{\mu^2} |\Sigma_1^\dagger \Sigma_2|^2 \,, \qquad (1.295)$$

including a free renormalization scale μ. The combination $\Sigma_1^\dagger \Sigma_2$ is a scalar. It is indeed gauge–invariant only under the diagonal subgroup of $SU(3)_1 \times SU(3)_2$, just like we discuss in case of the custodial $SU(2)_L \times SU(2)_R$ in Sect. 1.1.6. The combined gauge interactions g_1 and g_2 break the large symmetry group $SU(3) \times SU(3)$ of Eq. (1.294) to their diagonal subgroup $SU(3)_{\text{diag}}$. This happens as hard symmetry breaking via a loop–induced term in the Lagrangian.

Next, we translate Eq. (1.295) into the Higgs fields ϕ_j and see if it gives them a mass. This is easier if we re-organize the ϕ_j in a more symmetric manner; if we shift $\phi_j \to G \pm \phi$ the Goldstone modes are $SU(3)$ rotations common to Σ_1 and Σ_2 and lend longitudinal degrees of freedom to the massive gauge bosons of the gauged $SU(3)$ group. For the supposed Higgs mass term in the Lagrangian we find to leading order and neglecting commutators

$$\Sigma_1^\dagger \Sigma_2 = \begin{pmatrix} 0 & f \end{pmatrix} e^{\frac{i}{f}(\phi_1 \cdot \lambda)} e^{-\frac{i}{f}(\phi_2 \cdot \lambda)} \begin{pmatrix} 0 \\ f \end{pmatrix}$$

$$= \begin{pmatrix} 0 & f \end{pmatrix} e^{\frac{i}{f}(\phi \cdot \lambda)} e^{\frac{i}{f}(G \cdot \lambda)} e^{-\frac{i}{f}(G \cdot \lambda)} e^{\frac{i}{f}(\phi \cdot \lambda)} \begin{pmatrix} 0 \\ f \end{pmatrix} + \text{commutator terms}$$

$$\simeq \begin{pmatrix} 0 & f \end{pmatrix} e^{\frac{2i}{f}(\phi \cdot \lambda)} \begin{pmatrix} 0 \\ f \end{pmatrix}$$

$$= \begin{pmatrix} 0 & f \end{pmatrix} \left[\mathbb{1} + \frac{2i}{f} \begin{pmatrix} 0 & \phi \\ \phi^\dagger & 0 \end{pmatrix} + \frac{1}{2} \left(\frac{2i}{f}\right)^2 \begin{pmatrix} \phi\phi^\dagger & 0 \\ 0 & \phi^\dagger\phi \end{pmatrix} \right.$$

$$\left. + \frac{1}{6} \left(\frac{2i}{f}\right)^3 \begin{pmatrix} 0 & \phi\phi^\dagger\phi \\ \phi^\dagger\phi\phi^\dagger & 0 \end{pmatrix} + \frac{1}{24} \left(\frac{2i}{f}\right)^4 \begin{pmatrix} (\phi\phi^\dagger)^2 & 0 \\ 0 & (\phi^\dagger\phi)^2 \end{pmatrix} + \mathscr{O}\left(\frac{1}{f^5}\right) \right] \begin{pmatrix} 0 \\ f \end{pmatrix}$$

$$= f^2 - \frac{2}{f^2} \begin{pmatrix} 0 & f \end{pmatrix} \begin{pmatrix} 0 \\ \phi^\dagger\phi f \end{pmatrix} + \frac{2}{3f^4} \begin{pmatrix} 0 & f \end{pmatrix} \begin{pmatrix} 0 \\ (\phi^\dagger\phi)^2 f \end{pmatrix} + \mathscr{O}\left(\frac{1}{f^6}\right)$$

$$= f^2 - 2\phi^\dagger\phi + \frac{2}{3f^2}(\phi^\dagger\phi)^2 + \mathscr{O}\left(\frac{1}{f^4}\right) \,. \qquad (1.296)$$

After squaring this expression we find

$$|\Sigma_1^\dagger \Sigma_2|^2 \bigg|_{\text{gauge}} = f^4 - 4f^2 \phi^\dagger\phi + \frac{16}{3}(\phi^\dagger\phi)^2 + \mathscr{O}\left(\frac{1}{f^2}\right) \,. \qquad (1.297)$$

This combination of spontaneous symmetry breaking of the two $SU(3)$ symmetries at the scale f and explicit hard breaking to the diagonal $SU(3)$ the pseudo–Goldstone field ϕ develops a mass and a potential as powers of $|\Sigma_1^\dagger \Sigma_2|$. The mass scales for spontaneous symmetry breaking, f, and the hard breaking scale in Eq. (1.297) are linked by loop effects. For example, its mass term just combining the two above formulae reads

$$\boxed{\mathscr{L} \supset -\frac{g_1^2 g_2^2 f^2}{(2\pi)^2} \log \frac{\Lambda^2}{\mu^2} \, \phi^\dagger \phi} \quad \Leftrightarrow \quad \boxed{m_H \sim \frac{g^2 f}{2\pi} \gtrsim \frac{g^2 \Lambda}{8\pi^2} \sim \frac{\Lambda}{100}}.$$
(1.298)

This relation points to a *new physics energy scale* $f \sim 1\,\mathrm{TeV}$. Following the constraint given by Eq. (1.286) we do not expect $\log \Lambda/\mu$ to give a contribution to the Higgs mass beyond a factor of $\mathcal{O}(1)$. While this relation of scales indicates a suppression of g^2 instead of g, we do not collect additional factors $1/(4\pi)$, because we are still looking at one–loop diagrams.

The mechanism described above is called *collective symmetry breaking*. It is a convoluted way of spontaneously and explicitly breaking a global symmetry $SU(3)_1 \times SU(3)_2$ to our $SU(2)_L$, the latter by introducing gauge or Yukawa coupling terms in the Lagrangian. Of the two sets of Goldstones arising in the spontaneous breaking of each $SU(3)_{1,2} \to SU(2)_L$, denoted as ϕ_1 and ϕ_2, we use $(\phi_1 + \phi_2)/2$ to give the gauge bosons of one of the broken $SU(3)$ groups a mass around f. The remaining Goldstones $\phi = (\phi_1 - \phi_2)/2$ at this stage remain massless. They turn into pseudo–Goldstones and acquire a mass as well as a potential in the explicit breaking of the global $SU(3)_1 \times SU(3)_2$ symmetry into the gauged $SU(3)_{\mathrm{diag}}$.

The reason why this symmetry breaking is called 'collective' is that we need to break two symmetries explicitly to allow for mass and potential terms for the pseudo–Goldstone. Only breaking one of them leaves the respective other one as a global symmetry under which the Higgs fields transforms non–linearly. Because the original global symmetry group is explicitly broken the Higgs will develop mass and potential terms at the scale f, but doubly loop suppressed either via gauge boson or via fermion loops. This translates into a double Higgs mass suppression $g_1 g_2$ relative to f. Equation (1.298) tells us that we can write down a perturbative theory which is valid from $v \sim m_H$ to an ultraviolet cutoff around $100 \times m_H$.

1.9.3 Little Higgs Models

Collective symmetry breaking can be implemented in a wide variety of models. Our first example is based on the smallest useful extension of $SU(2)_L$, namely $SU(3)$. For decades people tried to implement a Goldstone Higgs in this symmetry structure and learned that to protect the Higgs mass a single broken $SU(3)$ symmetry is not sufficient. For the simplest little Higgs model or *Schmaltz model* we instead postulate a global $SU(3) \times SU(3)$ symmetry and break it down to $SU(2)_L$ the way

1.9 Alternatives and Extensions

we introduce it in Sect. 1.9.2. We can then express all mass scales in terms of the symmetry–breaking scale f. Starting from the ultraviolet the basic structure of our model in terms of its particle content in the gauge sector is

- For $E > 4\pi f$ our effective theory in E/f breaks down, so our theory is strongly interacting and/or needs a ultraviolet completion.
- Below that, the effective Lagrangian obeys a global and partly gauged $SU(3)_1 \times SU(3)_2$ symmetry with two gauge couplings $g_{1,2}$. Both couplings are attached to one set of $SU(3)$ gauge bosons, containing three $SU(2)$ gauge bosons plus complex fields $W'_\pm, W'_0$ with hypercharge $1/2$ and a singlet Z'.
- Through loop effects the combined gauge couplings explicitly break $SU(3)_1 \times SU(3)_2 \to SU(3)_{\text{diag}}$. The related Goldstone modes give masses of the order gf to the heavy $SU(3)$ gauge bosons.
- The other five broken generators of $SU(3)_1 \times SU(3)_2$ become Goldstone modes and the Standard Model Higgs doublet. Terms like $\Sigma_1^\dagger \Sigma_2$ give rise to a Higgs mass of the order $g^2 f/(2\pi)$.
- To introduce hypercharge $U(1)_Y$ we have to postulate another $U(1)_X$, which includes a heavy gauge boson mixing with the $SU(3)/SU(2)$ and the $SU(2)$ gauge bosons, to produce γ, Z, Z'. This will turn into a problem, because we lose custodial symmetry. For our discussion we ignore the $U(1)$ gauge bosons.

Until now we have not discussed any fermionic aspects of the little Higgs setup. However, to remove the leading quadratic divergence in Eq. (1.277) we obviously need to modify the *fermion sector* as well. For this purpose we enlarge the $SU(2)$ heavy–quark doublet Q to an $SU(3)$ triplet $\Psi = (t, b, T) \equiv (Q, T)$. The Yukawa couplings look like $\lambda \Sigma^\dagger \Psi t^c$, in analogy to the Standard Model, but with two right handed top singlets t_j^c which will combine to the Standard–Model and a heavy right handed top. We can compute this in terms of the physical fields,

$$\Sigma_j^\dagger \Psi = (0 \; f) \exp\left[\frac{i}{f}\begin{pmatrix} 0 & \phi \\ \phi^\dagger & 0 \end{pmatrix}\right]\begin{pmatrix} Q \\ T \end{pmatrix}$$

$$= (0 \; f)\left[\mathbb{1} + \frac{i}{f}\begin{pmatrix} 0 & \phi \\ \phi^\dagger & 0 \end{pmatrix} + \frac{1}{2}\left(\frac{i}{f}\right)^2 \begin{pmatrix} \phi\phi^\dagger & 0 \\ 0 & \phi^\dagger\phi \end{pmatrix} + \mathcal{O}\left(\frac{1}{f^3}\right)\right]\begin{pmatrix} Q \\ T \end{pmatrix}$$

$$= (0 \; f)\left[\begin{pmatrix} Q \\ T \end{pmatrix} + \frac{i}{f}\begin{pmatrix} \phi T \\ \phi^\dagger Q \end{pmatrix} - \frac{1}{2f^2}\begin{pmatrix} \phi\phi^\dagger Q \\ \phi^\dagger\phi T \end{pmatrix} + \mathcal{O}\left(\frac{1}{f^3}\right)\right]$$

$$= fT + i\phi^\dagger Q - \frac{1}{2f}\phi^\dagger\phi T + \mathcal{O}\left(\frac{1}{f^2}\right) . \tag{1.299}$$

Combining the two Yukawas with the simplification $\lambda_1 = \lambda_2 = \lambda$ gives us the leading terms

$$\mathcal{L} \supset \lambda f \left(1 - \frac{1}{2f^2}\phi^\dagger\phi\right) T T^c + \lambda \, \phi^\dagger Q t^c , \tag{1.300}$$

where we define the SM top quark as $t_2^c - t_1^2 = -i\sqrt{2}t^c$ and its orthogonal partner $t_1^c + t_2^c = \sqrt{2}T^c$.

According to Eq. (1.300) both top quarks contribute to the Higgs propagator and Higgs mass corrections, the Standard Model top through the usual *ttH* Yukawa coupling and the new heavy top particle through a four-point *TTHH* interaction. For the Feynman rule we will need to include an additional factor 2 in the *TTHH* coupling, stemming from the two permutations of the Higgs fields. The question becomes how these two diagrams cancel.

The scalar integrals involved we know; generally omitting a factor $1/(4\pi)^2$ the two-point function from the Standard Model top loop has a quadratic pole $B(0; m, m) \sim (\Lambda/m)^2$. Adding two fermion propagators with mass m_t and two Yukawa couplings λ gives a combined prefactor $i^4 \lambda^2 \Lambda^2 = \lambda^2 \Lambda^2$. The heavy top diagram gives a one-point function with the pole $A(m_T) \sim \Lambda^2$. Adding one fermion propagator with mass m_T and the coupling λ/f yields $i^2 \lambda/f \, m_T \Lambda^2 = -\lambda^2 \Lambda^2$. This hand–waving estimate illustrates how these two top quarks cancel each other's quadratic divergence for the Higgs mass. If we do this calculation more carefully, we find that for an $SU(3)$-invariant regulator the quadratic divergences cancel, and terms proportional to $\log m_t/m_T$ remain. Instead keeping the two λ_j separated we would find

$$m_T = \sqrt{\lambda_1^2 f_1^2 + \lambda_2^2 f_2^2} \sim \max_j(\lambda_j f_j) \qquad \lambda_t = \lambda_1 \lambda_2 \frac{\sqrt{f_1^2 + f_2^2}}{m_T}. \tag{1.301}$$

Following the same logic as for the gauge boson loop shown in Fig. 1.17 the combination of λ_1 and λ_2 breaks $SU(3)_1 \times SU(3)_2 \to SU(3)_{\text{diag}}$ explicitly. This turns the Higgs into a pseudo–Goldstone and allows contributions proportional to $\lambda_1 \lambda_2$ in the Higgs mass and potential.

To arrive at the Standard Model in the infrared limit we need to generate a *Higgs potential* $V = \mu^2 |\phi|^2 + \lambda |\phi|^2$. The two parameters are related via $\mu^2 = -\lambda v^2$. We already know that gauge boson loops generate such a potential, as shown in Eq. (1.297). Similarly, fermion loops in the Schmaltz model give

$$|\Sigma_1^\dagger \Sigma_2|^2 \bigg|_{\text{fermion}} = f^4 - 4f^2 \phi^\dagger \phi + \frac{14}{3}(\phi^\dagger \phi)^2 + \mathcal{O}\left(\frac{1}{f^2}\right)$$

$$\equiv f^4 + \mu^2 \phi^\dagger \phi + \lambda(\phi^\dagger \phi)^2 + \mathcal{O}\left(\frac{1}{f^2}\right)$$

$$\Rightarrow \qquad \left|\frac{\mu^2}{\lambda}\right|_{\text{fermion}} + \left|\frac{\mu^2}{\lambda}\right|_{\text{gauge}} \sim \frac{12 f^2}{14} + \frac{12 f^2}{16} \gg v^2. \tag{1.302}$$

1.9 Alternatives and Extensions

The self coupling λ is too small to give us anything like the Standard Model Higgs potential. There is no easy cure to this, but we can resort to ad–hoc introducing a tree level μ parameter with the proper size.

$$\mathscr{L} \supset \mu^2 \Sigma_1^\dagger \Sigma_2 = \mu^2 \left[f^2 - 2\,\phi^\dagger \phi + \mathscr{O}\left(\frac{1}{f^2}\right) \right] . \tag{1.303}$$

Roughly $\mu \sim v$ brings the Higgs potential terms to the correct value. Ironically, this ad-hoc terms gives the model its alternative name μ *model*. As a side remark, such a term also breaks the $U(1)$ symmetry linked to the 8th $SU(3)$ generators and gives the corresponding Goldstone a mass of the order v.

Now we can summarize the particle content of this first little Higgs model. Apart from the Standard Model particles and a protected light Higgs we find the new particle spectrum, still not including the $U(1)$ structure

$$\begin{aligned}
SU(3) \text{ gauge bosons } W'^{\pm}, W'^0 & \quad \text{with} \quad m_{W'} = \mathscr{O}(gf) \\
\text{singlet } Z' & \quad \text{with} \quad m_{Z'} = \mathscr{O}(gf) \\
\text{heavy top } T & \quad \text{with} \quad m_T = \mathscr{O}(\lambda_t f) .
\end{aligned} \tag{1.304}$$

The Schmaltz model discussed above has a few disadvantages. Among them is the missing $U(1)$ gauge group and the need for an ad-hoc μ term. From what we know about sigma models and collective symmetry breaking we can construct a second economic little–Higgs model, the *littlest Higgs* model. This time, we embed two gauge symmetries which overlap by the Standard Model Higgs doublet into one sigma field: it includes two copies of $SU(2)$ as part of the large global $SU(5)$ symmetry group. This will also have space to the $U(1)$ gauge group. Our enlarged symmetry group now has $(5^2 - 1) = 24$ generators T in the usual adjoint representation $SU(5)$. The Pauli matrices as one set of $SU(2)$ generators are arranged in the 5×5 matrix similarly to Eq. (1.283)

$$\begin{pmatrix} -SU(2)^* & 0_{2\times 3} \\ 0_{3\times 2} & 0_{3\times 3} \end{pmatrix} \qquad \begin{pmatrix} 0_{3\times 3} & 0_{2\times 3} \\ 0_{3\times 2} & SU(2) \end{pmatrix} . \tag{1.305}$$

This $SU(2)$ symmetry will need to stay unbroken when we break $SU(5)$. This double appearance looks like a double counting, so we will eventually get rid of half of these generators. Moreover, these unbroken $SU(2)$ generators only use half of the number of degrees of freedom which $SU(5)$ offers in each of these 2×2 sub-matrices.

Next, we need to identify those broken Goldstone modes $\hat{T}$ which form the Higgs field. In the $SU(5)$ generator matrix the Higgs doublet has to be arranged such that neither set of $SU(2)$ generators includes ϕ, so when we break $SU(5)$ into the $SU(2)$ subgroups the Higgs always stays a (pseudo–) Goldstone. We construct a pattern similar to Eq. (1.292) in the Schmaltz model when we only consider the upper left or lower right 3×3 matrices of the full $SU(5)$ group,

$$\Sigma = \exp\left[-\frac{i}{f}(w \cdot \hat{T})\right] \langle \Sigma \rangle \quad \text{with} \quad (w \cdot \hat{T}) = \begin{pmatrix} 0 & \phi^* & 0 \\ \phi^T & 0 & \phi^\dagger \\ 0 & \phi & 0 \end{pmatrix}. \quad (1.306)$$

Again, the four-fold appearance looks like we introduced too many Higgs fields, but the symmetry structure of the broken $SU(5)$ will ensure that they really all are the same field.

Spontaneously breaking the global $SU(5)$ symmetry by an appropriate vacuum expectation value $\langle \Sigma \rangle$ will eventually allow the ϕ doublet to develop a potential, including a mass and a self coupling. The Standard–Model $SU(2)_L$ generators should not be affected. We try

$$\langle \Sigma \rangle = \begin{pmatrix} 0 & 0 & \mathbb{1}_{2\times 2} \\ 0 & 1 & 0 \\ \mathbb{1}_{2\times 2} & 0 & 0 \end{pmatrix}. \quad (1.307)$$

This vacuum expectation value obviously breaks our global $SU(5)$ symmetry. What remains in the $\langle \Sigma \rangle$ background is an $SO(5)$ symmetry, generated by the antisymmetric tensor with $(4+3+2+1) = 10$ entries. This way 14 of the original 24 generators are broken and the multiple appearance of some of the Goldstone fields in Eqs. (1.305) and (1.306) is explained. Using commutation relations we can show that the Standard–Model $SU(2)_L$ generators in Eq. (1.305) are indeed unbroken. The corresponding unbroken $U(1)$ generators are the equally symmetric diagonals $\text{diag}(-3,-3,2,2,2)/10$ and $\text{diag}(-2,-2,-2,3,3)/10$.

To compute the spectrum of the littlest Higgs model based on breaking $SU(5) \to SO(5)$ through the vacuum expectation value shown in Eq. (1.307) we generalize Eq. (1.306) to include the complete set of Goldstones associated with the broken generators,

$$(w \cdot \hat{T}) = \begin{pmatrix} \chi_{2\times 2} & \phi^* & \kappa^\dagger_{2\times 2} \\ \phi^T & 0 & \phi^\dagger \\ \kappa_{2\times 2} & \phi & \chi^T_{2\times 2} \end{pmatrix} + \frac{\eta}{2\sqrt{5}} \begin{pmatrix} \mathbb{1} & 0 & 0 \\ 0 & -4 & 0 \\ 0 & 0 & \mathbb{1} \end{pmatrix}. \quad (1.308)$$

The form reflects the commutation property $\langle \Sigma \rangle \hat{T}^T = \hat{T} \langle \Sigma \rangle$, which links opposite corners of $w \cdot \hat{T}$. The χ field differs from the unbroken $SU(2)$ generators in Eq. (1.305) in the relative sign between the two appearances. They can be shown to also form hermitian traceless 2×2 matrices, which means that they can be written as a second triplet of $SU(2)$ fields. The combination of broken generators χ and χ^T and unbroken $SU(2)_L$ generators in Eq. (1.305) account for all degrees of freedom in those sub-matrices. The 2×2 matrix κ is not traceless, but complex symmetric, so instead of another set of $SU(2)$ gauge bosons they form a complex scalar triplet. The complex doublet ϕ will become the Standard Model Higgs doublet, and η is the usual real singlet. Together, these field indeed correspond to $3_\chi + 6_\kappa + 4_\phi + 1_\eta = 14$ broken Goldstones.

1.9 Alternatives and Extensions

Unless something happens the fields linked to the broken generators $\hat{T}$ can either turn into gauge boson mass terms of the order f or stay massless. In particular, χ will make one of the two sets of $SU(2)$ gauge bosons $W'^\pm, W'^0$ heavy, while η gets eaten by the B' field.

The trick in this littlest Higgs model is to mix the two $SU(2)$ groups in two opposite corners of the sigma field in Eqs. (1.305) and (1.308). We can for example introduce two gauge couplings g_1 and g_2, one for each corner. By construction, the combination with a relative minus sign between the upper–left and lower–right fields stays unbroken after spontaneously breaking $SU(5) \to SO(5)$, so this linear combination will give the Standard Model gauge bosons. Its orthogonal combination χ will become a set of heavy W' states. Introducing an $SU(2)$ mixing angle $\tan\theta = g_2/g_1$ we can define a rotation from the $SU(2)_1 \times SU(2)_2$ interaction basis to the mass basis after spontaneous symmetry breaking. Exactly the same works for the B fields corresponding to $U(1)_Y$,

$$\begin{pmatrix} W'^a \\ W^a \end{pmatrix} = \begin{pmatrix} -\cos\theta & \sin\theta \\ \sin\theta & \cos\theta \end{pmatrix} \begin{pmatrix} W_1^a \\ W_2^a \end{pmatrix} \qquad \begin{pmatrix} B' \\ B \end{pmatrix} = \begin{pmatrix} -\cos\theta' & \sin\theta' \\ \sin\theta' & \cos\theta' \end{pmatrix} \begin{pmatrix} B_1^a \\ B_2^a \end{pmatrix}. \tag{1.309}$$

Including all factors the heavy gauge bosons acquire the masses

$$m_{W'} = \frac{gf}{\sin(2\theta)} \qquad m_{B'} = \frac{g'f}{\sqrt{5}\sin(2\theta')}. \tag{1.310}$$

In our discussion of little Higgs models factors of two never really matter. However, in the case of $m_{B'}$ we see that for f in the TeV range the heavy $U(1)$ gauge boson is predicted to be very light, making the model experimentally very vulnerable.

Protecting the Higgs mass from quadratic divergences in the gauge sector of the littlest–Higgs model works similar to the Schmaltz model. Each of the two sets of $SU(2)_L$ generators in Eq. (1.305) corresponds to a 2×2 sub-matrix in one of the corners of the $SU(5)$ sigma field. If we break the global $SU(5)$ down to one of the two $SU(2)$ groups the Higgs doublet will be a broken generator of the global $SU(5)$ and therefore be massless. Unlike in the $SU(3) \times SU(3)$ setup the finite Higgs mass in the littlest Higgs model is not induced by gauge boson or fermion loops. It appears once we integrate out the heavy field κ in the presence of g_1 and g_2, communicated to the Higgs field ϕ via the *Coleman–Weinberg* mechanism introduced in Sect. 1.2.7. The resulting quartic Higgs term, only taking into account the $SU(2)$ couplings becomes

$$\mathscr{L} \supset -c \frac{g_1^2 g_2^2}{g_1^2 + g_2^2} |\phi^\dagger \phi|^2, \tag{1.311}$$

with an order-one constant c. Unlike in the Schmaltz model this value does not have to be too large; in the Coleman–Weinberg model we typically find

$$\frac{m_H^2}{\lambda} \sim \left(\frac{m_\kappa}{g}\right)^2 , \tag{1.312}$$

which we need to adjust to stay below the order f^2 which Eq. (1.302) gives for the Schmaltz model.

To protect the Higgs mass against the top loop we again extend the $SU(2)_L$ quark doublet to the triplet $\Psi = (t, b, T)$ and add a right handed singlet t'^c. Because we expect mixing between the two top singlets which will give us the Standard–Model and a heavy top quark we write two general Yukawa couplings for the Standard Model doublet and the additional heavy states. The first is mediated by the Σ field as

$$\mathscr{L} \supset \lambda_1 \, f \, \epsilon_{ijk} \Psi_i \, \Sigma_{j4} \Sigma_{k5} \, t_1^c + \lambda_2 \, f \, T t_2^c . \tag{1.313}$$

This form uses the 2×3 triplets from the upper–right corner of the Goldstone matrix in Eq. (1.308)

$$\Sigma_{jm} = \begin{pmatrix} \kappa^\dagger \\ \phi^\dagger \end{pmatrix} \qquad j = 1, 2, 3 \qquad m = 4, 5 . \tag{1.314}$$

These triplets represent the $SU(3)$ sub-matrix of the $SU(5)$ generators which requires the Higgs mass to be zero. This means that if we set $\lambda_2 = 0$ this Yukawa coupling as an anti–symmetric combination of three triplets is $SU(3)$ symmetric. The top–induced contributions to the Higgs mass will be proportional to $\lambda_1^2 \lambda_2^2$ and quadratic divergences are forbidden at one loop.

The two heavy quarks mix to the SM top quark and an additional heavy top with mass $m_T = \sqrt{\lambda_1^2 + \lambda_2^2}\, f$ where as before we assume $f = f_1 = f_2$. The actual top–Higgs coupling are of the order min λ_j for the three-point Higgs coupling to the Standard Model top and λ/f for the four-point Higgs coupling to a pair of heavy tops.

Looking at the complete set of $SU(5)$ generators in Eq. (1.308) we can collect the heavy spectrum of the littlest Higgs model,

$SU(2)_L \times U(1)_Y$ gauge bosons $B', W'^\pm, Z'$ with $m_{B',W',Z'} = \mathcal{O}(gf)$

heavy 'Higgs' triplet $\kappa = \begin{pmatrix} \kappa^{++} & \kappa^+ \\ \kappa^+ & \kappa^0 \end{pmatrix}$ with $m_\kappa = \mathcal{O}(gf)$

heavy top T with $m_T = \mathcal{O}(\lambda f) .$ (1.315)

From the B' and κ fields we expect a serious violation of the custodial $SU(2)$ symmetry. This means that electroweak precision data forces us to choose f unusually large, in conflict with the requirement in Eq. (1.312). Moreover, the Higgs triplet should not become too heavy to maintain the correct relative size of the Higgs mass and the Higgs self coupling. Such a Higgs triplet with a doubly charged

1.9 Alternatives and Extensions

Higgs boson has a smoking–gun signature at the LHC, namely its production in weak boson fusion $uu \to ddW^+W^+ \to ddH^{++}$.

While the littlest Higgs setup solves some of the issues of the Schmaltz model, it definitely has its problems linked to the Higgs triplet and the heavy $U(1)$ gauge boson. To not violate the custodial symmetry too badly it would be great to introduce some kind of Z_2 symmetry which allows only two heavy new particles in any vertex. The same symmetry would give us a stable lightest new particle as a weakly interacting dark matter candidate. All we need to do is define a quantum number with one value for all weak–scale Standard Model particles and another value for all particles with masses around f. Such a parity will be called T *parity*.

For the littlest Higgs, we would like to separate the additional heavy states, including the $SU(2)$ doublet, from our Standard Model Higgs and gauge bosons. The symmetry we want to introduce should multiply all heavy entries in the sigma field of Eq. (1.308) by a factor (-1). By multiplying out the matrices we show that there exists a matrix Ω such that

$$(w \cdot \hat{T}) \to \Omega^{-1} \left(w \cdot \hat{T}\right) \Omega \,. \tag{1.316}$$

One matrix Ω for which this works is

$$\begin{pmatrix} \chi & \phi^* & \kappa^\dagger \\ \phi^T & 0 & \phi^\dagger \\ \kappa & \phi & \chi^T \end{pmatrix} + \frac{\eta}{2\sqrt{5}} \begin{pmatrix} \mathbb{1} & 0 & 0 \\ 0 & -4 & 0 \\ 0 & 0 & \mathbb{1} \end{pmatrix}$$

$$\to i^2 \begin{pmatrix} \mathbb{1} & 0 & 0 \\ 0 & -1 & 0 \\ 0 & 0 & \mathbb{1} \end{pmatrix} \left[\begin{pmatrix} \chi & \phi^* & \kappa^\dagger \\ \phi^T & 0 & \phi^\dagger \\ \kappa & \phi & \chi^T \end{pmatrix} + \frac{\eta}{2\sqrt{5}} \begin{pmatrix} \mathbb{1} & 0 & 0 \\ 0 & -4 & 0 \\ 0 & 0 & \mathbb{1} \end{pmatrix} \right] \begin{pmatrix} \mathbb{1} & 0 & 0 \\ 0 & -1 & 0 \\ 0 & 0 & \mathbb{1} \end{pmatrix}$$

$$= -\begin{pmatrix} \mathbb{1} & 0 & 0 \\ 0 & -1 & 0 \\ 0 & 0 & \mathbb{1} \end{pmatrix} \left[\begin{pmatrix} \chi & -\phi^* & \kappa^\dagger \\ \phi^T & 0 & \phi^\dagger \\ \kappa & -\phi & \chi^T \end{pmatrix} + \frac{\eta}{2\sqrt{5}} \begin{pmatrix} \mathbb{1} & 0 & 0 \\ 0 & 4 & 0 \\ 0 & 0 & \mathbb{1} \end{pmatrix} \right]$$

$$= -\begin{pmatrix} \chi & -\phi^* & \kappa^\dagger \\ -\phi^T & 0 & -\phi^\dagger \\ \kappa & -\phi & \chi^T \end{pmatrix} - \frac{\eta}{2\sqrt{5}} \begin{pmatrix} \mathbb{1} & 0 & 0 \\ 0 & -4 & 0 \\ 0 & 0 & \mathbb{1} \end{pmatrix}$$

$$= \begin{pmatrix} -\chi & \phi^* & -\kappa^\dagger \\ \phi^T & 0 & \phi^\dagger \\ -\kappa & \phi & -\chi^T \end{pmatrix} - \frac{\eta}{2\sqrt{5}} \begin{pmatrix} \mathbb{1} & 0 & 0 \\ 0 & -4 & 0 \\ 0 & 0 & \mathbb{1} \end{pmatrix} \,. \tag{1.317}$$

This symmetry works perfectly for the additional gauge bosons, including the heavy scalars κ. For the massive twins of the $SU(2)_L$ gauge bosons we rely on the fact that in the special case $g_1 = g_2$ the Lagrangian involving $D_\mu \Sigma$ is symmetric under the exchange of the two $SU(2) \times U(1)$ groups. The eigenstates we can choose as $W_\pm = (W_1 \pm W_2)/\sqrt{2}$ and the same for the B fields. Of these two W_+, B_+ are Standard

Model gauge bosons, while W_-, B_- are heavy. Exchanging the indices $(1 \leftrightarrow 2)$ is an even transformation for W_+, while it is odd for W_-, again just as we want.

A problem arises when we assign such a quantum number to the heavy tops which form part of a triplet extending the usual Standard Model quark doublets. Getting worse, we have to be very careful to then implement T parity specifically taking care that it is not broken by anomalies. At this point, it turns our that we have to introduce additional fermions and the model rapidly loses its concise structure as the price for a better agreement with electroweak precision constraints.

In summary, it is fair to say that collective symmetry breaking is an attractive idea, based on a fundamental property like Goldstone's theorem. Already at the very beginning we notice that its ultraviolet completion will be strongly interacting, which some theorists would consider not attractive. Certainly, it is not clear how the measurement of an approximate gauge coupling unification would fit into such a picture. The same holds true for the fixed point arguments which we present in Sect. 1.2.5. What is more worrisome is that it appears to be hard to implement collective symmetry breaking in a compact model which is not obviously ruled out or inconsistent. This might well be a sign that protecting the Higgs mass through a pseudo–Goldstone property is not what is realized in Nature.

1.10 Higgs Inflation

Going beyond the weak scale and any energy scale we will probe with the LHC we can ask another question after discovering the first fundamental scalar particle in the Standard Model: can a Standard-Model-like Higgs boson be the scalar particle responsible for inflation?

One of the most pressing problems in cosmology is the question, why the cosmic microwave background radiation is so homogeneous while based on the usual evolution of the Universe different regions cannot be causally connected. A solution to this problem is to postulate an era of exponentially accelerated expansion of the Universe which would allow all these regions to actually know about each other.

We can trigger inflation through a scalar field χ in a potential $U(\chi)$. In the beginning, this field is located far away from its stable vacuum state. While moving towards the minimum of its potential it releases energy. Slow roll inflation can be linked to two physical conditions: on the one hand we require that in the equation of motion for the inflaton field χ we can neglect the kinetic term, which means

$$0 = \frac{\partial^2 \chi}{\partial t^2} + 3\hat{H}\frac{\partial \chi}{\partial t} + \frac{dU}{d\chi} \simeq 3\hat{H}\frac{\partial \chi}{\partial t} + \frac{dU}{d\chi} \; . \quad (1.318)$$

The second term is proportional to $\partial \chi/\partial t$ and therefore a friction term, called Hubble friction. To not confuse it with the Higgs field we denote the Hubble constant as $\hat{H}$. In the absence of the kinetic term the equation of state for the inflaton field χ can behave like $w \equiv p/\rho < -1/3$. Given the pressure p and the energy density ρ

1.10 Higgs Inflation

this is the condition for inflation. Negative values for w arise when the potential U dominates the energy of the inflaton and the change of the field value χ with time it small. Equivalently, we can require two parameters which describe the variation of the potential $U(\chi)$ to be small,

$$\epsilon = \frac{M_{\text{Planck}}^2}{2}\left(\frac{1}{U}\frac{dU}{d\chi}\right)^2 \ll 1 \qquad |\eta| = M_{\text{Planck}}^2 \left|\frac{1}{U}\frac{d^2U}{d\chi^2}\right| \ll 1. \tag{1.319}$$

The powers of M_{Planck} give the correct units. The two slow roll conditions in Eqs. (1.318) and (1.319) are equivalent, which means that for Higgs inflation we need to compute $U(\chi)$ with the appropriate inflaton field and test the conditions given in Eq. (1.319).

The starting point of any field theoretical explanation of inflation is the Einstein–Hilbert action for gravity,

$$\boxed{S = -\int d^4x \sqrt{-g}\, \frac{M_{\text{Planck}}^2}{2} R}, \tag{1.320}$$

with the Planck mass M_{Planck} as the only free parameter, the Ricci scalar with the graviton field, and no interactions between the gravitational sector and matter. In addition, we neglect the cosmological constant. If we combine the Higgs and gravitational sectors the minimal coupling of the two sectors is generated through the gravitational coupling to the energy–momentum tensor including all Standard Model particles. It turns out that for Higgs inflation we need an additional *non–minimal coupling* between the two sectors, so we start with the ansatz

$$\begin{aligned}S_J(H) &= \int d^4x \sqrt{-g}\left[-\frac{M^2}{2}R - \xi\frac{(v+H)^2}{2}R + \frac{1}{2}\partial_\mu H \partial^\mu H\right.\\&\quad \left.-\frac{\lambda}{4}((v+H)^2 - v^2)^2\right]\\&= \int d^4x \sqrt{-g}\left[-\frac{M^2}{2}R - \xi\frac{(v+H)^2}{2}R + \frac{1}{2}\partial_\mu H \partial^\mu H - \frac{\lambda}{4}(v+H)^4\right.\\&\quad \left. + \frac{\lambda v^2}{2}(v+H)^2 + \text{const}\right].\end{aligned} \tag{1.321}$$

This form of the Higgs potential in the second line corresponds Eq. (1.77) after inserting $\mu^2 = -\lambda v^2$ according to Eq. (1.86). The form of the *Einstein–Hilbert action* suggests that first of all the fundamental Planck mass M is replaced by an effective, observed Planck mass M_{Planck} in a scalar field background,

$$M_{\text{Planck}}^2 = M^2 + \xi(v+H)^2. \tag{1.322}$$

First, we assume $\xi v^2 \ll M^2$. For the specific case of Higgs inflation we in addition postulate $\xi \gg 1$, but with the original hierarchy still intact. The hierarchy between ξH and M will be discussed later.

The action in the Jordan frame given by Eq. (1.321) with the identification Eq. (1.322) is a little cumbersome to treat gravity problems. We can decouple the gravitational and Higgs sectors via a field re-definition into the Einstein frame and quote the result as

$$S_E(\chi) = \int d^4x \sqrt{-\hat{g}} \left[-\frac{M_{\text{Planck}}^2}{2} R + \frac{1}{2} \partial_\mu \chi \partial^\mu \chi \right.$$

$$\left. - \underbrace{\frac{\lambda}{4} \frac{M_{\text{Planck}}^4}{(M^2 + \xi(v+H(\chi))^2)^2} ((v+H(\chi))^2 - v^2)^2}_{\equiv U(\chi)} \right]$$

$$\hat{g}_{\mu\nu} = \frac{M^2 + \xi(v+H)^2}{M_{\text{Planck}}^2} g_{\mu\nu}$$

$$\frac{d\chi}{dH} = \left[\frac{1 + (\xi + 6\xi^2) \frac{(v+H)^2}{M_{\text{Planck}}^2}}{\left(1 + \xi \frac{(v+H)^2}{M_{\text{Planck}}^2}\right)^2} \right]^{1/2}. \quad (1.323)$$

The original Higgs potential in terms of H is replaced by the inflaton–Higgs scalar χ and its combined potential $U(\chi)$. The question is if this scalar field χ can explain inflation. After studying some basic features of this theory our main task will be to determine the value of ξ which would make such a model theoretically and experimentally feasible.

If we are interested in the evolution of the early universe the condition $\xi v^2 \ll M^2$ simplifies the above equations, but it does not imply anything for the hierarchy between the Higgs field values H and the fundamental Planck scale M. First, in the limit $\xi H^2, \xi^2 H^2 \ll M_{\text{Planck}}^2 \sim M^2$ we can solve the relation between the Higgs field H and its re-scaled counter part χ,

$$\frac{d\chi}{dH} = \left[1 + \mathcal{O}\left(\frac{\xi H^2}{M_{\text{Planck}}^2}\right) + \mathcal{O}\left(\frac{\xi^2 H^2}{M_{\text{Planck}}^2}\right) \right]^{1/2} \simeq 1 \quad \Leftrightarrow \quad \chi \simeq H. \quad (1.324)$$

In this derivation we already see that we have to deal with two additional energy scales, $M_{\text{Planck}}/\sqrt{\xi}$ and M_{Planck}/ξ. In this limit the potential for the re-scaled Higgs field χ in the Einstein frame becomes

1.10 Higgs Inflation

$$U(\chi) = \frac{\lambda}{4} \frac{M_{\text{Planck}}^4}{(M^2 + \xi(v+H)^2)^2} \, ((v+H)^2 - v^2)^2 \simeq \frac{\lambda}{4} \, ((v+\chi)^2 - v^2)^2 \,. \tag{1.325}$$

This is exactly the usual Higgs potential at low energies.

The opposite limit is $\xi H^2 \gg M_{\text{Planck}}^2 \sim M^2 \gg \xi v^2$. If we avoid Higgs field strength exceeding the Planck scale this condition implicitly assume $\xi \gg 1$ and therefore $\xi^2 H \gg \xi H^2$. We find

$$\frac{d\chi}{dH} = \left[\frac{6\xi^2 H^2/M_{\text{Planck}}^2}{\xi^2 H^4/M_{\text{Planck}}^4} + \mathcal{O}\left(\frac{M_{\text{Planck}}^2}{\xi H^2}\right) \right]^{1/2}$$

$$= \left[\frac{6 M_{\text{Planck}}^2}{H^2} + \mathcal{O}\left(\frac{M_{\text{Planck}}^2}{\xi H^2}\right) \right]^{1/2} \simeq \frac{\sqrt{6} M_{\text{Planck}}}{H}$$

$$\Leftrightarrow \quad \chi \simeq \sqrt{\frac{3}{2}} M_{\text{Planck}} \log \frac{\xi H^2}{M_{\text{Planck}}^2} \quad \Leftrightarrow \quad H^2 \simeq \frac{M_{\text{Planck}}^2}{\xi} e^{\sqrt{2}\chi/(\sqrt{3} M_{\text{Planck}})}. \tag{1.326}$$

The integration constants in this result are chosen appropriately. We can use this relation to compute the leading terms in the scalar potential for χ in the Einstein frame and for large Higgs field values,

$$U(\chi) \simeq \frac{\lambda}{4} \frac{M_{\text{Planck}}^4}{(M^2 + \xi H^2)^2} H^4 \simeq \frac{\lambda}{4} \frac{M_{\text{Planck}}^4 H^4}{\xi^2 H^4 \left(1 + \frac{M^2}{\xi H^2}\right)^2}$$

$$\simeq \frac{\lambda}{4} \frac{M_{\text{Planck}}^4}{\xi^2} \left(1 - 2 \frac{M_{\text{Planck}}^2}{\xi H^2}\right)$$

$$\boxed{U(\chi) \simeq \frac{\lambda}{4} \frac{M_{\text{Planck}}^4}{\xi^2} \left(1 - 2 e^{-\frac{\sqrt{2}\chi}{\sqrt{3} M_{\text{Planck}}}}\right)}. \tag{1.327}$$

We show $U(\chi)$ in Fig. 1.18. Following Eqs. (1.325) and (1.327) it resembles the usual Higgs potential at small values of H and χ and becomes flat at large field values. This means that we can indeed use the Higgs scalar as the inflaton, but we need to see what the slow roll conditions from Eq. (1.319) tell us about the model parameter ξ in the action introduced in Eq. (1.321). Following the discussion in Sect. 1.2.7 we can compute additional contributions to U from all Standard Model fields, but for our purpose the leading behavior is fine.

For this test we need to compute the first derivative of the potential U in the limit of large field values, because this is where we expect the field χ to act as the inflaton. For the first slow roll parameter ϵ we find

Fig. 1.18 Potential $U(\chi)$ as a function of the χ (Figure from Ref. [7])

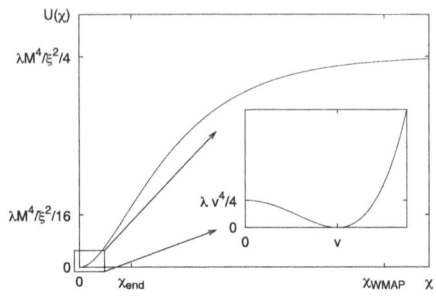

$$\frac{dU}{d\chi} = -\frac{\lambda}{2} \frac{M_{\text{Planck}}^4}{\xi^2} \frac{d}{d\chi} e^{-\frac{\sqrt{2}\chi}{\sqrt{3}M_{\text{Planck}}}} = \frac{\lambda}{\sqrt{6}} \frac{M_{\text{Planck}}^3}{\xi^2} e^{-\frac{\sqrt{2}\chi}{\sqrt{3}M_{\text{Planck}}}} = \frac{\lambda}{\sqrt{6}} \frac{M_{\text{Planck}}^5}{\xi^3 H^2}$$

$$\epsilon = \frac{M_{\text{Planck}}^2}{2} \left(\frac{\lambda^2}{6} \frac{M_{\text{Planck}}^{10}}{\xi^6 H^4} \right) \left(\frac{16}{\lambda^2} \frac{\xi^4}{M_{\text{Planck}}^8} \right)$$

$$= \frac{4}{3} \frac{M_{\text{Planck}}^4}{\xi^2 H^4} \ll 1 \quad \Leftrightarrow \quad H \gg \frac{M_{\text{Planck}}}{\sqrt{\xi}}. \qquad (1.328)$$

Similarly, the second parameter η comes out as

$$\frac{d^2U}{d\chi^2} = \frac{\lambda}{\sqrt{6}} \frac{M_{\text{Planck}}^3}{\xi^2} \frac{d}{d\chi} e^{-\frac{\sqrt{2}\chi}{\sqrt{3}M_{\text{Planck}}}} = -\frac{\lambda}{3} \frac{M_{\text{Planck}}^2}{\xi^2} e^{-\frac{\sqrt{2}\chi}{\sqrt{3}M_{\text{Planck}}}} = -\frac{\lambda}{3} \frac{M_{\text{Planck}}^4}{\xi^3 H^2}$$

$$|\eta| = M_{\text{Planck}}^2 \left(\frac{\lambda}{3} \frac{M_{\text{Planck}}^4}{\xi^3 H^2} \right) \left(\frac{4}{\lambda} \frac{\xi^2}{M_{\text{Planck}}^4} \right)$$

$$= \frac{4}{3} \frac{M_{\text{Planck}}^2}{\xi H^2} \ll 1 \quad \Leftrightarrow \quad H \gg \frac{M_{\text{Planck}}}{\sqrt{\xi}}. \qquad (1.329)$$

Both slow roll conditions are identical and correspond to the original condition we used to compute the potential $U(\chi)$ in the limit of large field values. The potential $U(\chi)$ leads to slow roll inflation, and since with time the Higgs field value becomes smaller inflation ends for $H \sim M_{\text{Planck}}/\sqrt{\xi}$ and correspondingly $\chi \sim M_{\text{Planck}}$.

After confirming that Higgs inflation has all the correct theoretical features we can confront it with data, most specifically with the different measurements of the cosmic microwave background. Without derivation we quote that during inflation the visible CMB modes crossed the horizon at field values $H \simeq 9 M_{\text{Planck}}/\sqrt{\xi}$, shortly before the condition $H \simeq M_{\text{Planck}}/\sqrt{\xi}$ for the end of inflation is finally reached. Experimentally, the normalization of these observed CMB modes requires

1.10 Higgs Inflation

$$\left(\frac{M_{\text{Planck}}}{36}\right)^4 \simeq \frac{U}{\epsilon} = \frac{\lambda}{4}\frac{M_{\text{Planck}}^4}{\xi^2}\frac{4}{3}\frac{\xi^2 H^4}{M_{\text{Planck}}^4} = \frac{\lambda}{3}H^4 \quad \text{using Eqs. (1.327) and (1.328)}$$

$$\simeq \frac{\lambda}{3}\left(\frac{9M_{\text{Planck}}}{\sqrt{\xi}}\right)^4 \quad \text{from CMB measurement}$$

$$\Leftrightarrow \quad \lambda = \frac{3}{324^4}\xi^2 \quad \Leftrightarrow \quad \boxed{\xi \simeq 60{,}000\sqrt{\lambda}}. \tag{1.330}$$

While at the electroweak scale we know that $\lambda \sim 1/8$ it becomes much smaller at high energies, so the actual value of ξ is tricky to extract with the Higgs data available right now. However, this measurement of ξ is in agreement with our original assumption. Additional measurements of the cosmic microwave background are in agreement with this setup. In particular, the Hubble scale $\sqrt{\lambda}M_{\text{Planck}}/\xi$ is small enough to avoid non–Gaussianities in the cosmic microwave background. This short introduction leaves us with a few open questions:

1. Higgs inflation is theoretically and experimentally feasible
2. The coupling strength $\xi \gg 1$ needs a good explanation
3. Adding the $\xi H^2 R$ coupling leads to unitarity violation in $HH \to HH$ scattering, so it requires an unknown ultraviolet completion
4. The cosmic microwave background decouples around $H \simeq M_{\text{Planck}}/10 \gg v$, so slow roll inflation and electroweak symmetry breaking are really described by separate potentials written as one function $U(\chi)$
5. Should the Higgs potential become unstable at high energies, the finite life time of the Universe can be used to save the Standard Model, but Higgs inflation will break down in the presence of an alternative global minimum.
6. In the standard setup of Higgs inflation the tensor-to-scalar ratio in the cosmis microwave background is predicted to be $r \sim 10\epsilon \sim 0.003$, which would be in disagreement with the BICEP2 measurement of $r \sim 0.2$.

Further Reading

At this point we are done with our brief review of the Higgs sector in the Standard Model and of contemporary Higgs phenomenology. From the discussions it should be clear that we have to move on to QCD, to understand the strengths and weaknesses of these searches and what distinguishes a good from a bad search channel.

Before moving on we mention a few papers where you can learn more about Higgs phenomenology at the LHC. Luckily, for Higgs searches there are several very complete and useful reviews available:

- You can look at the original articles by Higgs [23, 24] or Brout and Englert [17], but they are rather short and not really on top of the phenomenological aspects of the topic. Other papers for example by Guralnik, Hagen, Kibble [20] tend to be harder to read for phenomenologically interested students.

- Wolfgang Kilian's book [26] on the effective field theory approach to the Higgs mechanism is what the corresponding sections in these lecture notes are largely based on it. The underlying symmetry structure, including the custodial symmetry is nicely described in Scott Willenbrock's TASI lectures [38].
- If you are interested in a comprehensive overview of Higgs physics as an effective field theory with a special focus on higher-dimensional operators in linear and non-linear sigma models there is a really useful paper from Spain [8].
- Abdelhak Djouadi's compilation on 'absolutely everything we know about Higgs phenomenology' is indeed exhaustive. It has two volumes, one on the Standard Model [15] and another one on its minimal supersymmetric extension [16]
- For more experimental details you might want to have a look at Karl Jakobs' and Volker Büscher's review of LHC Higgs physics [9].
- As a theory view on LHC Higgs physics, mostly focused on gluon fusion production and its QCD aspects, there is Michael Spira's classic [35]. This is where you can find more information on the low energy theorem. Michael and his collaborators also published a set of lecture notes on Higgs physics [19].
- As always, there is a TASI lecture on the topic. TASI lecture notes are generally what you should look for when you are interested in an area of high energy physics. Dave Rainwater did not only write his thesis on Higgs searches in weak boson fusion [31], he also left us all he knows about Higgs phenomenology at the LHC in his TASI notes [32].
- Tao Han wrote a very comprehensible set of TASI lecture notes on basic LHC phenomenology, in case you need to catch up on this [22].
- For some information on electroweak precision data and the ρ parameter, there are James Wells' TASI lectures [36].
- If you are interested in Higgs production in association with a W or Z boson and the way to observe boosted $H \to b\bar{b}$ decays you need to read the original paper [11]. The same is true for the $t\bar{t}H$ analysis.
- For cut rules and scalar integrals the best writeup I know is Wim Beenakker's PhD thesis. Unfortunately, I am not sure where to get it from except from the author by request.
- If you are getting interested in fixed points in RGE analyses you can look at Christoph Wetterich's original paper [37] or a nice introductory review by Barbara Schrempp and Michael Wimmer [34].
- My discussion on technicolor largely follows the extensive review by Chris Hill and Elisabeth Simmons [25].
- A really nice writeup which my little Higgs discussion is based on is Martin Schmaltz' and David Tucker-Smith's review article [33].
- For more information on Higgs inflation you can start with a nice set of TASI lectures on inflation [4] and then dive into a specific review of Higgs inflation by one of its inventors [7].

References

1. G. Aad et al., [ATLAS Collaboration], Observation of a new particle in the search for the Standard Model Higgs boson with the ATLAS detector at the LHC. Phys. Lett. B **716**, 1 (2012)
2. S. Asai et al., Prospects for the search for a standard model Higgs boson in ATLAS using vector boson fusion. Eur. Phys. J. C **32S2**, 19 (2004)
3. A. Banfi, G.P. Salam, G. Zanderighi, NLL+NNLO predictions for jet-veto efficiencies in Higgs-boson and Drell-Yan production. JHEP **1206**, 159 (2012)
4. D. Baumann, TASI lectures on inflation. arXiv:0907.5424 [hep-th]
5. U. Baur, T. Plehn, D.L. Rainwater, Determining the Higgs boson selfcoupling at hadron colliders. Phys. Rev. D **67**, 033003 (2003)
6. G.L. Bayatian et al., [CMS Collaboration], CMS technical design report, volume II: physics performance. J. Phys. G **34**, 995 (2007)
7. F. Bezrukov, The Higgs field as an inflaton. Class. Quantum Gravity **30**, 214001 (2013). [arXiv:1307.0708 [hep-ph]]
8. I. Brivio et al., Disentangling a dynamical Higgs. arXiv:1311.1823 [hep-ph]
9. V. Büscher, K. Jakobs, Higgs boson searches at hadron colliders. Int. J. Mod. Phys. A **20**, 2523 (2005)
10. D. Buttazzo, G. Degrassi, P.P. Giardino, G.F. Giudice, F. Sala, A. Salvio, A. Strumia, Investigating the near-criticality of the Higgs boson. arXiv:1307.3536 [hep-ph]
11. J.M. Butterworth, A.R. Davison, M. Rubin, G.P. Salam, Jet substructure as a new Higgs search channel at the LHC. Phys. Rev. Lett. **100**, 242001 (2008)
12. N. Cabibbo, L. Maiani, G. Parisi, R. Petronzio, Bounds on the fermions and Higgs boson masses in grand unified theories. Nucl. Phys. B **158**, 295 (1979)
13. J.M. Campbell, J.W. Huston, W.J. Stirling, Hard interactions of quarks and gluons: a primer for LHC physics. Rep. Prog. Phys. **70**, 89 (2007)
14. S. Chatrchyan et al., [CMS Collaboration], Observation of a new boson at a mass of 125 GeV with the CMS experiment at the LHC. Phys. Lett. B **716**, 30 (2012)
15. A. Djouadi, The anatomy of electro-weak symmetry breaking. I: the Higgs boson in the standard model. Phys. Rep. **457**, 1 (2008)
16. A. Djouadi, The anatomy of electro-weak symmetry breaking. II: the Higgs bosons in the minimal supersymmetric model. Phys. Rep. **459**, 1 (2008)
17. F. Englert, R. Brout, Broken symmetry and the mass of gauge vector mesons. Phys. Rev. Lett. **13**, 321 (1964)
18. C. Englert, D. Goncalves-Netto, K. Mawatari, T. Plehn, Higgs quantum numbers in weak boson fusion. JHEP **1301**, 148 (2013)
19. M. Gomez-Bock, M. Mondragon, M. Mühlleitner, M. Spira, P.M. Zerwas, Concepts of electroweak symmetry breaking and Higgs physics. arXiv:0712.2419 [hep-ph]
20. G.S. Guralnik, C.R. Hagen, T.W.B. Kibble, Global conservation laws and massless particles. Phys. Rev. Lett. **13**, 585 (1964)
21. T. Hahn, S. Heinemeyer, W. Hollik, H. Rzehak, G. Weiglein, FeynHiggs: a program for the calculation of MSSM Higgs-boson observables – version 2.6.5. Comput. Phys. Commun. **180**, 1426 (2009)
22. T. Han, Collider phenomenology: basic knowledge and techniques. arXiv:hep-ph/0508097
23. P.W. Higgs, Broken symmetries, massless particles and gauge fields. Phys. Lett. **12**, 132 (1964)
24. P.W. Higgs, Broken symmetries and the masses of gauge bosons. Phys. Rev. Lett. **13**, 508 (1964)
25. C.T. Hill, E.H. Simmons, Strong dynamics and electroweak symmetry breaking. Phys. Rep. **381**, 235 (2003)
26. W. Kilian, *Electroweak Symmetry Breaking: The Bottom-Up Approach*. Springer Tracts in Modern Physics, vol. 198 (Springer, New York, 2003), p. 1.
27. M. Klute, R. Lafaye, T. Plehn, M. Rauch, D. Zerwas, Measuring Higgs couplings from LHC data. Phys. Rev. Lett. **109**, 101801 (2012)

28. M. Lindner, Implications of triviality for the standard model. Z. Phys. C **31**, 295 (1986)
29. T. Plehn, M. Rauch, Higgs Couplings after the discovery. Europhys. Lett. **100**, 11002 (2012)
30. T. Plehn, D. Rainwater, D. Zeppenfeld, Determining the structure of Higgs couplings at the LHC. Phys. Rev. Lett. **88**, 051801 (2002)
31. D.L. Rainwater, Intermediate-mass Higgs searches in weak boson fusion. arXiv:hep-ph/9908378
32. D. Rainwater, Searching for the Higgs boson. arXiv:hep-ph/0702124
33. M. Schmaltz, D. Tucker-Smith, Little Higgs review. Ann. Rev. Nucl. Part. Sci. **55**, 229 (2005). [hep-ph/0502182]
34. B. Schrempp, M. Wimmer, Top quark and Higgs boson masses: interplay between infrared and ultraviolet physics. Prog. Part. Nucl. Phys. **37**, 1 (1996)
35. M. Spira, QCD effects in Higgs physics. Fortschr. Phys. **46**, 203 (1998)
36. J.D. Wells, TASI lecture notes: introduction to precision electroweak analysis. arXiv:hep-ph/0512342
37. C. Wetterich, Gauge hierarchy due to strong interactions? Phys. Lett. B **104**, 269 (1981)
38. S. Willenbrock, Symmetries of the standard model. arXiv:hep-ph/0410370

Chapter 2
QCD

Just as Sect. 1 is not meant to be a complete introduction to electroweak symmetry breaking but is aimed at introducing the aspects of Higgs physics most relevant to the LHC this section cannot cover the entire field of QCD. Instead, we will focus on QCD as it impacts LHC physics, like for example the Higgs searches discussed in the first part of the lecture.

In Sect. 2.1 we will introduce the most important process at the LHC, the Drell–Yan process or lepton pair production. This process will lead us through all of the introduction into QCD. Ultraviolet divergences and renormalization we will only mention in passing, to get an idea how much of the treatment of ultraviolet and infrared divergences works the same way. After discussing in detail infrared divergences in Sects. 2.3–2.5 we will spend some time on modern approaches on combining QCD matrix element calculations at leading order and next–to–leading order in perturbative QCD with parton showers. This last part is fairly close to current research with the technical details changing rapidly. Therefore, we will rely on toy models to illustrate the different approaches.

2.1 Drell–Yan Process

Most text books on QCD start from a very simple class of QCD processes, called deep inelastic scattering. These are processes with the HERA initial state $e^{\pm} p$. The problem with this approach is that in the LHC era we would like to instead understand processes of the kind $pp \to W+$jets, $pp \to H+$jets, $pp \to t\bar{t}+$jets, or the production of new particles with or without jets. These kind of signal and background processes and their relevance in an LHC analysis we already mentioned in Sect. 1.5.4.

From a QCD perspective such processes are very complex, so we need to step back a little and start with a simple question: we know how to compute the production rate and distributions for photon or Z production for example at LEP,

$e^+e^- \to \gamma, Z \to \ell^+\ell^-$. What is then the production rate for the same final state at the LHC, how do we account for quarks inside the protons, and what are the best-suited kinematic variables to use at a hadron collider?

2.1.1 Gauge Boson Production

The simplest question we can ask at the LHC is: how do we compute the production of a single weak gauge boson? This process we refer to as the *Drell–Yan production* process, in spite of producing neither Drell nor Yan at the LHC. In our first attempts we will explicitly not care about additional jets, so if we assume the proton to consist of quarks and gluons and simply compute the process $q\bar{q} \to \gamma, Z$ under the assumption that the quarks are partons inside protons. Gluons do not couple to electroweak gauge bosons, so we only have to consider valence quark vs sea antiquark scattering in the initial state. Modulo the $SU(2)_L$ and $U(1)_Y$ charges which describe the $Zf\bar{f}$ and $\gamma f\bar{f}$ couplings in the Feynman rules

$$\boxed{-i\gamma^\mu (\ell \mathbb{P}_L + r \mathbb{P}_R)} \quad \text{with} \quad \ell = \frac{e}{s_w c_w} (T_3 - 2Qs_w^2) \quad r = \ell \Big|_{T_3=0} \quad (Zf\bar{f})$$

$$\ell = r = Qe \qquad (\gamma f\bar{f}),$$
(2.1)

with $T_3 = \pm 1/2$, the matrix element and the squared matrix element for the partonic process

$$q\bar{q} \to \gamma, Z \qquad (2.2)$$

will be the same as the corresponding matrix element squared for $e^+e^- \to \gamma, Z$, with an additional color factor. The general amplitude for *massless fermions* is

$$\mathcal{M} = -i\bar{v}(k_2)\gamma^\mu (\ell \mathbb{P}_L + r \mathbb{P}_R) u(k_1)\epsilon_\mu . \qquad (2.3)$$

At the LHC massless fermions are a good approximation for all particles except for the top quark. For the bottom quark we need to be careful with some aspects of this approximation, but the first two generations of quarks and all leptons are usually assumed to be massless in LHC simulations. Once we will arrive at infrared divergences in LHC cross sections we will specifically discuss ways of regulating them without introducing masses.

Squaring the matrix element in Eq. (2.3) means adding the same structure once again, just walking through the Feynman diagram in the opposite direction. Luckily, we do not have to care about factors of $(-i)$ since we are only interested in the absolute value squared. Because the chiral projectors $\mathbb{P}_{L,R} = (\mathbb{1} \mp \gamma_5)/2$ defined in Eq. (1.14) are real and $\gamma_5^T = \gamma_5$ is symmetric, the left and right handed gauge

2.1 Drell–Yan Process

boson vertices described by the Feynman rules in Eq. (2.1) do not change under transposition. For the production of a massive Z boson on or off its mass shell we obtain

$$|\mathcal{M}|^2 = \sum_{\text{spin,pol,color}} \bar{u}(k_1)\gamma^\nu (\ell \mathbb{P}_L + r \mathbb{P}_R) v(k_2) \, \bar{v}(k_2)\gamma^\mu$$

$$\times (\ell \mathbb{P}_L + r \mathbb{P}_R) u(k_1) \, \epsilon_\mu \epsilon_\nu^* \qquad \text{incoming (anti-) quark } k_{1,2}$$

$$= N_c \, \text{Tr}\,[\slashed{k}_1 \gamma^\nu (\ell \mathbb{P}_L + r \mathbb{P}_R) \slashed{k}_2 \gamma^\mu (\ell \mathbb{P}_L + r \mathbb{P}_R)]$$

$$\times \left(-g_{\mu\nu} + \frac{q_\mu q_\nu}{m_Z^2} \right) \qquad \text{unitary gauge with } q = -k_1 - k_2$$

$$= N_c \, \text{Tr}\,[\slashed{k}_1 \gamma^\nu (\ell \mathbb{P}_L + r \mathbb{P}_R)(\ell \mathbb{P}_L + r \mathbb{P}_R) \slashed{k}_2 \gamma^\mu]$$

$$\times \left(-g_{\mu\nu} + \frac{q_\mu q_\nu}{m_Z^2} \right) \qquad \text{with } \{\gamma_\mu, \gamma_5\} = 0$$

$$= N_c \, \text{Tr}\,\left[\slashed{k}_1 \gamma^\nu \left(\ell^2 \frac{1}{2} + r^2 \frac{1}{2} \right) \slashed{k}_2 \gamma^\mu \right]$$

$$\times \left(-g_{\mu\nu} + \frac{q_\mu q_\nu}{m_Z^2} \right) \qquad \text{symmetric polarization sum}$$

$$= \frac{N_c}{2} (\ell^2 + r^2) \, \text{Tr}\,[\slashed{k}_1 \gamma^\nu \slashed{k}_2 \gamma^\mu] \left(-g_{\mu\nu} + \frac{q_\mu q_\nu}{m_Z^2} \right)$$

$$= 2 N_c (\ell^2 + r^2) [k_1^\mu k_2^\nu + k_1^\nu k_2^\mu - (k_1 k_2) g^{\mu\nu}] \left(-g_{\mu\nu} + \frac{q_\mu q_\nu}{m_Z^2} \right)$$

$$= 2 N_c (\ell^2 + r^2) \Bigg[-2(k_1 k_2) + 4(k_1 k_2)$$

$$+ 2 \frac{(-k_1 k_2)^2}{m_Z^2} - \frac{(k_1 k_2) q^2}{m_Z^2} \Bigg] \qquad \text{with } (qk_1) = -(k_1 k_2)$$

$$= 2 N_c (\ell^2 + r^2) \left[2(k_1 k_2) + \frac{q^4}{2 m_Z^2} - \frac{q^4}{2 m_Z^2} \right] \qquad \text{with } q^2 = (k_1 + k_2)^2$$

$$= 2 N_c (\ell^2 + r^2) q^2 \qquad (2.4)$$

The color factor N_c accounts for the number of $SU(3)$ states which can be combined to form a color singlet like the Z.

An interesting aspect coming out of our calculation is that the $1/m_Z$-dependent terms in the polarization sum do not contribute—as far as the matrix element squared is concerned the Z boson could as well be transverse. This reflects the fact that the Goldstone modes do not couple to massless fermions, just like the Higgs boson. This means that not only the matrix element squared for the on–shell Z case corresponds to $q^2 = m_Z^2$ but also that the on–shell photon case is given by $q^2 \to 0$.

The apparently vanishing matrix element in this limit has to be combined with the phase space definition to give a finite result.

What is still missing is an averaging factor for initial–state spins and colors, only the sum is included in Eq. (2.4). For incoming electrons as well as incoming quarks this factor K_{ij} includes $1/4$ for the spins. Since we do not observe color in the initial state, and the color structure of the incoming $q\bar{q}$ pair has no impact on the Z-production matrix element, we also average over the color. This gives us another factor $1/N_c^2$ for the averaged matrix element, which altogether becomes

$$K_{ij} = \frac{1}{4N_c^2} \,. \qquad (2.5)$$

In spite of our specific case in Eq. (2.4) looking that way, matrix elements we compute from our Feynman rules are not automatically numbers with mass unit zero.

If for the partonic invariant mass of the two quarks we introduce the *Mandelstam variable* $s = (k_2 + k_2)^2 = 2(k_1 k_2)$, so momentum conservation for on–shell Z production implies $s = q^2 = m_Z^2$. In four space–time dimensions (this detail will become important later) we can compute a *total cross section* from the matrix element squared, for example as given in Eq. (2.4), as

$$\boxed{s \left. \frac{d\sigma}{dy} \right|_{2 \to 1} = \frac{\pi}{(4\pi)^2} K_{ij} \, (1 - \tau) \, |\mathcal{M}|^2} \qquad \tau = \frac{m_Z^2}{s} \qquad y = \frac{1 - \cos\theta}{2} \,.$$

$$(2.6)$$

The scattering angle θ enters through the definition of $y = 0 \ldots 1$. The mass of the final state appears in τ, with $\tau = 0$ for a massless photon. It would be replaced to include m_W or the Higgs mass or the mass of a Kaluza–Klein graviton if needed. At the production threshold of an on–shell particle the phase space opens in the limit $\tau \to 1$, slowly increasing the cross section above threshold $\tau < 1$.

We know that such a heavy gauge boson we do not actually observe at colliders. What we should really calculate is the production for example of a pair of fermions through an s-channel Z and γ, where the Z might or might not be on its mass shell. The matrix element for this process we can derive from the same Feynman rules in Eq. (2.1), now for an incoming fermion k_1, incoming anti–fermion k_2, outgoing fermion p_1 and outgoing anti–fermion p_2. To make it easy to switch particles between initial and final states, we can define all momenta as incoming, so momentum conservation means $k_1 + k_2 + p_1 + p_2 = 0$. The additional *Mandelstam variables* we need to describe this $(2 \to 2)$ process are $t = (k_1 + p_1)^2 < 0$ and $u = (k_1 + p_2)^2 < 0$, as usually with $s + t + u = 0$ for massless final-state particles. The $(2 \to 2)$ matrix element for the two sets of incoming and outgoing fermions becomes

2.1 Drell–Yan Process

$$\mathcal{M} = (-i)^2\, \bar{u}(p_1)\gamma^\nu \left(\ell' \mathbb{P}_L + r' \mathbb{P}_R\right) v(p_2)\, \bar{v}(k_2)\gamma^\mu \left(\ell \mathbb{P}_L + r \mathbb{P}_R\right)$$

$$\times u(k_1)\, \frac{i}{q^2 - m_Z^2} \left(-g_{\mu\nu} + \frac{q_\mu q_\nu}{m_Z^2}\right). \tag{2.7}$$

The coupling to the gauge bosons are ℓ and r for the incoming quarks and ℓ' and r' for the outgoing leptons. The chiral projectors are defined in Eq. (1.14). When we combine the four different spinors and their momenta correctly the matrix element squared factorizes into twice the trace we have computed before. The corresponding picture is two fermion currents interacting with each other through a gauge boson. All we have to do is combine the traces properly. If the incoming trace in the matrix element and its conjugate includes the indices μ and ρ and the outgoing trace the indices ν and σ, the Z bosons link μ and ν as well as ρ and σ.

To make the results a little more compact we compute this process for a *massless photon* instead of the Z boson, i.e. for the physical scenario where the initial-state fermions do not have enough energy to excite the intermediate Z boson. The specific features of an intermediate massive Z boson we postpone to Sect. 2.1.2. The assumption of a massless photon simplifies the couplings to $(\ell^2 + r^2) = 2Q^2 e^2$ and the polarization sums to $-g_{\mu\nu}$ and $-g_{\rho\sigma}$:

$$|\mathcal{M}|^2 = 4N_c\, (2Q^2 e^2)\, (2Q'^2 e^2)\, \frac{1}{q^4} \left[k_1^\mu k_2^\rho + k_1^\rho k_2^\mu - (k_1 k_2) g^{\mu\rho}\right]$$

$$(-g_{\mu\nu}) \left[p_1^\nu p_2^\sigma + p_1^\sigma p_2^\nu - (p_1 p_2) g^{\nu\sigma}\right] (-g_{\rho\sigma})$$

$$= 16 N_c\, Q^2 Q'^2 e^4\, \frac{1}{q^4} \left[k_1^\mu k_{2\sigma} + k_{1\sigma} k_2^\mu - (k_1 k_2) g_\sigma^\mu\right]$$

$$\left[p_{1\mu} p_2^\sigma + p_1^\sigma p_{2\mu} - (p_1 p_2) g_\mu^\sigma\right]$$

$$= 16 N_c\, Q^2 Q'^2 e^4\, \frac{1}{q^4} \left[2(k_1 p_1)(k_2 p_2) + 2(k_1 p_2)(k_2 p_1) - 2(k_1 k_2)(p_1 p_2)\right.$$

$$\left. -2(k_1 k_2)(p_1 p_2) + 4(k_1 k_2)(p_1 p_2)\right]$$

$$= 32 N_c\, Q^2 Q'^2 e^4\, \frac{1}{q^4} \left[(k_1 p_1)(k_2 p_2) + (k_1 p_2)(k_2 p_1)\right]$$

$$= 32 N_c\, Q^2 Q'^2 e^4\, \frac{1}{s^2} \left[\frac{t^2}{4} + \frac{u^2}{4}\right]$$

$$= 8 N_c\, Q^2 Q'^2 e^4\, \frac{1}{s^2} \left[s^2 + 2st + 2t^2\right]$$

$$= 8 N_c\, Q^2 Q'^2 e^4 \left[1 + 2\frac{t}{s} + 2\frac{t^2}{s^2}\right]. \tag{2.8}$$

We can briefly check if this number is indeed positive, using the definition of the Mandelstam variable t for massless external particles in terms of the polar angle

$t = s(-1 + \cos\theta)/2 = -s \cdots 0$: the upper phase space boundary $t = 0$ inserted into the brackets in Eq. (2.8) gives $[\cdots] = 1$, just as the lower boundary $t = -s$ with $[\cdots] = 1 - 2 + 2 = 1$. For the central value $t = -s/2$ the minimum value of the brackets is $[\cdots] = 1 - 1 + 0.5 = 0.5$.

The azimuthal angle ϕ plays no role at colliders, unless you want to compute gravitational effects on Higgs production at ATLAS and CMS. Any LHC Monte Carlo will either random-generate a reference angle ϕ for the partonic process or pick one and keep it fixed.

The two-particle phase space integration for massless particles then gives us

$$\boxed{s^2 \frac{d\sigma}{dt}\bigg|_{2\to 2} = \frac{\pi}{(4\pi)^2} K_{ij} |\mathcal{M}|^2} \qquad t = \frac{s}{2}(-1 + \cos\theta) \ . \qquad (2.9)$$

For our Drell–Yan process we then find the differential cross section in four space-time dimensions, using $\alpha = e^2/(4\pi)$

$$\begin{aligned}
\frac{d\sigma}{dt} &= \frac{1}{s^2} \frac{\pi}{(4\pi)^2} \frac{1}{4N_c} 8 \, Q^2 Q'^2 (4\pi\alpha)^2 \left(1 + 2\frac{t}{s} + 2\frac{t^2}{s^2}\right) \\
&= \frac{1}{s^2} \frac{2\pi\alpha^2}{N_c} Q^2 Q'^2 \left(1 + 2\frac{t}{s} + 2\frac{t^2}{s^2}\right) \ ,
\end{aligned} \qquad (2.10)$$

which we can integrate over the polar angle or the Mandelstam variable t to compute the total cross section

$$\begin{aligned}
\sigma &= \frac{1}{s^2} \frac{2\pi\alpha^2}{N_c} Q^2 Q'^2 \int_{-s}^0 dt \left(1 + 2\frac{t}{s} + 2\frac{t^2}{s^2}\right) \\
&= \frac{1}{s^2} \frac{2\pi\alpha^2}{N_c} Q^2 Q'^2 \left[t + \frac{t^2}{s} + \frac{2t^3}{3s^2}\right]_{-s}^0 \\
&= \frac{1}{s^2} \frac{2\pi\alpha^2}{N_c} Q^2 Q'^2 \left(s - \frac{s^2}{s} + \frac{2s^3}{3s^2}\right) \\
&= \frac{1}{s} \frac{2\pi\alpha^2}{N_c} Q^2 Q'^2 \frac{2}{3} \quad \Rightarrow \quad \boxed{\sigma(q\bar{q} \to \ell^+ \ell^-)\bigg|_{\text{QED}} = \frac{4\pi\alpha^2}{3N_c s} Q_\ell^2 Q_q^2}
\end{aligned} \qquad (2.11)$$

As a side remark—in the history of QCD, the same process but read *right-to-left* played a crucial role, namely the production rate of quarks in $e^+ e^-$ scattering. For small enough energies we can neglect the Z exchange contribution. At leading order we can then compute the corresponding production cross sections for muon pairs and for quark pairs in $e^+ e^-$ collisions. Moving the quarks into the final state means that we do not average of the color in the initial state, but sum over all possible color

2.1 Drell–Yan Process

combinations, which in Eq. (2.9) gives us an averaging factor $K_{ij} = 1/4$. Everything else stays the same as for the Drell–Yan process

$$\boxed{R \equiv \frac{\sigma(e^+e^- \to \text{hadrons})}{\sigma(e^+e^- \to \ell^+\ell^-)}} = \frac{\sum_{\text{quarks}} \frac{4\pi\alpha^2 N_c}{3s} Q_e^2 Q_q^2}{\frac{4\pi\alpha^2}{3s} Q_e^2 Q_\ell^2} = N_c \left(3\frac{1}{9} + 2\frac{4}{9}\right) = \frac{11 N_c}{9},$$

(2.12)

for example for five quark flavors where the top quark is too heavy to be produced at the given e^+e^- collider energy. For those interested in the details we did take one short cut: hadrons are also produced in the hadronic decays of $e^+e^- \to \tau^+\tau^-$ which we strictly speaking need to subtract. This way, R as a function of the collider energy is a beautiful measurement of the weak and color charges of the quarks in QCD.

2.1.2 Massive Intermediate States

At hadron colliders we cannot tune the energies of the incoming partons. This means that for any particle we will always observe a mix of on–shell and off–shell production, depending on the structure of the matrix element and the distribution of partons inside the proton. For hadron collider analyses this has profound consequences: unlike at an e^+e^- collider we have to base all measurements on reconstructed final–state particles. For studies of particles decaying to jets this generally limits the possible precision at hadron colliders to energy scales above Λ_{QCD}. Therefore, before we move on to describing incoming quarks inside protons we should briefly consider the second Feynman diagram contributing to the Drell–Yan production rate in Eq. (2.11), the on–shell or off–shell Z boson

$$|\mathcal{M}|^2 = |\mathcal{M}_\gamma + \mathcal{M}_Z|^2 = |\mathcal{M}_\gamma|^2 + |\mathcal{M}_Z|^2 + 2\,\text{Re}\,\mathcal{M}_Z \mathcal{M}_\gamma. \quad (2.13)$$

Interference occurs in phase space regions where for both intermediate states the invariant masses of the muon pair are the same.

For the photon the on–shell pole is not a problem. It has zero mass, which means that we hit the pole $1/q^2$ in the matrix element squared only in the limit of zero incoming energy. Strictly speaking we never hit it, because the energy of the incoming particles has to be large enough to produce the final–state particles with their tiny but finite masses and with some kind of momentum driving them through the detector.

A problem arises when we consider the intermediate Z boson. In that case, the propagator contributes as $|\mathcal{M}|^2 \propto 1/(s - m_Z^2)^2$ which diverges on the mass shell. Before we can ask what such a pole means for LHC simulations we have to recall how we deal with it in field theory. There, we encounter the same issue when we solve for example the *Klein–Gordon equation*. The Green function for a field obeying this equation is the inverse of the Klein–Gordon operator

$$(\Box + m^2) G(x - x') = \delta^4(x - x') . \tag{2.14}$$

Fourier transforming $G(x - x')$ into momentum space we find

$$G(x - x') = \int \frac{d^4q}{(2\pi)^4} e^{-iq\cdot(x-x')} \tilde{G}(q)$$

$$(\Box + m^2) G(x - x') = \int \frac{d^4q}{(2\pi)^4} (\Box + m^2) e^{-iq\cdot(x-x')} \tilde{G}(q)$$

$$= \int \frac{d^4q}{(2\pi)^4} ((iq)^2 + m^2) e^{-iq\cdot(x-x')} \tilde{G}(q)$$

$$= \int \frac{d^4q}{(2\pi)^4} e^{-iq\cdot(x-x')} (-q^2 + m^2) \tilde{G}(q) \stackrel{!}{=} \delta^4(x - x')$$

$$= \int \frac{d^4q}{(2\pi)^4} e^{-iq\cdot(x-x')}$$

$$\Leftrightarrow \quad (-q^2 + m^2) \tilde{G}(q) = 1 \quad \Leftrightarrow \quad \tilde{G}(q) = -\frac{1}{q^2 - m^2} . \tag{2.15}$$

The problem with the Green function in momentum space is that as an inverse it is not defined for $q^2 = m^2$. We usually avoid this problem by slightly shifting this pole following the *Feynman* $i\epsilon$ prescription to $m^2 \to m^2 - i\epsilon$, or equivalently deforming our integration contours appropriately. The sign of this infinitesimal shift we need to understand because it will become relevant for phenomenology when we introduce an actual finite decay width of intermediate states.

In the Feynman $i\epsilon$ prescription the sign is crucial to correctly complete the q_0 integration of the Fourier transform in the complex plane

$$\int_{-\infty}^{\infty} dq_0 \frac{e^{-iq_0 x_0}}{q^2 - m^2 + i\epsilon}$$

$$= (\theta(x_0) + \theta(-x_0)) \int_{-\infty}^{\infty} dq_0 \frac{e^{-iq_0 x_0}}{q_0^2 - (\omega^2 - i\epsilon)} \quad \text{with } \omega^2 = \vec{q}^2 + m^2$$

$$= (\theta(x_0) + \theta(-x_0)) \int_{-\infty}^{\infty} dq_0 \frac{e^{-iq_0 x_0}}{(q_0 - \sqrt{\omega^2 - i\epsilon})(q_0 + \sqrt{\omega^2 - i\epsilon})} \tag{2.16}$$

$$= \left(\theta(x_0) \oint_{C_2} + \theta(-x_0) \oint_{C_1}\right) dq_0 \frac{e^{-iq_0 x_0}}{(q_0 - \omega(1 - i\epsilon'))(q_0 + \omega(1 - i\epsilon'))} \quad \text{with } \epsilon' = \frac{\epsilon}{2\omega^2}$$

2.1 Drell–Yan Process

In the last step we have closed the integration contour along the real q_0 axis in the complex q_0 plane. Because the integrand has to vanish for large q_0, we have to make sure the exponent $-ix_0 i \operatorname{Im} q_0 = x_0 \operatorname{Im} q_0$ is negative. For $x_0 > 0$ this means $\operatorname{Im} q_0 < 0$ and vice versa. This argument forces C_1 to close for positive and C_2 for negative imaginary parts in the complex q_0 plane.

The contour integrals we can solve using Cauchy's formula, keeping in mind that the integrand has two poles at $q_0 = \pm \omega(1 - i\epsilon')$. They lie in the upper (lower) half plane for negative (positive) real parts of q_0. The contour C_1 through the upper half plane includes the pole at $q_0 \sim -\omega$ while the contour C_2 includes the pole at $q_0 \sim \omega$, all assuming $\omega > 0$:

$$\int_{-\infty}^{\infty} dq_0 \, \frac{e^{-iq_0 x_0}}{q^2 - m^2 + i\epsilon} = 2\pi i \left[\theta(x_0) \frac{(-1)e^{-i\omega x_0}}{\omega + \omega(1 - i\epsilon')} + \theta(-x_0) \frac{e^{i\omega x_0}}{-\omega - \omega(1 - i\epsilon')} \right]$$

$$\stackrel{\epsilon' \to 0}{=} -i\frac{\pi}{\omega} \left[\theta(x_0) e^{-i\omega x_0} + \theta(-x_0) e^{i\omega x_0} \right] . \quad (2.17)$$

The factor (-1) in the C_2 integration arises because Cauchy's integration formula requires us to integrate counter–clockwise, while going from negative to positive $\operatorname{Re} q_0$ the contour C_2 is defined clockwise. Using this result we can complete the four-dimensional Fourier transform from Eq. (2.15)

$$G(x)$$
$$= \int d^4q \, e^{-i(q \cdot x)} \tilde{G}(q)$$
$$= \int d^4q \, \frac{e^{-i(q \cdot x)}}{q^2 - m^2 + i\epsilon}$$
$$= -i\pi \int d^3\vec{q} \, e^{i\vec{q} \cdot \vec{x}} \, \frac{1}{\omega} \left[\theta(x_0) e^{-i\omega x_0} + \theta(-x_0) e^{i\omega x_0} \right]$$
$$= -i\pi \int d^4q \, e^{i\vec{q} \cdot \vec{x}} \, \frac{1}{\omega} \left[\theta(x_0) e^{-iq_0 x_0} \delta(q_0 - \omega) + \theta(-x_0) e^{-iq_0 x_0} \delta(q_0 + \omega) \right]$$
$$= -i\pi \int d^4q \, e^{-i(q \cdot x)} \, \frac{1}{\omega} \left[\theta(x_0) \delta(\omega - q_0) + \theta(-x_0) \delta(\omega + q_0) \right] \quad \text{with } \delta(x) = \delta(-x)$$
$$= -i\pi \int d^4q \, e^{-i(q \cdot x)} \, \frac{1}{\omega} \, 2\omega \left[\theta(x_0) \delta(\omega^2 - q_0^2) + \theta(-x_0) \delta(\omega^2 - q_0^2) \right]$$
$$= -2\pi i \int d^4q \, e^{-i(q \cdot x)} \, [\theta(x_0) + \theta(-x_0)] \, \delta(q_0^2 - \omega^2)$$
$$= -2\pi i \int d^4q \, e^{-i(q \cdot x)} \, [\theta(x_0) + \theta(-x_0)] \, \delta(q^2 - m^2) \quad \text{with } q_0^2 - \omega^2 = q^2 - m^2 .$$

$$(2.18)$$

This is exactly the usual decomposition of the propagator function $\Delta_F(x) = \theta(x_0)\Delta^+(x) + \theta(-x_0)\Delta^-(x)$ into positive and negative energy contributions.

Let us briefly recapitulate what would have happened if we instead had chosen the Feynman parameter $\epsilon < 0$. We summarize all steps leading to the propagator function in Eq. (2.18) in Table 2.1. For the wrong sign of $i\epsilon$ the two poles in the complex q_0 plane would be mirrored by the real axis. The solution with $\text{Re}\, q_0 > 0$ would sit in the quadrant with $\text{Im}\, q_0 > 0$ and the second pole at a negative real and imaginary part. To be able to close the integration path in the upper half plane in the mathematically positive direction the real pole would have to be matched up with $\theta(-x_0)$. The residue in the Cauchy integral would now include a factor $+1/(2\omega)$. At the end, the two poles would give the same result as for the correct sign of $i\epsilon$, except with a wrong over–all sign.

When we are interested in the kinematic distributions of on–shell massive states the situation is a little different. Measurements of differential distributions for example at LEP include information on the *physical width* of the decaying particle, which means we cannot simply apply the Feynman $i\epsilon$ prescription as if we were dealing with an asymptotic stable state. From the same couplings governing the Z decay, the Z propagator receives corrections, for example including fermion loops:

Such one-particle irreducible diagrams can occur in the same propagator repeatedly. Schematically written as a scalar they are of the form

$$\frac{i}{q^2 - m_0^2 + i\epsilon} + \frac{i}{q^2 - m_0^2 + i\epsilon}(-iM^2)\frac{i}{q^2 - m_0^2 + i\epsilon}$$

$$+ \frac{i}{q^2 - m_0^2 + i\epsilon}(-iM^2)\frac{i}{q^2 - m_0^2 + i\epsilon}(-iM^2)\frac{i}{q^2 - m_0^2 + i\epsilon} + \cdots$$

$$= \frac{i}{q^2 - m_0^2 + i\epsilon}\sum_{n=0}^{\infty}\left(\frac{M^2}{q^2 - m_0^2 + i\epsilon}\right)^n$$

$$= \frac{i}{q^2 - m_0^2 + i\epsilon}\frac{1}{1 - \dfrac{M^2}{q^2 - m_0^2 + i\epsilon}} \quad \text{summing the geometric series}$$

$$= \frac{i}{q^2 - m_0^2 + i\epsilon - M^2}. \tag{2.19}$$

2.1 Drell–Yan Process

Table 2.1 Contributions to the propagator function Eq. (2.18) for both signs of $i\epsilon$

	$\dfrac{1}{q^2-m^2+i\epsilon}$		$\dfrac{1}{q^2-m^2-i\epsilon}$	
Pole	$q_0=\omega(1-i\epsilon)$	$q_0=-\omega(1-i\epsilon)$	$q_0=\omega(1+i\epsilon)$	$q_0=-\omega(1+i\epsilon)$
Complex quadrant	$(+,-)$	$(-,+)$	$(+,+)$	$(-,-)$
Convergence: $x_0\mathrm{Im}q_0<0$	$x_0>0$	$x_0<0$	$x_0<0$	$x_0>0$
Part of real axis	$\theta(x_0)$	$\theta(-x_0)$	$\theta(-x_0)$	$\theta(x_0)$
Closed contour	$\mathrm{Im}\,q_0<0$	$\mathrm{Im}\,q_0>0$	$\mathrm{Im}\,q_0>0$	$\mathrm{Im}\,q_0<0$
Direction of contour	-1	$+1$	$+1$	-1
Residue	$+\dfrac{1}{2\omega}$	$-\dfrac{1}{2\omega}$	$+\dfrac{1}{2\omega}$	$-\dfrac{1}{2\omega}$
Fourier exponent	$e^{-i\omega x_0}$	$e^{+i\omega x_0}$	$e^{+i\omega x_0}$	$e^{-i\omega x_0}$
All combined	$-\dfrac{e^{-i\omega x_0}}{2\omega}\theta(x_0)$	$-\dfrac{e^{+i\omega x_0}}{2\omega}\theta(-x_0)$	$+\dfrac{e^{+i\omega x_0}}{2\omega}\theta(-x_0)$	$+\dfrac{e^{-i\omega x_0}}{2\omega}\theta(x_0)$

We denote the loop as M^2 for reasons which will become obvious later. Requiring that the residue of the propagator be unity at the pole we *renormalize* the wave function and the mass in the corresponding process. For example for a massive scalar or gauge boson with a real correction $M^2(q^2)$ this reads

$$\boxed{\dfrac{i}{q^2-m_0^2-M^2(q^2)} = \dfrac{iZ}{q^2-m^2} \quad \text{for } q^2 \sim m^2}, \qquad (2.20)$$

including a renormalized mass m and a wave function renormalization constant Z.

The important step in our argument is that in analogy to the effective ggH coupling discussed in Sect. 1.5.1 the one-loop correction M^2 depends on the momentum flowing through the propagator. Above a certain threshold it can develop an imaginary part because the momentum flowing through the diagram is large enough to produce on-shell states in the loop. Just as for the ggH coupling such *absorptive parts* appear when a real decay like $Z \to \ell^+\ell^-$ becomes kinematically allowed. After splitting $M^2(q^2)$ into its real and imaginary parts we know what to do with the real part: the solution to $q^2 - m_0^2 - \mathrm{Re}M^2(q^2) \stackrel{!}{=} 0$ defines the renormalized particle mass $q^2 = m^2$ and the wave function renormalization Z. The imaginary part looks like the Feynman $i\epsilon$ term discussed before

$$\dfrac{i}{q^2-m_0^2+i\epsilon-\mathrm{Re}M^2(q^2)-i\mathrm{Im}M^2} = \dfrac{iZ}{q^2-m^2+i\epsilon-iZ\mathrm{Im}M^2}$$

$$\equiv \dfrac{iZ}{q^2-m^2+im\Gamma}$$

$$\Leftrightarrow \Gamma = -\dfrac{Z}{m}\,\mathrm{Im}\,M^2(q^2=m^2), \qquad (2.21)$$

for $\epsilon \to 0$ and finite $\Gamma \neq 0$. We can illustrate the link between the element squared M^2 of a self energy and the partial width by remembering one way to compute scalar integrals or one-loop amplitudes by gluing them together using tree level amplitudes. Schematically written, the Cutkosky cutting rule discussed in Sect. 1.5.1 tells us Im $M^2 \sim M^2|_\text{cut} \equiv \Gamma$. Cutting the one-loop bubble diagram at the one possible place is nothing but squaring the two tree level matrix element for the decay $Z \to \ell^+ \ell^-$. One thing that we need to keep track of, apart from the additional factor m due to dimensional analysis, is the sign of the $im\Gamma$ term which just like the $i\epsilon$ prescription is fixed by causality.

Going back to the Drell–Yan process $q\bar{q} \to \ell^+ \ell^-$ we now know that for massive unstable particles the Feynman epsilon which we need to define the Green function for internal states acquires a finite value, proportional to the total width of the unstable particle. This definition of a propagator of an unstable particle in the s-channel is what we need for the second Feynman diagram contributing to the Drell–Yan process: $q\bar{q} \to Z^* \to \ell^+ \ell^-$. The resulting shape of the propagator squared is a *Breit–Wigner propagator*

$$\boxed{\sigma(q\bar{q} \to Z \to \ell^+ \ell^-) \propto \left| \frac{1}{s - m_Z^2 + i m_Z \Gamma_Z} \right|^2 = \frac{1}{(s - m_Z^2)^2 + m_Z^2 \Gamma_Z^2}} \,. \quad (2.22)$$

When taking everything into account, the $(2 \to 2)$ production cross section also includes the squared matrix element for the decay $Z \to \ell^+ \ell^-$ in the numerator. In the *narrow width approximation*, the $(2 \to 2)$ matrix element factorizes into the production process times the branching ratio for $Z \to \ell^+ \ell^-$, simply by definition of the Breit–Wigner or Lorentz or Cauchy distribution

$$\lim_{\Gamma \to 0} \frac{\Gamma_{Z,\ell\ell}}{(s - m_Z^2)^2 + m_Z^2 \Gamma_{Z,\text{tot}}^2} = \Gamma_{Z,\ell\ell} \frac{\pi}{\Gamma_{Z,\text{tot}}} \delta(s - m_Z^2) \equiv \pi \, \text{BR}(Z \to \ell\ell) \, \delta(s - m_Z^2) \,. \quad (2.23)$$

The additional factor π will be absorbed in the different one-particle and two-particle phase space definitions. We immediately see that this narrow width approximation is only exact for scalar particles. It does not keep information about the structure of the matrix element, e.g. when a non–trivial structure of the numerator gives us the spin and angular correlations between the production and decay processes.

Because of the γ-Z interference we will always simulate lepton pair production using the full on–shell and off–shell $(2 \to 2)$ process. For example for top pair production with three-body decays this is less clear. Sometimes, we will simulate the production and the decay independently and rely on the limit $\Gamma \to 0$. In that case it makes sense to nevertheless require a Breit–Wigner shape for the momenta of the supposedly on–shell top quarks. For top mass measurements we do, however, have to take into account off–shell effects and QCD effects linking the decay and production sides of the full Feynman diagrams.

2.1 Drell–Yan Process

Equation (2.23) uses a mathematical relation we might want to remember for life, and that is the definition of the one-dimensional *Dirac delta distribution* in three ways and including all factors of 2 and π

$$\delta(x) = \int \frac{dq}{2\pi} e^{-ixq} = \lim_{\sigma \to 0} \frac{1}{\sigma \sqrt{\pi}} e^{-x^2/\sigma^2} = \lim_{\Gamma \to 0} \frac{1}{\pi} \frac{\Gamma}{x^2 + \Gamma^2} . \quad (2.24)$$

The second distribution is a *Gaussian* and the third one we would refer to as a *Breit–Wigner* shape while most other people call it a Cauchy distribution.

Now, we know everything necessary to compute all Feynman diagrams contributing to muon pair production at a hadron collider. Strictly speaking, the two amplitudes interfere, so we end up with three distinct contributions: γ exchange, Z exchange and the $\gamma - Z$ interference terms. They have the properties

- For small energies the γ contribution dominates and can be linked to the R parameter.
- On the Z pole the rate is regularized by the Z width and Z contribution dominates over the photon.
- In the tails of the Breit–Wigner distribution we expect $Z - \gamma$ interference. For $m_{\ell\ell} > 120$ GeV the γ and Z contributions at the LHC are roughly equal in size.
- For large energies we are again dominated by the photon channel.
- Quantum effects allow unstable particles like the Z to decay off–shell, defining a Breit–Wigner propagator.
- In the limit of vanishing width the Z contribution factorizes into $\sigma \cdot$ BR.

2.1.3 Parton Densities

At the end of Sect. 2.1.1 the discussion of different energy regimes for R experimentally makes sense—at an e^+e^- collider we can tune the energy of the initial state. At hadron colliders the situation is very different. The energy distribution of incoming quarks as parts of the colliding protons has to be taken into account. We first assume that quarks move collinearly with the surrounding proton such that at the LHC incoming partons have zero p_T. Under that condition we can define a probability distribution for finding a parton just depending on the respective fraction of the proton's momentum. For this momentum fraction $x = 0 \cdots 1$ the *parton density function* (pdf) is written as $f_i(x)$, where i denotes the different partons in the proton, for our purposes u, d, c, s, g and, depending on the details, b. All incoming partons we assume to be massless.

In contrast to so-called structure functions a pdf is not an observable. It is a distribution in the mathematical sense, which means it has to produce reasonable results when we integrate it together with a test function. Different parton densities have very different behavior—for the valence quarks (uud) they peak somewhere around $x \lesssim 1/3$, while the gluon pdf is small at $x \sim 1$ and grows very rapidly towards small x. For some typical part of the relevant parameter space ($x = 10^{-3} \cdots 10^{-1}$) the gluon density roughly scales like $f_g(x) \propto x^{-2}$. Towards smaller x values it becomes even steeper. This steep gluon distribution was initially not expected and means that for small enough x LHC processes will dominantly be gluon fusion processes.

While we cannot actually compute parton distribution functions $f_i(x)$ as a function of the momentum fraction x there are a few predictions we can make based on symmetries and properties of the hadrons. Such arguments for example lead to *sum rules*:

The parton distributions inside an antiproton are linked to those inside a proton through the CP symmetry, which is an exact symmetry of QCD. Therefore, we know that

$$f_q^{\bar{p}}(x) = f_{\bar{q}}(x) \qquad f_{\bar{q}}^{\bar{p}}(x) = f_q(x) \qquad f_g^{\bar{p}}(x) = f_g(x) \qquad (2.25)$$

for all values of x.

If the proton consists of three valence quarks uud, plus quantum fluctuations from the vacuum which can either involve gluons or quark–antiquark pairs, the contribution from the sea quarks has to be symmetric in quarks and antiquarks. The expectation values for the signed numbers of up and down quarks inside a proton have to fulfill

$$\langle N_u \rangle = \int_0^1 dx \, (f_u(x) - f_{\bar{u}}(x)) = 2 \qquad \langle N_d \rangle = \int_0^1 dx \, (f_d(x) - f_{\bar{d}}(x)) = 1 \,.$$
(2.26)

Similarly, the total momentum of the proton has to consist of sum of all parton momenta. We can write this as the expectation value of $\sum x_i$

$$\left\langle \sum x_i \right\rangle = \int_0^1 dx \, x \left(\sum_q f_q(x) + \sum_{\bar{q}} f_{\bar{q}}(x) + f_g(x) \right) = 1 \qquad (2.27)$$

What makes this prediction interesting is that we can compute the same sum only taking into account the measured quark and antiquark parton densities. We find that the momentum sum rule only comes to 1/2. Half of the proton momentum is then carried by gluons.

Given the correct definition and normalization of the pdf we can now compute the *hadronic cross section* from its partonic counterpart, like the QED result in Eq. (2.11), as

$$\boxed{\sigma_{\text{tot}} = \int_0^1 dx_1 \int_0^1 dx_2 \sum_{ij} f_i(x_1) \, f_j(x_2) \, \hat{\sigma}_{ij}(x_1 x_2 S)} \,, \qquad (2.28)$$

where i, j are the incoming partons with the momentum factions $x_{i,j}$. The partonic energy of the scattering process is $s = x_1 x_2 S$ with the LHC proton energy of eventually $\sqrt{S} = 14$ TeV. The partonic cross section $\hat{\sigma}$ corresponds to the cross sections σ computed for example in Eq. (2.11). It has to include all the necessary θ and δ functions for energy–momentum conservation. When we express a general n–particle cross section $\hat{\sigma}$ including the phase space integration, the x_i integrations and the phase space integrations can of course be interchanged, but Jacobians will make life hard. In Sect. 2.1.5 we will discuss an easier way to compute kinematic distributions instead of from the fully integrated total rate in Eq. (2.28).

2.1.4 Hadron Collider Kinematics

Hadron colliders have a particular kinematic feature in that event by event we do not know the longitudinal velocity of the initial state, i.e. the relative longitudinal boost from the laboratory frame to the partonic center of mass. This sensitivity to longitudinal boosts is reflected in the choice of kinematic variables. The first thing we consider is the projection of all momenta onto the transverse plane. These transverse components are trivially invariant under longitudinal boosts because the two are orthogonal to each other.

In addition, for the production of a single electroweak gauge boson we remember that the produced particle does not have any momentum transverse to the beam direction. This reflects the fact that the incoming quarks are collinear with the protons and hence have zero transverse momentum. Such a gauge boson not recoiling against anything else cannot develop a finite transverse momentum. Of course, once we decay this gauge boson for example into a pair of muons, each muon will have transverse momentum, only their vector sum will be zero:

$$\sum_{\text{final state}} \vec{p}_{T,j} = \vec{0} . \qquad (2.29)$$

This is a relation between two-dimensional, not three dimensional vectors. For more than one particle in the final state we define an azimuthal angle in the transverse plane transverse. While differences of azimuthal angles are observables, the over–all angle is a symmetry of the detector as well as of our physics.

In addition to the transverse plane we need to parameterize the longitudinal momenta in a way which makes it easy to implement longitudinal boosts. In Eq. (2.28) we integrate over the two momentum fractions $x_{1,2}$ and can at best determine their product $x_1 x_2 = s/S$ from the final–state kinematics. Our task is to replace both, x_1 and x_2 with a more physical variable which should be well behaved under longitudinal boosts.

A longitudinal boost for example from the rest frame of a massive particle reads

$$\begin{pmatrix} E \\ p_L \end{pmatrix} = \exp\left[y \begin{pmatrix} 0 & 1 \\ 1 & 0 \end{pmatrix}\right] \begin{pmatrix} m \\ 0 \end{pmatrix}$$

$$= \left[\mathbb{1} + y \begin{pmatrix} 0 & 1 \\ 1 & 0 \end{pmatrix} + \frac{y^2}{2}\mathbb{1} + \frac{y^3}{6}\begin{pmatrix} 0 & 1 \\ 1 & 0 \end{pmatrix} \cdots \right] \begin{pmatrix} m \\ 0 \end{pmatrix}$$

$$= \left[\mathbb{1} \sum_{j \text{ even}} \frac{y^j}{j!} + \begin{pmatrix} 0 & 1 \\ 1 & 0 \end{pmatrix} \sum_{j \text{ odd}} \frac{y^j}{j!}\right] \begin{pmatrix} m \\ 0 \end{pmatrix}$$

$$= \left[\mathbb{1} \cosh y + \begin{pmatrix} 0 & 1 \\ 1 & 0 \end{pmatrix} \sinh y\right] \begin{pmatrix} m \\ 0 \end{pmatrix} = m \begin{pmatrix} \cosh y \\ \sinh y \end{pmatrix}. \qquad (2.30)$$

We can re-write the *rapidity y* defined above in a way which allows us to compute it from the four-momentum for example in the LHC lab frame

$$\frac{1}{2} \log \frac{E + p_L}{E - p_L} = \frac{1}{2} \log \frac{\cosh y + \sinh y}{\cosh y - \sinh y} = \frac{1}{2} \log \frac{e^y}{e^{-y}} = y. \qquad (2.31)$$

We can explicitly check that the rapidity is indeed additive by applying a second longitudinal boost to (E, p_L) in Eq. (2.30)

$$\begin{pmatrix} E' \\ p'_L \end{pmatrix} = \exp\left[y' \begin{pmatrix} 0 & 1 \\ 1 & 0 \end{pmatrix}\right] \begin{pmatrix} E \\ p_L \end{pmatrix} = \left[\mathbb{1} \cosh y' + \begin{pmatrix} 0 & 1 \\ 1 & 0 \end{pmatrix} \sinh y'\right] \begin{pmatrix} E \\ p_L \end{pmatrix}$$

$$= \begin{pmatrix} E \cosh y' + p_L \sinh y' \\ p_L \cosh y' + E \sinh y' \end{pmatrix}, \qquad (2.32)$$

which gives for the combined rapidity, following its extraction in Eq. (2.31)

$$\frac{1}{2} \log \frac{E' + p'_L}{E' - p'_L} = \frac{1}{2} \log \frac{(E + p_L)(\cosh y' + \sinh y')}{(E - p_L)(\cosh y' - \sinh y')}$$

$$= \frac{1}{2} \log \frac{E + p_L}{E - p_L} + y' = y + y'. \qquad (2.33)$$

Two successive boosts with rapidities y and y' can be combined into a single boost by $y + y'$. This combination of several longitudinal boosts is important in the case of massless particles. They do not have a rest frame, which means we can only boost them from one finite-momentum frame to the other. For such massless particles we

2.1 Drell–Yan Process

can simplify the formula for the rapidity Eq. (2.31), in terms of the polar angle θ. We use that for massless particles $E = |\vec{p}|$, giving us

$$y = \frac{1}{2} \log \frac{E + p_L}{E - p_L} = \frac{1}{2} \log \frac{|\vec{p}| + p_L}{|\vec{p}| - p_L} = \frac{1}{2} \log \frac{1 + \cos\theta}{1 - \cos\theta}$$

$$= \frac{1}{2} \log \frac{1}{\tan^2 \frac{\theta}{2}} = -\log \tan \frac{\theta}{2} \equiv \eta \qquad (2.34)$$

This *pseudo-rapidity* η is more handy, but coincides with the actual rapidity only for massless particles. To get an idea about the experimental setup at the LHC— in CMS and ATLAS we can observe different particles to polar angles of between 10 and 1.3°, corresponding to maximum pseudo-rapidities of 2.5–4.5. Because this is numerically about the same range as the range of the *azimuthal angle* $[0, \pi]$ we define a distance measure inside the detector

$$(\Delta R)^2 = (\Delta y)^2 + (\Delta \phi)^2$$

$$= (\Delta \eta)^2 + (\Delta \phi)^2 \qquad \text{massless particles}$$

$$= \left(\log \frac{\tan \frac{\theta + \Delta \theta}{2}}{\tan \frac{\theta}{2}} \right)^2 + (\Delta \phi)^2$$

$$= \frac{(\Delta \theta)^2}{\sin^2 \theta} + (\Delta \phi)^2 + \mathcal{O}((\Delta \theta)^3) \qquad (2.35)$$

The angle θ is the polar angle of one of the two particles considered and in our leading approximation can be chosen as each of them without changing Eq. (2.35).

Still for the case of single gauge boson production we can express the final–state kinematics in terms of two parameters, the invariant mass of the final–state particle q^2 and its rapidity. We already know that the transverse momentum of a single particle in the final state is zero. The two incoming and approximately massless protons have the momenta

$$p_1 = (E, 0, 0, E) \qquad p_2 = (E, 0, 0, -E) \qquad S = (2E)^2 . \qquad (2.36)$$

For the momentum of the final–state gauge boson in terms of the parton momentum fractions this means in combination with Eq. (2.30)

$$q = x_1 p_1 + x_2 p_2 = E \begin{pmatrix} x_1 + x_2 \\ 0 \\ 0 \\ x_1 - x_2 \end{pmatrix} \stackrel{!}{=} \sqrt{q^2} \begin{pmatrix} \cosh y \\ 0 \\ 0 \\ \sinh y \end{pmatrix} = 2E\sqrt{x_1 x_2} \begin{pmatrix} \cosh y \\ 0 \\ 0 \\ \sinh y \end{pmatrix}$$

$$\Leftrightarrow \quad \cosh y = \frac{x_1 + x_2}{2\sqrt{x_1 x_2}} = \frac{1}{2}\left(\sqrt{\frac{x_1}{x_2}} + \sqrt{\frac{x_2}{x_1}}\right)$$

$$\Leftrightarrow \quad e^y = \sqrt{\frac{x_1}{x_2}} \ . \tag{2.37}$$

This result can be combined with $x_1 x_2 = q^2/S$ to obtain

$$x_1 = \sqrt{\frac{q^2}{S}} e^y \qquad x_2 = \sqrt{\frac{q^2}{S}} e^{-y} \ . \tag{2.38}$$

These relations allow us to for example compute the hadronic *total cross section* for lepton pair production in QED

$$\boxed{\sigma(pp \to \ell^+ \ell^-)\bigg|_{\text{QED}} = \frac{4\pi \alpha^2 Q_\ell^2}{3 N_c} \int_0^1 dx_1 dx_2 \sum_j Q_j^2 \, f_j(x_1) \, f_{\bar{j}}(x_2) \, \frac{1}{q^2} \ ,}$$

$$\tag{2.39}$$

instead in terms of the hadronic phase space variables $x_{1,2}$ in terms of the kinematic final–state observables q^2 and y. Remember that the partonic or quark–antiquark cross section $\hat{\sigma}$ is already integrated over the (symmetric) azimuthal angle ϕ and the polar angle Mandelstam variable t. The transverse momentum of the two leptons is therefore fixed by momentum conservation.

The Jacobian for this change of variables reads

$$\frac{\partial(q^2, y)}{\partial(x_1, x_2)} = \begin{vmatrix} x_2 S & x_1 S \\ 1/(2x_1) & -1/(2x_2) \end{vmatrix} = S = \frac{q^2}{x_1 x_2} \ , \tag{2.40}$$

which inserted into Eq. (2.39) gives us

$$\sigma(pp \to \ell^+ \ell^-)\bigg|_{\text{QED}} = \frac{4\pi \alpha^2 Q_\ell^2}{3 N_c} \int dq^2 dy \, \frac{x_1 x_2}{q^2} \frac{1}{q^2} \sum_j Q_j^2 \, f_j(x_1) f_{\bar{j}}(x_2)$$

$$= \frac{4\pi \alpha^2 Q_\ell^2}{3 N_c} \int dq^2 dy \, \frac{1}{q^4} \sum_j Q_j^2 \, x_1 f_j(x_1) \, x_2 f_{\bar{j}}(x_2) \ .$$

$$\tag{2.41}$$

2.1 Drell–Yan Process

In contrast to the original form of the integration over the hadronic phase space this form reflects the kinematic observables. For the Drell–Yan process at leading order the q^2 distribution is the same as $m_{\ell\ell}^2$, one of the most interesting distributions to study because of different contributions from the photon, the Z boson, or extra dimensional gravitons. On the other hand, the rapidity integral still suffers from the fact that at hadron colliders we do not know the longitudinal kinematics of the initial state and therefore have to integrate over it.

2.1.5 Phase Space Integration

In the previous example we have computed the simple two-dimensional distribution, by leaving out the double integration in Eq. (2.41)

$$\left.\frac{d\sigma(pp \to \ell^+\ell^-)}{dq^2 dy}\right|_{\text{QED}} = \frac{4\pi\alpha^2 Q_\ell^2}{3N_c q^4} \sum_j Q_j^2 \, x_1 f_j(x_1) \, x_2 f_{\bar{j}}(x_2) \,. \qquad (2.42)$$

We can numerically evaluate this expression and compare it to experiment. However, the rapidity y and the momentum transfer q^2 of the $\ell^+\ell^-$ pair are by no means the only distribution we would like to look at. Moreover, we have to integrate numerically over the parton densities $f(x)$, so we will have to rely on numerical integration tools no matter what we are doing. Looking at a simple ($2 \to 2$) process we can write the total cross section as

$$\sigma_{\text{tot}} = \int d\phi \int d\cos\theta \int dx_1 \int dx_2 \, F_{\text{PS}} \, |\mathcal{M}|^2 = \int_0^1 dy_1 \cdots dy_4 \, J_{\text{PS}}(\vec{y}) \, |\mathcal{M}|^2 \,, \qquad (2.43)$$

with an appropriate function F_{PS}. In the second step we have re-written the phase space integral as an integral over the four-dimensional unit cube, implicitly defining the appropriate Jacobian. Like any integral we can numerically evaluate this phase space integral by binning the variable we integrate over

$$\int_0^1 dy \, f(y) \quad \longrightarrow \quad \sum_j (\Delta y)_j \, f(y_j) \sim \Delta y \sum_j f(y_j) \,. \qquad (2.44)$$

Without any loss of generality we assume that the integration boundaries are $0 \cdots 1$. We can divide the integration variable y into a discrete set of points y_j, for example equidistant in y or as a chain of random numbers $y_j \in [0, 1]$. In the latter case we need to keep track of the bin widths $(\Delta y)_j$. When we extend the integral to d dimensions we can in principle divide each axis into bins and compute the functional values for this grid. For not equidistant bins generated by random numbers we again keep track of the associated phase space volume for each random number vector. Once we know these phase space weights for each phase space point there is no reason to consider the set of random numbers as in any way linked to the d axes. All we need is a chain of random points with an associated phase space weight and their transition matrix element, to integrate over the phase space in complete analogy to Eq. (2.44).

The obvious question is how such random numbers can be chosen in a smart way. However, before we discuss how to best evaluate such an integral numerically, let us first illustrate how this integral is much more useful than just to provide the total cross section. If we are interested in the distribution of an observable, like for example the distribution of the transverse momentum of a muon in the Drell–Yan process, we need to compute $d\sigma/dp_T$ as a function of p_T. In terms of Eq. (2.43) any physical y_1 distribution is given by

$$\sigma = \int dy_1 \cdots dy_d \; f(\vec{y}) = \int dy_1 \; \frac{d\sigma}{dy_1}$$

$$\left.\frac{d\sigma}{dy_1}\right|_{y_1^0} = \int dy_2 \cdots dy_d \; f(y_1^0) = \int dy_1 \cdots dy_d \; f(\vec{y}) \, \delta(y_1 - y_1^0) \, . \quad (2.45)$$

We can compute this distribution numerically in two ways: one way corresponds to the first line in Eq. (2.45) and means evaluating the $y_2 \cdots y_d$ integrations and leaving out the y_1 integration. The result will be a function of y_1 which we then evaluate at different points y_1^0.

The second and much more efficient option corresponds to the second line of Eq. (2.45), with the delta distribution defined for discretized y_1. First, we define an array with the size given by the number of bins in the y_1 integration. Then, for each y_1 value of the complete $y_1 \cdots y_d$ integration we decide where the value y_1 goes in this array and add $f(\vec{y})$ to the corresponding column. Finally, we print these columns as a function of y_1 to see the distribution. This set of columns is referred to as a *histogram* and can be produced using publicly available software. This histogram approach does not sound like much, but imagine we want to compute a distribution $d\sigma/dp_T$, where $p_T(\vec{y})$ is a complicated function of the integration variables and kinematic phase space cuts. We then simply evaluate

$$\frac{d\sigma}{dp_T} = \int dy_1 \cdots dy_d \; f(\vec{y}) \, \delta\left(p_T(\vec{y}) - p_T^0\right) \quad (2.46)$$

numerically and read off the p_T distribution as a side product of the calculation of the total rate. Histograms mean that computing a total cross section numerically we can trivially extract all distributions in the same process.

The procedure outlined above has an interesting interpretation. Imagine we do the entire phase space integration numerically. Just like computing the interesting observables we can compute the momenta of all external particles. These momenta are not all independent, because of energy–momentum conservation, but this can be taken care of. The tool which translates the vector of integration variables $\vec{y}$ into

2.1 Drell–Yan Process

the external momenta is called a *phase space generator*. Because the phase space is not uniquely defined in terms of the integration variables, the phase space generator also returns the Jacobian J_{PS}, called the phase space weight. If we think of the integration as an integration over the unit cube, this weight needs to be combined with the matrix element squared $|\mathcal{M}|^2$. Once we compute the unique phase space configuration $(k_1, k_2, p_1 \cdots)_j$ corresponding to the vector $\vec{y}_j$, the combined weight $W = J_{PS} |\mathcal{M}|^2$ is the probability that this configuration will appear at the LHC. This means we do not only integrate over the phase space, we really simulate LHC events. The only complication is that the probability of a given configuration is not only given by the frequency with which it appears, but also by the explicit weight. So when we run our numerical integration through the phase space generator and histogram all the distributions we are interested in we generate *weighted events*. These events, which consist of the momenta of all external particles and the weight W, we can for example store in a big file.

This simulation is not yet what experimentalists want—they want to represent the probability of a certain configuration appearing only by its frequency and not by an additional event weight. Experimentally measured events do not come with a variable weight, either they are recorded or they are not. This means we have to unweight the events by translating the event weight into frequency.

There are two ways to do that. On the one hand, we can look at the minimum event weight and express all other events in relative probability to this event. Translating this relative event weight into a frequency means replacing an event with the relative weight W_j/W_{min} by W_j/W_{min} unit-weight events in the same phase space point. The problem with this method is that we are really dealing with a binned phase space, so we would not know how to distribute these events in and around the given bin.

Alternatively, we can translate the weight of each event into a probability to keep it or drop it. Because such a probability has to be limited from above we start from to the maximum weight W_{max} and compute the ratio $W_j/W_{max} \epsilon [0, 1]$ for each event. We then generate a flat random number $r \epsilon [0, 1]$ and only keep an event if $r < W_j/W_{max}$. This way, we keep an event with a large weight $W_j/W_{max} \sim 1$ for almost all values of r, while events with small weights are more likely to drop out. The challenge in this translation is that we always lose events. If it was not for the experimentalists we would hardly use such *unweighted events*, but they have good reasons to want such unweighted events which feed best through detector simulations.

The last comment is that if the phase space configuration $(k_1, k_2, p_1 \cdots)_j$ can be measured, its weight W_j better be positive. This is not trivial once we go beyond leading order. There, we need to add several contributions to produce a physical event, like for example different n-particle final states. There is no guarantee for each of them to be positive. Instead, we ensure that after adding up all contributions and after integrating over any kind of unphysical degrees of freedom we might have introduced, the probability of a physics configuration is positive. From this point of view negative values for parton densities $f(x) < 0$ are in principle not problematic, as long as we always keep a positive hadronic rate $d\sigma_{pp \to X} > 0$.

Going back to the numerical phase space integration for many particles, it faces two problems. First, the partonic phase space for n on–shell particles in the final state has $3(n+2) - 3$ dimensions. If we divide each of these directions in 100 bins, the number of phase space points we need to evaluate for a $(2 \to 4)$ process is $100^{15} = 10^{30}$, which is not realistic.

To integrate over a large number of dimensions we use *Monte Carlo integration*. In this approach we first replace the binned directions in phase space by a chain of random numbers Y_j, which can be organized in any number of dimensions. In one dimension it replaces the equidistant bins in the direction y. Because the distance between these new random numbers is not constant, each random number will come with yet another weight. The probability of finding $Y_j \epsilon [y, y+dy]$ is given by a smartly chosen function $p_Y(y)$. Integrating a function $g(y)$ now returns an expectation value of g evaluated over the chain Y_j,

$$\langle g(Y) \rangle = \int_0^1 dy \, p_Y(y) \, g(y) \quad \longrightarrow \quad \frac{1}{N_Y} \sum_{j=1}^{N_Y} g(Y_j) \,. \qquad (2.47)$$

First of all, we can immediately generalize this approach to any number of d dimensions, just by organizing the random numbers Y_j in one large chain instead of a d-dimensional array. Second, in Eq. (2.43) we are interested in an integral over g, which means that we should rewrite the integration as

$$\int_0^1 d^d y \, f(y) = \int_0^1 d^d y \, \frac{f(y)}{p_Y(y)} \, p_Y(y) = \left\langle \frac{f(Y)}{p_Y(Y)} \right\rangle \quad \longrightarrow \quad \frac{1}{N_Y} \sum_j \frac{f(Y_j)}{p_Y(Y_j)} \,. \qquad (2.48)$$

To compute the integral we now average over all values of f/p_Y along the random number chain Y_j. In the ideal case where we exactly know the form of the integrand and can map it into our random numbers, the error of the numerical integration will be zero. So what we have to find is a way to encode $f(Y_j)$ into $p_Y(Y_j)$. This task is called *importance sampling* and you can find some documentation for example on the standard implementation VEGAS to look at the details.

Technically, VEGAS will call the function which computes the weight $W = J_{PS} |\mathcal{M}|^2$ for a number of phase space points and average over these points, but including another weight factor W_{MC} representing the importance sampling. If we want to extract distributions via histograms we have to add the total weight $W = W_{MC} J_{PS} |\mathcal{M}|^2$ to the columns.

The second numerical challenge is that the matrix elements for interesting processes are by no means flat. We would therefore like to help our adaptive or importance sampling Monte Carlo by defining the integration variables such that the integrand becomes as flat as possible. For example for the integration over the partonic momentum fraction we know that the integrand usual falls off as $1/x$. In that situation we can substitute

$$\int_\delta dx \frac{C}{x} = \int_{\log \delta} d\log x \left(\frac{d\log x}{dx}\right)^{-1} \frac{C}{x} = \int_{\log \delta} d\log x \, C , \qquad (2.49)$$

to obtain a flat integrand. There exists an even more impressive and relevant example: intermediate particles with Breit–Wigner propagators squared are particularly painful to integrate over the momentum $s = p^2$ flowing through it

$$P(s,m) = \frac{1}{(s-m^2)^2 + m^2\Gamma^2} . \qquad (2.50)$$

For example, a Standard Model Higgs boson with a mass of 126 GeV has a width around 0.005 GeV, which means that the integration over the invariant mass of the Higgs decay products $\sqrt{s}$ requires a relative resolution of 10^{-5}. Since this is unlikely to be achievable what we should really do is find a substitution which produces the inverse Breit–Wigner as a Jacobian and leads to a flat integrand—et voilá

$$\int ds \frac{C}{(s-m^2)^2 + m^2\Gamma^2} = \int dz \left(\frac{dz}{ds}\right)^{-1} \frac{C}{(s-m^2)^2 + m^2\Gamma^2}$$
$$= \int dz \frac{(s-m^2)^2 + m^2\Gamma^2}{m\Gamma} \frac{C}{(s-m^2)^2 + m^2\Gamma^2}$$
$$= \frac{1}{m\Gamma} \int dz \, C \quad \text{with} \quad \tan z = \frac{s-m^2}{m\Gamma} . \qquad (2.51)$$

This is the most useful *phase space mapping* in LHC physics. Of course, any adaptive Monte Carlo will eventually converge on such an integrand, but a well-chosen set of integration parameters will speed up simulations very significantly.

2.2 Ultraviolet Divergences

From general field theory we know that when we are interested for example in cross section prediction with higher precision we need to compute further terms in its perturbative series in α_s. This computation will lead to ultraviolet divergences which can be absorbed into counter terms for any parameter in the Lagrangian. The crucial feature is that for a renormalizable theory like our Standard Model the number of counter terms is finite, which means once we know all parameters including their counter terms our theory becomes predictive.

In Sect. 2.3 we will see that in QCD processes we also encounter another kind of divergences. They arise from the infrared momentum regime. Infrared divergences is what this lecture is really going to be about, but before dealing with them it is very instructive to see what happens to the much better understood ultraviolet divergences. In Sect. 2.2.1 we will review how such ultraviolet divergences arise and how they are removed. In Sect. 2.2.2 we will review how running parameters appear in this procedure, i.e. how scale dependence is linked to the appearance of divergences. Finally, in Sect. 2.2.3 we will interpret the use of running parameters physically and see that in perturbation theory they resum classes of logarithms to all orders in perturbation theory. Later in Sect. 2.3 we will follow exactly the same steps for infrared divergences and develop some crucial features of hadron collider physics.

2.2.1 Counter Terms

Renormalization as the proper treatment of ultraviolet divergences is one of the most important things to understand about field theories; you can find more detailed discussions in any book on advanced field theory. The particular aspect of renormalization which will guide us through this section is the appearance of the renormalization scale.

In perturbation theory, scales automatically arise from the regularization of infrared or ultraviolet divergences. We can see this by writing down a simple scalar loop integral, with to two virtual scalar propagators with masses $m_{1,2}$ and an external momentum p flowing through a diagram, similar to those summed in Sect. 2.1.2

$$B(p^2; m_1, m_2) \equiv \int \frac{d^4q}{16\pi^2} \frac{1}{q^2 - m_1^2} \frac{1}{(q+p)^2 - m_2^2} \,. \tag{2.52}$$

Such two-point functions appear for example in the gluon self energy with virtual gluons, with massless ghost scalars, with a Dirac trace in the numerator for quarks, and with massive scalars for supersymmetric scalar quarks. In those cases the two masses are identical $m_1 = m_2$. The integration measure $1/(16\pi^2)$ is dictated by the Feynman rule for the integration over loop momenta. Counting powers of q in Eq. (2.52) we see that the integrand is not suppressed by powers of $1/q$ in the ultraviolet, so it is logarithmically divergent and we have to regularize it. Regularizing means expressing the divergence in a well–defined manner or scheme, allowing us to get rid of it by renormalization.

One regularization scheme is to introduce a cutoff into the momentum integral Λ, for example through the so-called Pauli—Villars regularization. Because the ultraviolet behavior of the integrand or integral cannot depend on any parameter living at a small energy scales, the parameterization of the ultraviolet divergence in Eq. (2.52) cannot involve the mass m or the external momentum p^2. The scalar two-point function has mass dimension zero, so its divergence has to be proportional to $\log(\Lambda/\mu_R)$ with a dimensionless prefactor and some scale μ_R^2 which is an artifact of the regularization of such a Feynman diagram.

2.2 Ultraviolet Divergences

A more elegant regularization scheme is *dimensional regularization*. It is designed not to break gauge invariance and naively seems to not introduce a mass scale μ_R. When we shift the momentum integration from 4 to $4-2\epsilon$ dimensions and use analytic continuation in the number of space–time dimensions to renormalize the theory, a *renormalization scale* μ_R nevertheless appears once we ensure the two-point function and with it observables like cross sections keep their correct mass dimension

$$\int \frac{d^4 q}{16\pi^2} \cdots \longrightarrow \mu_R^{2\epsilon} \int \frac{d^{4-2\epsilon} q}{16\pi^2} \cdots = \frac{i\mu_R^{2\epsilon}}{(4\pi)^2} \left[\frac{C_{-1}}{\epsilon} + C_0 + C_1 \epsilon + \mathcal{O}(\epsilon^2) \right]. \tag{2.53}$$

At the end, the scale μ_R might become irrelevant and drop out after renormalization and analytic continuation, but to be on the safe side we keep it. The constants C_i in the series in $1/\epsilon$ depend on the loop integral we are considering. To regularize the ultraviolet divergence we assume $\epsilon > 0$ and find mathematically well defined poles $1/\epsilon$. Defining scalar integrals with the integration measure $1/(i\pi^2)$ will make for example C_{-1} come out as of the order $\mathcal{O}(1)$. This is the reason we usually find factors $1/(4\pi)^2 = \pi^2/(2\pi)^4$ in front of the loop integrals.

The poles in $1/\epsilon$ will cancel with the universal *counter terms* once we renormalize the theory. Counter terms we include by shifting parameters in the Lagrangian and the leading order matrix element. They cancel the poles in the combined leading order and virtual one-loop prediction

$$|\mathcal{M}_{\text{LO}}(g) + \mathcal{M}_{\text{virt}}|^2 = |\mathcal{M}_{\text{LO}}(g)|^2 + 2\,\text{Re}\,\mathcal{M}_{\text{LO}}(g)\mathcal{M}_{\text{virt}} + \cdots$$

$$\to |\mathcal{M}_{\text{LO}}(g + \delta g)|^2 + 2\,\text{Re}\,\mathcal{M}_{\text{LO}}(g)\mathcal{M}_{\text{virt}} + \cdots$$

$$\text{with} \quad g \to g^{\text{bare}} = g + \delta g \quad \text{and} \quad \delta g \propto \alpha_s/\epsilon. \tag{2.54}$$

The dots indicate higher orders in α_s, for example absorbing the δg corrections in the leading order and virtual interference. As we can see in Eq. (2.54) the counter terms do not come with a factor $\mu_R^{2\epsilon}$ in front. Therefore, while the poles $1/\epsilon$ cancel just fine, the scale factor $\mu_R^{2\epsilon}$ will not be matched between the actual ultraviolet divergence and the counter term.

We can keep track of the renormalization scale best by expanding the prefactor of the regularized but not yet renormalized integral in Eq. (2.53) in a Taylor series in ϵ, no question asked about convergence radii

$$\mu_R^{2\epsilon} \left[\frac{C_{-1}}{\epsilon} + C_0 + \mathcal{O}(\epsilon) \right] = e^{2\epsilon \log \mu_R} \left[\frac{C_{-1}}{\epsilon} + C_0 + \mathcal{O}(\epsilon) \right]$$

$$= \left[1 + 2\epsilon \log \mu_R + \mathcal{O}(\epsilon^2) \right] \left[\frac{C_{-1}}{\epsilon} + C_0 + \mathcal{O}(\epsilon) \right]$$

$$= \frac{C_{-1}}{\epsilon} + C_0 + C_{-1} \log \mu_R^2 + \mathcal{O}(\epsilon)$$

$$\to \frac{C_{-1}}{\epsilon} + C_0 + C_{-1} \log \frac{\mu_R^2}{M^2} + \mathcal{O}(\epsilon) \, . \tag{2.55}$$

In the last step we correct by hand for the fact that $\log \mu_R^2$ with a mass dimension inside the logarithm cannot appear in our calculations. From somewhere else in our calculation the logarithm will be matched with a $\log M^2$ where M^2 is the typical mass or energy scale in our process. This little argument shows that also in dimensional regularization we introduce a mass scale μ_R which appears as $\log(\mu_R^2/M^2)$ in the renormalized expression for our observables. There is no way of removing ultraviolet divergences without introducing some kind of renormalization scale.

In Eq. (2.55) there appear two contributions to a given observable, the expected C_0 and the renormalization-induced C_{-1}. Because the factors C_{-1} are linked to the counter terms in the theory we can often guess them without actually computing the loop integral, which is very useful in cases where they numerically dominate.

Counter terms as they schematically appear in Eq. (2.54) are not uniquely defined. They need to include a given divergence to return finite observables, but we are free to add any finite contribution we want. This opens many ways to define a counter term for example based on physical processes where counter terms do not only cancel the pole but also finite contributions at a given order in perturbation theory. Needless to say, such schemes do not automatically work universally. An example for such a *physical renormalization scheme* is the on-shell scheme for masses, where we define a counter term such that external on-shell particles do not receive any corrections to their masses. For the top mass this means that we replace the leading order mass with the bare mass, for which we then insert the expression in terms of the renormalized mass and the counter term

$$m_t^{\text{bare}} = m_t + \delta m_t$$

$$= m_t + m_t \frac{\alpha_s C_F}{4\pi} \left(3 \left(-\frac{1}{\epsilon} + \gamma_E - \log(4\pi) - \log \frac{\mu_R^2}{M^2} \right) - 4 + 3 \log \frac{m_t^2}{M^2} \right)$$

$$\equiv m_t + m_t \frac{\alpha_s C_F}{4\pi} \left(-\frac{3}{\tilde{\epsilon}} - 4 + 3 \log \frac{m_t^2}{M^2} \right)$$

$$\Leftrightarrow \quad \frac{1}{\tilde{\epsilon}\left(\frac{\mu_R}{M}\right)} \equiv \frac{1}{\epsilon} - \gamma_E + \log \frac{4\pi \mu_R^2}{M^2} \, , \tag{2.56}$$

with the color factor $C_F = (N^2 - 1)/(2N)$. The convenient scale dependent pole $1/\tilde{\epsilon}$ includes the universal additional terms like the Euler gamma function and the scaling logarithm. This logarithm is the big problem in this universality argument, since we need to introduce the arbitrary energy scale M to separate the universal

2.2 Ultraviolet Divergences

logarithm of the renormalization scale and the parameter-dependent logarithm of the physical process.

A theoretical problem with this *on–shell renormalization scheme* is that it is not gauge invariant. On the other hand, it describes for example the kinematic features of top pair production at hadron colliders in a stable perturbation series. This means that once we define a more appropriate scheme for heavy particle masses in collider production mechanisms it better be numerically close to the pole mass. For the computation of total cross sections at hadron colliders or the production thresholds at e^+e^- colliders the pole mass is not well suited at all, but as we will see in Sect. 3 this is not where we expect to measure particle masses at the LHC, so we should do fine with something very similar to the pole mass.

Another example for a process dependent renormalization scheme is the mixing of γ and Z propagators. There we choose the counter term of the weak mixing angle such that an on–shell Z boson cannot oscillate into a photon, and vice versa. We can generalize this scheme for mixing scalars as they for example appear in supersymmetry, but it is not gauge invariant with respect to the weak gauge symmetries of the Standard Model either. For QCD corrections, on the other hand, it is the most convenient scheme keeping all exchange symmetries of the two scalars.

To finalize this discussion of process dependent mass renormalization we quote the result for a scalar supersymmetric quark, a squark, where in the on–shell scheme we find

$$m_{\tilde{q}}^{\text{bare}} = m_{\tilde{q}} + \delta m_{\tilde{q}} = m_{\tilde{q}} + m_{\tilde{q}}$$

$$\frac{\alpha_s C_F}{4\pi} \left(-\frac{2r}{\tilde{\epsilon}} - 1 - 3r - (1-2r)\log r - (1-r)^2 \log\left|\frac{1}{r} - 1\right| - 2r \log \frac{m_{\tilde{q}}^2}{M^2} \right). \quad (2.57)$$

with $r = m_{\tilde{g}}^2/m_{\tilde{q}}^2$. The interesting aspect of this squark mass counter term is that it also depends on the gluino mass, not just the squark mass itself. The reason why QCD counter terms tend to depend only on the renormalized quantity itself is that the gluon is massless. In the limit of vanishing gluino contribution the squark mass counter term is again only proportional to the squark mass itself

$$\left. m_{\tilde{q}}^{\text{bare}} \right|_{m_{\tilde{g}}=0} = m_{\tilde{q}} + \delta m_{\tilde{q}} = m_{\tilde{q}} + m_{\tilde{q}} \frac{\alpha_s C_F}{4\pi} \left(-\frac{1}{\tilde{\epsilon}} - 3 + \log \frac{m_{\tilde{q}}^2}{M^2} \right). \quad (2.58)$$

Taking the limit of Eq. (2.57) to derive Eq. (2.58) is computationally not trivial, though.

One common feature of all mass counter terms listed above is $\delta m \propto m$, which means that we actually encounter a multiplicative renormalization

$$m^{\text{bare}} = Z_m m = (1 + \delta Z_m) m = \left(1 + \frac{\delta m}{m}\right) m = m + \delta m, \quad (2.59)$$

with $\delta Z_m = \delta m/m$ linking the two ways of writing the mass counter term. This form implies that particles with zero mass will not obtain a finite mass through renormalization. If we remember that chiral symmetry protects a Lagrangian from acquiring fermion masses this means that on–shell renormalization does not break this symmetry. A massless theory cannot become massive by mass renormalization. Regularization and renormalization schemes which do not break symmetries of the Lagrangian are ideal.

When we introduce counter terms in general field theory we usually choose a slightly more model independent scheme—we define a renormalization point. This is the energy scale at which the counter terms cancels all higher order contributions, divergent as well as finite. The best known example is the electric charge which we renormalize in the *Thomson limit* of zero momentum transfer through the photon propagator

$$e \to e^{\text{bare}} = e + \delta e \,. \tag{2.60}$$

Looking back at δm_t as defined in Eq. (2.56) we also see a way to define a completely general counter term: if dimensional regularization, i.e. the introduction of $4 - 2\epsilon$ dimensions does not break any of the symmetries of our Lagrangian, like Lorentz symmetry or gauge symmetries, we can simply subtract the ultraviolet pole and nothing else. The only question is: do we subtract $1/\epsilon$ in the MS scheme or do we subtract $1/\bar{\epsilon}$ in the $\overline{MS}$ *scheme*. In the $\overline{MS}$ scheme the counter term is then scale dependent.

Carefully counting, there are three scales present in such a scheme. First, there is the physical scale in the process. In our case of a top self energy this is for example the top mass m_t appearing in the matrix element for the process $pp \to t\bar{t}$. Next, there is the renormalization scale μ_R, a reference scale which is part of the definition of any counter term. And last but not least, there is the scale M separating the counter term from the process dependent result, which we can choose however we want, but which as we will see implies a running of the counter term. The role of this scale M will become clear when we go through the example of the running strong coupling α_s. Of course, we would prefer to choose all three scales the same, but in a complex physical process this might not always be possible. For example, any massive $(2 \to 3)$ production process naturally involves several external physical scales.

Just a side remark for completeness: a one loop integral which has no intrinsic mass scale is the two-point function with zero mass in the loop and zero momentum flowing through the integral: $B(p^2 = 0; 0, 0)$. It appears for example in the self energy corrections of external quarks and gluons. Based on dimensional arguments this integral has to vanish altogether. On the other hand, we know that like any massive two-point function it has to be ultraviolet divergent $B \sim 1/\epsilon_{\text{UV}}$ because setting all internal and external mass scales to zero is nothing special from an ultraviolet point of view. This can only work if the scalar integral also has an infrared divergence appearing in dimensional regularization. We can then write the entire massless two-point function as

2.2 Ultraviolet Divergences

$$B(p^2 = 0; 0, 0) = \int \frac{d^4q}{16\pi^2} \frac{1}{q^2} \frac{1}{(q+p)^2} = \frac{i\pi^2}{16\pi^2} \left(\frac{1}{\epsilon_{UV}} - \frac{1}{\epsilon_{IR}} \right), \quad (2.61)$$

keeping track of the divergent contributions from the infrared and the ultraviolet regimes. For this particular integral they precisely cancel, so the result for $B(0; 0, 0)$ is zero, but setting it to zero too early will spoil any ultraviolet and infrared finiteness test. Treating the two divergences strictly separately and dealing with them one after the other also ensures that for ultraviolet divergences we can choose $\epsilon > 0$ while for infrared divergences we require $\epsilon < 0$.

2.2.2 Running Strong Coupling

To get an idea what these different scales which appear in the process of renormalization mean let us compute such a scale dependent parameter, namely the *running strong coupling* $\alpha_s(\mu_R^2)$. The Drell–Yan process is one of the very few relevant processes at hadron colliders where the strong coupling does not appear at tree level, so we cannot use it as our toy process this time. Another simple process where we can study this coupling is bottom pair production at the LHC, where at some energy range we will be dominated by valence quarks: $q\bar{q} \to b\bar{b}$. The only Feynman diagram is an *s*-channel off–shell gluon with a momentum flow $p^2 \equiv s$.

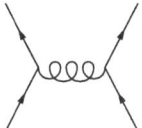

At next–to–leading order this gluon propagator will be corrected by self energy loops, where the gluon splits into two quarks or gluons and re-combines before it produces the two final–state bottoms. Let us for now assume that all quarks are massless. The Feynman diagrams for the gluon self energy include a quark look, a gluon loop, and the ghost loop which removes the unphysical degrees of freedom of the gluon inside the loop.

The gluon self energy correction or *vacuum polarization*, as propagator corrections to gauge bosons are usually labelled, will be a scalar. This way, all fermion lines close in the Feynman diagram and the Dirac trace is computed inside the loop. In color space the self energy will (hopefully) be diagonal, just like the gluon propagator itself, so we can ignore the color indices for now. In unitary gauge the gluon propagator is proportional to the transverse tensor $T^{\mu\nu} = g^{\mu\nu} - p^\nu p^\mu / p^2$.

As mentioned in the context of the effective gluon–Higgs coupling, the same should be true for the gluon self energy, which we therefore write as $\Pi^{\mu\nu} \equiv \Pi \, T^{\mu\nu}$. Unlike for two different external momenta $k_1 \neq k_2$ shown in Eq. (1.187) the case with only one external momentum gives us the useful simple relations

$$T^{\mu\nu} g_\nu^\rho = \left(g^{\mu\nu} - \frac{p^\mu p^\nu}{p^2} \right) g_\nu^\rho = T^{\mu\rho}$$

$$T^{\mu\nu} T_\nu^\rho = \left(g^{\mu\nu} - \frac{p^\mu p^\nu}{p^2} \right) \left(g_\nu^\rho - \frac{p_\nu p^\rho}{p^2} \right) = g^{\mu\rho} - 2 \frac{p^\mu p^\rho}{p^2} + p^2 \frac{p^\mu p^\rho}{p^4} = T^{\mu\rho} \, . \tag{2.62}$$

Including the gluon, quark, and ghost loops the regularized gluon self energy with a momentum flow p^2 through the propagator reads

$$-\frac{1}{p^2} \Pi \left(\frac{\mu_R^2}{p^2} \right) = \frac{\alpha_s}{4\pi} \left(-\frac{1}{\tilde{\epsilon}} + \log \frac{p^2}{M^2} \right) \left(\frac{13}{6} N_c - \frac{2}{3} n_f \right) + \mathcal{O}(\log m_t^2)$$

$$\equiv \alpha_s \left(-\frac{1}{\tilde{\epsilon}} + \log \frac{p^2}{M^2} \right) b_0 + \mathcal{O}(\log m_t^2)$$

$$\text{with} \quad \boxed{b_0 = \frac{1}{4\pi} \left(\frac{11}{3} N_c - \frac{2}{3} n_f \right)} . \tag{2.63}$$

The minus sign arises from the factors i in the propagators, as shown in Eq. (2.19). The number of fermions coupling to the gluons is n_f. From the comments on $B(p^2; 0, 0)$ we could guess that the loop integrals will only give a logarithm $\log p^2$ which is then matched by the logarithm $\log M^2$ implicitly included in the definition of $\tilde{\epsilon}$. The factor b_0 arises from the one-loop corrections to the gluon self energy, i.e. from diagrams which include one additional factor α_s. Strictly speaking, this form is the first term in a *perturbative series* in the strong coupling $\alpha_s = g_s^2/(4\pi)$. Later on, we will indicate where additional higher order corrections would enter. For later we keep in mind that for a sufficiently small number of quark generations the sign of b_0 is positive.

In the second step of Eq. (2.63) we have sneaked in additional contributions to the renormalization of the strong coupling from the other one-loop diagrams in the process, replacing the factor 13/6 by a factor 11/3. This is related to the fact that there are actually three types of divergent virtual gluon diagrams in the physical process $q\bar{q} \to b\bar{b}$: the external quark self energies with renormalization factors $Z_f^{1/2}$, the internal gluon self energy Z_A, and the vertex corrections Z_{Aff}. The only physical parameters we can renormalize in this process are the strong coupling and, if finite, the bottom mass. Wave function renormalization constants are not physical, but vertex renormalization terms are. The entire divergence in our $q\bar{q} \to b\bar{b}$ process which needs to be absorbed in the strong coupling Z_g is given by the combination

2.2 Ultraviolet Divergences

$$Z_{Aff} = Z_g Z_A^{1/2} Z_f \qquad \Leftrightarrow \qquad \frac{Z_{Aff}}{Z_A^{1/2} Z_f} \equiv Z_g \,. \qquad (2.64)$$

We can check this definition of Z_g by comparing all vertices in which the strong coupling g_s appears, namely the gluon coupling to quarks, ghosts as well as the triple and quartic gluon vertex. All of them need to have the same divergence structure

$$\frac{Z_{Aff}}{Z_A^{1/2} Z_f} \stackrel{!}{=} \frac{Z_{A\eta\eta}}{Z_A^{1/2} Z_\eta} \stackrel{!}{=} \frac{Z_{3A}}{Z_A^{3/2}} \stackrel{!}{=} \sqrt{\frac{Z_{4A}}{Z_A^2}} \,. \qquad (2.65)$$

If we had done the same calculation in QED and looked for a running electric charge, we would have found that the vacuum polarization diagrams for the photon do account for the entire counter term of the electric charge. The other two renormalization constants Z_{Aff} and Z_f cancel because of gauge invariance.

In contrast to QED, the strong coupling diverges in the Thomson limit because QCD is confined towards large distances and weakly coupled at small distances. Lacking a well enough motivated reference point we are lead to renormalize α_s in the $\overline{\text{MS}}$ scheme. From Eq. (2.63) we know that the ultraviolet pole which needs to be cancelled by the counter term is proportional to the function b_0

$$g_s^{\text{bare}} = Z_g g_s = \left(1 + \delta Z_g\right) g_s = \left(1 + \frac{\delta g_s}{g_s}\right) g_s$$

$$\Rightarrow \quad (g_s^2)^{\text{bare}} = (Z_g g_s)^2 = \left(1 + \frac{\delta g_s}{g_s}\right)^2 g_s^2 = \left(1 + 2\frac{\delta g_s}{g_s}\right) g_s^2 = \left(1 + \frac{\delta g_s^2}{g_s^2}\right) g_s^2$$

$$\Rightarrow \quad \alpha_s^{\text{bare}} = \left(1 + \frac{\delta \alpha_s}{\alpha_s}\right) \alpha_s \stackrel{!}{=} \left(1 - \left.\frac{\Pi}{p^2}\right|_{\text{pole}}\right)$$

$$\alpha_s(M^2) \stackrel{\text{Eq. (2.63)}}{=} \left(1 - \frac{\alpha_s}{\tilde{\epsilon}\left(\frac{\mu_R}{M}\right)} b_0\right) \alpha_s(M^2) \,. \qquad (2.66)$$

Only in the last step we have explicitly included the scale dependence of the counter term. Because the bare coupling does not depend on any scales, this means that α_s depends on the unphysical scale M. Similar to the top mass renormalization scheme we can switch to a more physical scheme for the strong coupling as well: we can absorb also the finite contributions of $\Pi(\mu_R^2/p^2)$ into the strong coupling by simply identifying $M^2 = p^2$. Based again on Eq. (2.63) this implies

$$\alpha_s^{\text{bare}} = \alpha_s(p^2) \left(1 - \frac{\alpha_s(p^2) b_0}{\tilde{\epsilon}} + \alpha_s(p^2) b_0 \log \frac{p^2}{M^2}\right) \,. \qquad (2.67)$$

On the right hand side α_s is consistently evaluated as a function of the physical scale p^2. This formula defines a running coupling $\alpha_s(p^2)$, because the definition of the coupling now has to account for a possible shift between the original argument p^2 and the scale M^2 coming out of the $\overline{\text{MS}}$ scheme. Since according to Eqs. (2.66) and (2.67) the bare strong coupling can be expressed in terms of $\alpha_s(M^2)$ as well as in terms of $\alpha_s(p^2)$ we can link the two scales through

$$\alpha_s(M^2) = \alpha_s(p^2) + \alpha_s^2(p^2) b_0 \log \frac{p^2}{M^2} = \alpha_s(p^2) \left(1 + \alpha_s(p^2) b_0 \log \frac{p^2}{M^2} \right)$$

$$\Leftrightarrow \quad \frac{d\alpha_s(p^2)}{d \log p^2} = -\alpha_s^2(p^2) b_0 + \mathcal{O}(\alpha_s^3) \,. \tag{2.68}$$

To the given loop order the argument of the strong coupling squared in this formula can be neglected—its effect is of higher order. We nevertheless keep the argument as a higher order effect to later distinguish different approaches to the running coupling. From Eq. (2.63) we know that $b_0 > 0$, which means that towards larger scales the strong coupling has a negative slope. The ultraviolet limit of the strong coupling is zero. This makes QCD an *asymptotically free* theory. We can compute the function b_0 in general models by simply adding all contributions of strongly interacting particles in this loop

$$b_0 = -\frac{1}{12\pi} \sum_{\text{colored states}} D_j \, T_{R,j} \,, \tag{2.69}$$

where we need to know some kind of counting factor D_j which is -11 for a vector boson (gluon), $+4$ for a Dirac fermion (quark), $+2$ for a Majorana fermion (gluino), $+1$ for a complex scalar (squark) and $+1/2$ for a real scalar. Note that this sign is not given by the fermionic or bosonic nature of the particle in the loop. The color charges are $T_R = 1/2$ for the fundamental representation of $SU(3)$ and $C_A = N_c$ for the adjoint representation. The masses of the loop particles are not relevant in this approximation because we are only interested in the ultraviolet regime of QCD where all particles can be regarded massless. When we really model the running of α_s we need to take into account threshold effects of heavy particles, because particles can only contribute to the running of α_s at scales above their mass scale. This is why the R ratio computed in Eq. (2.12) is so interesting once we vary the energy of the incoming electron–positron pair.

We can do even better than this fixed order in perturbation theory: while the correction to α_s in Eq. (2.67) is perturbatively suppressed by the usual factor $\alpha_s/(4\pi)$ it includes a logarithm of a ratio of scales which does not need to be small. Instead of simply including these gluon self energy corrections at a given order in perturbation theory we can instead include chains of one-loop diagrams with Π appearing many times in the off–shell gluon propagator. This series of Feynman diagrams is identical to the one we sum for the mass renormalization in Eq. (2.19). It means we replace the off–shell gluon propagator by

2.2 Ultraviolet Divergences

$$\frac{T^{\mu\nu}}{p^2} \to \frac{T^{\mu\nu}}{p^2} + \left(\frac{T}{p^2} \cdot (-T\,\Pi) \cdot \frac{T}{p^2}\right)^{\mu\nu}$$

$$+ \left(\frac{T}{p^2} \cdot (-T\,\Pi) \cdot \frac{T}{p^2} \cdot (-T\,\Pi) \cdot \frac{T}{p^2}\right)^{\mu\nu} + \cdots$$

$$= \frac{T^{\mu\nu}}{p^2} \sum_{j=0}^{\infty} \left(-\frac{\Pi}{p^2}\right)^j = \frac{T^{\mu\nu}}{p^2} \frac{1}{1 + \Pi/p^2}, \qquad (2.70)$$

schematically written without the factors i. To avoid indices we abbreviate $T^{\mu\nu} T_\nu^\rho = T \cdot T$ which make sense because of $(T \cdot T \cdot T)^{\mu\nu} = T^{\mu\rho} T_\rho^\sigma T_\sigma^\nu = T^{\mu\nu}$. This resummation of the logarithm which appears in the next–to–leading order corrections to α_s moves the finite shift in α_s shown in Eqs. (2.63) and (2.67) into the denominator, while we assume that the pole will be properly taken care off in any of the schemes we discuss

$$\alpha_s^{\text{bare}} = \alpha_s(M^2) - \frac{\alpha_s^2 b_0}{\tilde{\epsilon}} \equiv \frac{\alpha_s(p^2)}{1 - \alpha_s(p^2)\, b_0\, \log\dfrac{p^2}{M^2}} - \frac{\alpha_s^2 b_0}{\tilde{\epsilon}}. \qquad (2.71)$$

Just as in the case without resummation, we can use this complete formula to relate the values of α_s at two reference points, i.e. we consider it a *renormalization group equation* (RGE) which evolves physical parameters from one scale to another in analogy to the fixed order version in Eq. (2.68)

$$\frac{1}{\alpha_s(M^2)} = \frac{1}{\alpha_s(p^2)}\left(1 - \alpha_s(p^2)\, b_0\, \log\frac{p^2}{M^2}\right) = \frac{1}{\alpha_s(p^2)} - b_0\, \log\frac{p^2}{M^2} + \mathcal{O}(\alpha_s). \qquad (2.72)$$

The factor α_s inside the parentheses we can again evaluate at either of the two scales, the difference is a higher order effect. If we keep it at p^2 we see that the expression in Eq. (2.72) is different from the un-resummed version in Eq. (2.67). If we ignore this higher order effect the two formulas become equivalent after switching p^2 and M^2. Resumming the vacuum expectation bubbles only differs from the un-resummed result once we include some next–to–leading order contribution. When we differentiate $\alpha_s(p^2)$ with respect to the momentum transfer p^2 we find, using the relation $d/dx(1/\alpha_s) = -1/\alpha_s^2\, d\alpha_s/dx$

$$\frac{1}{\alpha_s} \frac{d\alpha_s}{d\log p^2} = -\alpha_s \frac{d}{d\log p^2}\frac{1}{\alpha_s} = -\alpha_s\, b_0 + \mathcal{O}(\alpha_s^2) \qquad \text{or}$$

$$\boxed{p^2 \frac{d\alpha_s}{dp^2} \equiv \frac{d\alpha_s}{d\log p^2} = \beta = -\alpha_s^2 \sum_{n=0} b_n \alpha_s^n}. \qquad (2.73)$$

This is the famous running of the strong coupling constant including all higher order terms b_n.

In the running of the strong coupling constant we relate the different values of α_s through multiplicative factors of the kind

$$\left(1 \pm \alpha_s(p^2) b_0 \log \frac{p^2}{M^2}\right). \tag{2.74}$$

Such factors appear in the un-resummed computation of Eq. (2.68) as well as in Eq. (2.71) after resummation. Because they are multiplicative, these factors can move into the denominator, where we need to ensure that they do not vanish. Dependent on the sign of b_0 this becomes a problem for large scale ratios $|\alpha_s \log p^2/M^2| > 1$, where it leads to the Landau pole. We discuss it in detail for the Higgs self coupling in Sect. 1.2.4. For the strong coupling with $b_0 > 0$ and large coupling values at small scales $p^2 \ll M^2$ the combination $(1 + \alpha_s b_0 \log p^2/M^2)$ can indeed vanish and become a problem. For the opposite case of a coupling with $b_0 < 0$ and large coupling values at large scales $p^2 \gg M^2$ the same combination $(1 + \alpha_s b_0 \log p^2/M^2)$ crosses zero.

It is customary to replace the renormalization point of α_s in Eq. (2.71) with a reference scale defined by the Landau pole. At one loop order this reads

$$1 + \alpha_s b_0 \log \frac{\Lambda_{QCD}^2}{M^2} \stackrel{!}{=} 0 \quad \Leftrightarrow \quad \log \frac{\Lambda_{QCD}^2}{M^2} = -\frac{1}{\alpha_s(M^2) b_0}$$

$$\Leftrightarrow \log \frac{p^2}{M^2} = \log \frac{p^2}{\Lambda_{QCD}^2} - \frac{1}{\alpha_s(M^2) b_0}$$

$$\frac{1}{\alpha_s(p^2)} \stackrel{\text{Eq. (2.72)}}{=} \frac{1}{\alpha_s(M^2)} + b_0 \log \frac{p^2}{M^2} \tag{2.75}$$

$$= \frac{1}{\alpha_s(M^2)} + b_0 \log \frac{p^2}{\Lambda_{QCD}^2} - \frac{1}{\alpha_s(M^2)} = b_0 \log \frac{p^2}{\Lambda_{QCD}^2}$$

$$\Leftrightarrow \boxed{\alpha_s(p^2) = \frac{1}{b_0 \log \frac{p^2}{\Lambda_{QCD}^2}}}.$$

This scheme can be generalized to any order in perturbative QCD and is not that different from the Thomson limit renormalization scheme of QED, except that with the introduction of Λ_{QCD} we are choosing a reference point which is particularly hard to compute perturbatively. One thing that is interesting in the way we introduce Λ_{QCD} is the fact that we introduce a scale into our theory without ever setting it. All we did was renormalize a coupling which becomes strong at large energies and search for the mass scale of this strong interaction. This trick is called *dimensional transmutation*.

2.2 Ultraviolet Divergences

In terms of language, there is a little bit of *confusion* between field theorists and phenomenologists: up to now we have introduced the renormalization scale μ_R as the renormalization point, for example of the strong coupling constant. In the $\overline{\text{MS}}$ scheme, the subtraction of $1/\tilde{\epsilon}$ shifts the scale dependence of the strong coupling to M^2 and moves the logarithm $\log M^2/\Lambda_{\text{QCD}}^2$ into the definition of the renormalized parameter. This is what we will from now on call the renormalization scale in the phenomenological sense, i.e. the argument we evaluate α_s at. Throughout this section we will keep the symbol M for this renormalization scale in the $\overline{\text{MS}}$ scheme, but from Sect. 2.3 on we will shift back to μ_R instead of M as the argument of the running coupling, to be consistent with the literature.

2.2.3 Resumming Scaling Logarithms

In the last Sect. 2.2.2 we have introduced the running strong coupling in a fairly abstract manner. For example, we did not link the resummation of diagrams and the running of α_s in Eqs. (2.68) and (2.73) to physics. In what way does the resummation of the one-loop diagrams for the s-channel gluon improve our prediction of the bottom pair production rate at the LHC?

To illustrate those effects we best look at a simple observable which depends on just one energy scale p^2. The first observable coming to mind is again the Drell–Yan cross section $\sigma(q\bar{q} \to \mu^+\mu^-)$, but since we are not really sure what to do with the parton densities which are included in the actual hadronic observable, we better use an observable at an e^+e^- collider. Something that will work and includes α_s at least in the one-loop corrections is the R parameter defined in Eq. (2.12)

$$R = \frac{\sigma(e^+e^- \to \text{hadrons})}{\sigma(e^+e^- \to \mu^+\mu^-)} = N_c \sum_{\text{quarks}} Q_q^2 = \frac{11 N_c}{9}. \tag{2.76}$$

The numerical value at leading order assumes five quarks. Including higher order corrections we can express the result in a power series in the renormalized strong coupling α_s. In the $\overline{\text{MS}}$ scheme we subtract $1/\tilde{\epsilon}(\mu_R/M)$ and in general include a scale dependence on M in the individual prefactors r_n

$$R\left(\frac{p^2}{M^2}, \alpha_s\right) = \sum_{n=0}^{\infty} r_n\left(\frac{p^2}{M^2}\right) \alpha_s^n(M^2) \qquad r_0 = \frac{11 N_c}{9}. \tag{2.77}$$

The r_n we can assume to be dimensionless—if they are not, we can scale R appropriately using p^2. This implies that the r_n only depend on ratios of two scales, the externally fixed p^2 on the one hand and the artificial M^2 on the other.

At the same time we know that R is an observable, which means that including all orders in perturbation theory it cannot depend on any artificial scale choice M. Writing this dependence as a total derivative and setting it to zero we find an equation which would be called a *Callan–Symanzik equation* if instead of the running coupling we had included a running mass

$$0 \stackrel{!}{=} M^2 \frac{d}{dM^2} R\left(\frac{p^2}{M^2}, \alpha_s(M^2)\right) = M^2 \left[\frac{\partial}{\partial M^2} + \frac{\partial \alpha_s}{\partial M^2} \frac{\partial}{\partial \alpha_s}\right] R\left(\frac{p^2}{M^2}, \alpha_s\right)$$

$$= \left[M^2 \frac{\partial}{\partial M^2} + \beta \frac{\partial}{\partial \alpha_s}\right] \sum_{n=0} r_n\left(\frac{p^2}{M^2}\right) \alpha_s^n$$

$$= \sum_{n=1} M^2 \frac{\partial r_n}{\partial M^2} \alpha_s^n + \sum_{n=1} \beta\, r_n\, n \alpha_s^{n-1} \qquad \text{with} \quad r_0 = \frac{11 N_c}{9} = \text{const}$$

$$= M^2 \sum_{n=1} \frac{\partial r_n}{\partial M^2} \alpha_s^n - \sum_{n=1} \sum_{m=0} n r_n \alpha_s^{n+m+1} b_m \qquad \text{with} \quad \beta = -\alpha_s^2 \sum_{m=0} b_m \alpha_s^m$$

$$= M^2 \frac{\partial r_1}{\partial M^2} \alpha_s + \left(M^2 \frac{\partial r_2}{\partial M^2} - r_1 b_0\right) \alpha_s^2$$

$$+ \left(M^2 \frac{\partial r_3}{\partial M^2} - 2 r_2 b_0 - r_1 b_1\right) \alpha_s^3 + \mathcal{O}(\alpha_s^4) \,. \tag{2.78}$$

In the second line we have to remember that the M dependence of α_s is already included in the appearance of β, so α_s should be considered a variable by itself. This perturbative series in α_s has to vanish in each order of perturbation theory. The non-trivial structure, namely the mix of r_n derivatives and the perturbative terms in the β function we can read off the α_s^3 term in Eq. (2.78): first, we have the appropriate NNNLO corrections r_3. Next, we have one loop in the gluon propagator b_0 and two loops for example in the vertex r_2. And finally, we need the two-loop diagram for the gluon propagator b_1 and a one-loop vertex correction r_1. The kind–of–Callan–Symanzik equation (2.78) requires

$$\frac{\partial r_1}{\partial \log M^2/p^2} = 0$$

$$\frac{\partial r_2}{\partial \log M^2/p^2} = r_1 b_0$$

$$\frac{\partial r_3}{\partial \log M^2/p^2} = r_1 b_1 + 2 r_2(M^2) b_0$$

$$\cdots \tag{2.79}$$

The dependence on the argument M^2 vanishes for r_0 and r_1. Keeping in mind that there will be integration constants c_n and that another, in our case, unique momentum scale p^2 has to cancel the mass units inside $\log M^2$ we find

2.2 Ultraviolet Divergences

$$r_0 = c_0 = \frac{11 N_c}{9}$$

$$r_1 = c_1$$

$$r_2 = c_2 + r_1 b_0 \log \frac{M^2}{p^2} = c_2 + c_1 b_0 \log \frac{M^2}{p^2}$$

$$r_3 = \int d \log \frac{M'^2}{p^2} \left(c_1 b_1 + 2 \left(c_2 + c_1 b_0 \log \frac{M'^2}{p^2} \right) b_0 \right)$$

$$= c_3 + (c_1 b_1 + 2 c_2 b_0) \log \frac{M^2}{p^2} + c_1 b_0^2 \log^2 \frac{M^2}{p^2}$$

$$\cdots \tag{2.80}$$

This chain of r_n values looks like we should interpret the apparent fixed-order perturbative series for R in Eq. (2.77) as a series which implicitly includes terms of the order $\log^{n-1} M^2/p^2$ in each r_n. They can become problematic if this logarithm becomes large enough to spoil the fast convergence in terms of $\alpha_s \sim 0.1$, evaluating the observable R at scales far away from the scale choice for the strong coupling constant M.

Instead of the series in r_n we can use the conditions in Eq. (2.80) to express R in terms of the c_n and collect the logarithms appearing with each c_n. The geometric series we then resum to

$$R = \sum_n r_n \left(\frac{p^2}{M^2} \right) \alpha_s^n(M^2)$$

$$= c_0 + c_1 \left(1 + \alpha_s(M^2) b_0 \log \frac{M^2}{p^2} + \alpha_s^2(M^2) b_0^2 \log^2 \frac{M^2}{p^2} + \cdots \right) \alpha_s(M^2)$$

$$+ c_2 \left(1 + 2 \alpha_s(M^2) b_0 \log \frac{M^2}{p^2} + \cdots \right) \alpha_s^2(M^2) + \cdots$$

$$= c_0 + c_1 \frac{\alpha_s(M^2)}{1 - \alpha_s(M^2) b_0 \log \frac{M^2}{p^2}} + c_2 \left(\frac{\alpha_s(M^2)}{1 - \alpha_s(M^2) b_0 \log \frac{M^2}{p^2}} \right)^2 + \cdots$$

$$\equiv \sum c_n \alpha_s^n(p^2) . \tag{2.81}$$

In the original ansatz α_s is always evaluated at the scale M^2. In the last step we use Eq. (2.72) with flipped arguments p^2 and M^2, derived from the resummation of the vacuum polarization bubbles. In contrast to the r_n integration constants the c_n are by definition independent of p^2/M^2 and therefore more suitable as a perturbative series in the presence of potentially large logarithms. Note that the un-resummed version of the running coupling in Eq. (2.67) would not give the correct result, so Eq. (2.81) only holds for resummed vacuum polarization bubbles.

This re-organization of the perturbation series for R can be interpreted as *resumming all logarithms* of the kind $\log M^2/p^2$ in the new organization of the perturbative series and absorbing them into the running strong coupling evaluated at the scale p^2. All scale dependence in the perturbative series for the dimensionless observable R is moved into α_s, so possibly large logarithms $\log M^2/p^2$ have disappeared. In Eq. (2.81) we also see that this series in c_n will never lead to a scale-invariant result when we include a finite order in perturbation theory. Some higher–order factors c_n are known, for example inserting $N_c = 3$ and five quark flavors just as we assume in Eq. (2.76)

$$R = \frac{11}{3}\left(1 + \frac{\alpha_s(p^2)}{\pi} + 1.4\left(\frac{\alpha_s(p^2)}{\pi}\right)^2 - 12\left(\frac{\alpha_s(p^2)}{\pi}\right)^3 + \mathcal{O}\left(\frac{\alpha_s(p^2)}{\pi}\right)^4\right). \tag{2.82}$$

This alternating series with increasing perturbative prefactors seems to indicate the asymptotic instead of convergent behavior of perturbative QCD. At the bottom mass scale the relevant coupling factor is only $\alpha_s(m_b^2)/\pi \sim 1/14$, so a further increase of the c_n would become dangerous. However, a detailed look into the calculation shows that the dominant contributions to c_n arise from the analytic continuation of logarithms, which are large finite terms for example from $\text{Re}(\log^2(-E^2)) = \log^2 E^2 + \pi^2$. In the literature such π^2 terms arising from the analytic continuation of loop integrals are often phrased in terms of $\zeta_2 = \pi^2/6$.

Before moving on we collect the logic of the argument given in this section: when we regularize an ultraviolet divergence we automatically introduce a reference scale μ_R. Naively, this could be an ultraviolet cutoff scale, but even the seemingly scale invariant dimensional regularization in the conformal limit of our field theory cannot avoid the introduction of a scale. There are several ways of dealing with such a scale: first, we can renormalize our parameter at a reference point. Secondly, we can define a running parameter and this way absorb the scale logarithm into the $\overline{\text{MS}}$ counter term. In that case introducing Λ_{QCD} leaves us with a compact form of the running coupling $\alpha_s(M^2; \Lambda_{\text{QCD}})$.

Strictly speaking, at each order in perturbation theory the scale dependence should vanish together with the ultraviolet poles, as long as there is only one scale affecting a given observable. However, defining the running strong coupling we sum one-loop vacuum polarization graphs. Even when we compute an observable at a given loop order, we implicitly include higher order contributions. They lead to a dependence of our perturbative result on the artificial scale M^2, which phenomenologists refer to as renormalization scale dependence.

Using the R ratio we see what our definition of the running coupling means in terms of resumming logarithms: reorganizing our perturbative series to get rid of the ultraviolet divergence $\alpha_s(p^2)$ resums the scale logarithms $\log p^2/M^2$ to all orders in perturbation theory. We will need this picture once we introduce infrared divergences in the following section.

2.3 Infrared Divergences

After this brief excursion into ultraviolet divergences and renormalization we can return to the original example, the Drell–Yan process. Last, we wrote down the hadronic cross sections in terms of parton distributions at leading order in Eq. (2.39). At this stage parton distributions (pdfs) in the proton are only functions of the collinear momentum fraction of the partons inside the proton about which from a theory point of view we only know a set of sum rules.

The perturbative question we need to ask for $\mu^+\mu^-$ production at the LHC is: what happens if together with the two leptons we produce additional jets which for one reason or another we do not observe in the detector. Such jets could for example come from the radiation of a gluon from the initial-state quarks. In Sect. 2.3.1 we will study the kinematics of radiating such jets and specify the infrared divergences this leads to. In Sects. 2.3.2 and 2.3.3 we will show that these divergences have a generic structure and can be absorbed into a re-definition of the parton densities, similar to an ultraviolet renormalization of a Lagrangian parameter. In Sects. 2.3.4 and 2.3.5 we will again follow the example of the ultraviolet divergences and specify what absorbing these divergences means in terms logarithms appearing in QCD calculations.

Throughout this writeup we will use the terms *jets and final state partons* synonymously. This is not really correct once we include jet algorithms and hadronization. On the other hand, in Sect. 3.1.2 we will see that the purpose of a jet algorithm is to take us from some kind of energy deposition in the calorimeter to the parton radiated in the hard process. The two should therefore be closely related.

2.3.1 Single Jet Radiation

Let us get back to the radiation of additional partons in the Drell–Yan process. We can start for example by computing the cross section for the partonic process $q\bar{q} \to Zg$. However, this partonic process involves renormalization of ultraviolet divergences as well as loop diagrams which we have to include before we can say anything reasonable, i.e. ultraviolet and infrared finite.

To make life easier and still learn about the structure of collinear infrared divergences we instead look at the crossed process

It should behave similar to any other $(2 \to 2)$ jet radiation, except that it has a different incoming state than the leading order Drell–Yan process and hence does not involve virtual corrections. This means we do not have to deal with ultraviolet divergences and renormalization, and can concentrate on parton or jet radiation from the initial state. Moreover, let us go back to Z production instead of a photon, to avoid confusion with additional massless particles in the final state.

The amplitude for this $(2 \to 2)$ process is—modulo charges and averaging factors, but including all Mandelstam variables

$$|\mathcal{M}|^2 \sim -\frac{t}{s} - \frac{s^2 - 2m_Z^2(s+t-m_Z^2)}{st} . \tag{2.83}$$

As discussed in Sect. 2.1.1, the Mandelstam variable t for one massless final–state particle can be expressed as $t = -s(1-\tau)y$ in terms of the rescaled gluon emission angle $y = (1-\cos\theta)/2$ and $\tau = m_Z^2/s$. Similarly, we obtain $u = -s(1-\tau)(1-y)$, so as a first check we can confirm that $t + u = -s(1-\tau) = -s + m_Z^2$. The collinear limit when the gluon is radiated in the beam direction is given by $y \to 0$, corresponding to negative $t \to 0$ with finite $u = -s + m_Z^2$. In this limit the matrix element can also be written as

$$|\mathcal{M}|^2 \sim \frac{s^2 - 2sm_Z^2 + 2m_Z^4}{s(s-m_Z^2)} \frac{1}{y} + \mathcal{O}(y) . \tag{2.84}$$

This expression is divergent for collinear gluon radiation or gluon splitting, i.e. for small angles y. We can translate this $1/y$ divergence for example into the transverse momentum of the gluon or Z

$$sp_T^2 = tu = s^2(1-\tau)^2 \, y(1-y) = (s-m_Z^2)^2 y + \mathcal{O}(y^2) \tag{2.85}$$

In the collinear limit our matrix element squared in Eq. (2.84) becomes

$$\boxed{|\mathcal{M}|^2 \sim \frac{s^2 - 2sm_Z^2 + 2m_Z^4}{s^2} \frac{(s-m_Z^2)}{p_T^2} + \mathcal{O}(p_T^0)} . \tag{2.86}$$

The matrix element for the tree level process $qg \to Zq$ has a leading divergence proportional to $1/p_T^2$. To compute the total cross section for this process we need to integrate the matrix element over the entire two-particle phase space. Starting from Eq. (2.41) and using the appropriate Jacobian this integration can be written in terms of the reduced angle y. Approximating the matrix element as C'/y or C/p_T^2, we then integrate

2.3 Infrared Divergences

$$\int_{y^{\min}}^{y^{\max}} dy \frac{C'}{y} = \int_{p_T^{\min}}^{p_T^{\max}} dp_T^2 \frac{C}{p_T^2} = 2\int_{p_T^{\min}}^{p_T^{\max}} dp_T\, p_T\, \frac{C}{p_T^2} \simeq 2C \int_{p_T^{\min}}^{p_T^{\max}} dp_T \frac{1}{p_T}$$

$$= 2C \, \log \frac{p_T^{\max}}{p_T^{\min}} \qquad (2.87)$$

The form C/p_T^2 for the matrix element is of course only valid in the collinear limit; in the non–collinear phase space C is not a constant. However, Eq. (2.87) describes well the collinear divergence arising from quark radiation at the LHC.

Next, we follow the same strategy as for the ultraviolet divergence. First, we regularize the divergence for example using dimensional regularization. Then, we find a well–defined way to get rid of it. Dimensional regularization means writing the two-particle phase space in $n = 4 - 2\epsilon$ dimensions. Just for reference, the complete formula in terms of the angular variable y reads

$$s\,\frac{d\sigma}{dy} = \frac{\pi(4\pi)^{-2+\epsilon}}{\Gamma(1-\epsilon)} \left(\frac{\mu_F^2}{m_Z^2}\right)^\epsilon \frac{\tau^\epsilon (1-\tau)^{1-2\epsilon}}{y^\epsilon(1-y)^\epsilon} |\mathcal{M}|^2 \sim \left(\frac{\mu_F^2}{m_Z^2}\right)^\epsilon \frac{|\mathcal{M}|^2}{y^\epsilon(1-y)^\epsilon}. \qquad (2.88)$$

In the second step we only keep the factors we are interested in. The additional factor $1/y^\epsilon$ regularizes the integral at $y \to 0$, as long as $\epsilon < 0$ by slightly increasing the suppression of the integrand in the infrared regime. This means that for infrared divergences we can as well choose $n = 4 + 2\epsilon$ space–time dimensions with $\epsilon > 0$. After integrating the leading collinear divergence $1/y^{1+\epsilon}$ we are left with a pole $1/(-\epsilon)$. This regularization procedure is symmetric in $y \leftrightarrow (1-y)$. What is important to notice is again the appearance of a scale $\mu_F^{2\epsilon}$ with the n-dimensional integral. This scale arises from the infrared regularization of the phase space integral and is referred to as *factorization scale*. The actual removal of the infrared pole—corresponding to the renormalization in the ultraviolet case—is called *mass factorization* and works exactly the same way as renormalizing a parameter: in a well–defined scheme we simply subtract the pole from the fixed-order matrix element squared.

2.3.2 Parton Splitting

From the discussion of the process $qg \to Zq$ we can at least hope that after taking care of all other infrared and ultraviolet divergences the collinear structure of the process $q\bar{q} \to Zg$ will be similar. In this section we will show that we can indeed write all collinear divergences in a universal form, independent of the hard process which we choose as the Drell–Yan process. In the collinear limit, the radiation of additional partons or the splitting into additional partons will be described by universal *splitting functions*.

Fig. 2.1 Splitting of one gluon into two gluons (Figure from Ref. [7])

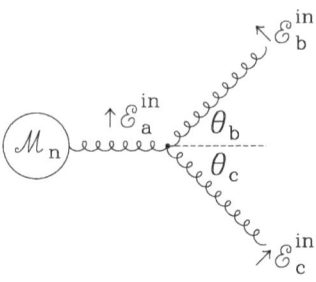

Infrared divergences occur for massless particles in the initial or final state, so we need to go through all ways incoming or outgoing gluons and quark can split into each other. The description of the factorized phase space, with which we will start, is common to all these different channels. The first and at the LHC most important case is the splitting of one gluon into two, shown in Fig. 2.1. The two daughter gluons are close to mass shell while the mother has to have a finite positive invariant mass $p_a^2 \gg p_b^2, p_c^2$. We again assign the direction of the momenta as $p_a = -p_b - p_c$, which means we have to take care of minus signs in the particle energies. The kinematics of this approximately collinear process we can describe in terms of the energy fractions z and $1-z$ defined as

$$z = \frac{|E_b|}{|E_a|} = 1 - \frac{|E_c|}{|E_a|} \qquad p_a^2 = (-p_b - p_c)^2 = 2(p_b p_c) = 2z(1-z)(1-\cos\theta)$$

$$E_a^2 = z(1-z)E_a^2\theta^2 + \mathcal{O}(\theta^4)$$

$$\Leftrightarrow \qquad \theta \equiv \theta_b + \theta_c \simeq \frac{1}{|E_a|}\sqrt{\frac{p_a^2}{z(1-z)}}, \qquad (2.89)$$

in the collinear limit and in terms of the opening angle θ between $\vec{p}_b$ and $\vec{p}_c$. Because $p_a^2 > 0$ we call this final-state splitting configuration *time-like branching*. For this configuration we can write down the so-called *Sudakov decomposition* of the four-momenta

$$-p_a = p_b + p_c = (-zp_a + \beta n + p_T) + (-(1-z)p_a - \beta n - p_T). \qquad (2.90)$$

It defines an arbitrary unit four-vector n, a component orthogonal to the mother momentum and n, i.e. $p_a (p_a p_T) = 0 = (np_T)$, and a free factor β. This way, we can specify n such that it defines the direction of the p_b–p_c decay plane. In this decomposition we can set only one invariant mass to zero, for example that of a radiated gluon $p_c^2 = 0$. The second final state will have a finite invariant mass $p_b^2 \neq 0$.

2.3 Infrared Divergences

As specific choice for the three reference four-vectors is

$$p_a = \begin{pmatrix} |E_a| \\ 0 \\ 0 \\ p_{a,3} \end{pmatrix} = |E_a| \begin{pmatrix} 1 \\ 0 \\ 0 \\ 1 + \mathcal{O}(\theta) \end{pmatrix} \quad n = \begin{pmatrix} 1 \\ 0 \\ 0 \\ -1 \end{pmatrix} \quad p_T = \begin{pmatrix} 0 \\ p_{T,1} \\ p_{T,2} \\ 0 \end{pmatrix}. \quad (2.91)$$

Relative to $\vec{p}_a$ we can split the opening angle θ for massless partons according to Fig. 2.1

$$\theta = \theta_b + \theta_c \quad \text{and} \quad \frac{\theta_b}{\theta_c} = \frac{p_T}{|E_b|} \left(\frac{p_T}{|E_c|} \right)^{-1} = \frac{1-z}{z} \quad \Leftrightarrow \quad \theta = \frac{\theta_b}{1-z} = \frac{\theta_c}{z}. \quad (2.92)$$

The momentum choice in Eq. (2.91) has the additional feature that $n^2 = 0$, which allows us to extract β from the momentum parameterization shown in Eq. (2.90) and the additional condition that $p_c^2 = 0$

$$\begin{aligned}
p_c^2 &= (-(1-z)p_a - \beta n - p_T)^2 \\
&= (1-z)^2 p_a^2 + p_T^2 + 2\beta(1-z)(np_a) \\
&= (1-z)^2 p_a^2 + p_T^2 + 4\beta(1-z)|E_a|(1+\mathcal{O}(\theta)) \stackrel{!}{=} 0 \Leftrightarrow \beta \simeq -\frac{p_T^2 + (1-z)^2 p_a^2}{4(1-z)|E_a|}.
\end{aligned} \quad (2.93)$$

Using this specific phase space parameterization we can divide an $(n+1)$-particle process into an n-particle process and a *splitting process* of quarks and gluons. First, this requires us to split the $(n+1)$-particle phase space alone into an n-particle phase space and the collinear splitting. The general $(n+1)$-particle phase space separating off the n-particle contribution

$$\begin{aligned}
d\Phi_{n+1} &= \cdots \frac{d^3 \vec{p}_b}{2(2\pi)^3 |E_b|} \frac{d^3 \vec{p}_c}{2(2\pi)^3 |E_c|} \\
&= \cdots \frac{d^3 \vec{p}_a}{2(2\pi)^3 |E_a|} \frac{d^3 \vec{p}_c}{2(2\pi)^3 |E_c|} \frac{|E_a|}{|E_b|} \quad \text{at fixed } p_a \\
&= d\Phi_n \frac{dp_{c,3} dp_T p_T d\phi}{2(2\pi)^3 |E_c|} \frac{1}{z} \\
&= d\Phi_n \frac{dp_{c,3} dp_T^2 d\phi}{4(2\pi)^3 |E_c|} \frac{1}{z}
\end{aligned} \quad (2.94)$$

is best expressed in terms of the energy fraction z and the azimuthal angle ϕ. In other words, separating the $(n+1)$-particle space into an n-particle phase space and a $(1 \to 2)$ splitting phase space is possible without any approximation, and all we have to take care of is the correct prefactors in the new parameterization.

Our next task is to translate the phase space parameters $p_{c,3}$ and p_T^2 appearing in Eq. (2.94) into z and p_a^2. Starting from Eq. (2.90) for $p_{c,3}$ with the third components of p_a and p_T given by Eq. (2.91) we insert β from Eq. (2.93) and obtain

$$\begin{aligned}
\frac{dp_{c,3}}{dz} &= \frac{d}{dz}\left[-(1-z)|E_a|(1+\mathcal{O}(\theta)) + \beta\right] \\
&= \frac{d}{dz}\left[-(1-z)|E_a|(1+\mathcal{O}(\theta)) - \frac{p_T^2 + (1-z)^2 p_a^2}{4(1-z)|E_a|}\right] \\
&= |E_a|(1+\mathcal{O}(\theta)) - \frac{p_T^2}{4(1-z)^2 E_a} + \frac{p_a^2}{4|E_a|} \\
&= \frac{|E_c|}{1-z}(1+\mathcal{O}(\theta)) - \frac{\theta^2 z^2 E_c^2}{4(1-z)^2 E_a} + \frac{z(1-z)E_a^2\theta^2 + \mathcal{O}(\theta^4)}{4|E_a|}
\end{aligned}$$

$$\text{using Eqs. (2.89) and (2.92)}$$

$$= \frac{|E_c|}{1-z} + \mathcal{O}(\theta) \quad \Leftrightarrow \quad \frac{dp_{c,3}}{|E_c|} \simeq \frac{dz}{1-z}. \tag{2.95}$$

In addition to substituting $dp_{c,3}$ by dz in Eq. (2.94) we also replace dp_T^2 with dp_a^2 according to

$$\frac{p_T^2}{p_a^2} = \frac{E_b^2 \theta_b^2}{z(1-z)E_a^2\theta^2} = \frac{z^2 E_a^2(1-z)^2\theta^2}{z(1-z)E_a^2\theta^2} = z(1-z) \quad \Leftrightarrow \quad dp_T^2 = z(1-z)dp_a^2. \tag{2.96}$$

This gives us the final result for the separated *collinear phase space*

$$\boxed{d\Phi_{n+1} = d\Phi_n \, \frac{dz\, dp_a^2\, d\phi}{4(2\pi)^3} = d\Phi_n \, \frac{dz\, dp_a^2}{4(2\pi)^2}}, \tag{2.97}$$

where in the second step we assume an azimuthal symmetry.

Adding the transition matrix elements to this factorization of the phase space and ignoring the initial–state flux factor which is common to both processes we can now postulate a full factorization for one collinear emission and in the collinear approximation

2.3 Infrared Divergences

$$d\sigma_{n+1} = \overline{|\mathscr{M}_{n+1}|^2}\, d\Phi_{n+1}$$

$$= \overline{|\mathscr{M}_{n+1}|^2}\, d\Phi_n \frac{dp_a^2\, dz}{4(2\pi)^2}\, (1 + \mathscr{O}(\theta))$$

$$\simeq \frac{2g_s^2}{p_a^2} \hat{P}(z)\, \overline{|\mathscr{M}_n|^2}\, d\Phi_n \frac{dp_a^2\, dz}{16\pi^2} \quad \text{assuming} \quad \boxed{\overline{|\mathscr{M}_{n+1}|^2} \simeq \frac{2g_s^2}{p_a^2} \hat{P}(z)\, \overline{|\mathscr{M}_n|^2}}\,.$$

(2.98)

This last step is an assumption. We will proceed to show it step by step by constructing the appropriate *splitting kernels* $\hat{P}(z)$ for all different quark and gluon configurations. If Eq. (2.98) holds true this means that we can compute the $(n+1)$ particle amplitude squared from the n-particle case convoluted with the appropriate splitting kernel. Using $d\sigma_n \sim \overline{|\mathscr{M}_n|^2}\, d\Phi_n$ and $g_s^2 = 4\pi\alpha_s$ we can write this relation in its most common form

$$\boxed{\sigma_{n+1} \simeq \int \sigma_n\, \frac{dp_a^2}{p_a^2}\, dz\, \frac{\alpha_s}{2\pi}\, \hat{P}(z)}\,. \tag{2.99}$$

Reminding ourselves that relations of the kind $\overline{|\mathscr{M}_{n+1}|^2} = p\overline{|\mathscr{M}_n|^2}$ can typically be summed, for example for the case of successive soft photon radiation in QED, we see that Eq. (2.99) is not the final answer. It does not include the necessary phase space factor $1/n!$ from identical bosons in the final state which leads to the simple exponentiation.

As the first parton splitting in QCD we study a *gluon splitting into two gluons*, shown in Fig. 2.1. To compute its transition amplitude we write down all gluon momenta and polarizations in a specific frame. With respect to the scattering plane opened by $\vec{p}_b$ and $\vec{p}_c$ all three gluons have two transverse polarizations, one in the plane, $\epsilon^{\|}$, and one perpendicular to it, $\epsilon^{\perp}$. In the limit of small scattering angles, the three parallel as well as the three perpendicular polarization vectors are aligned. The perpendicular polarizations are also orthogonal to all three gluon momenta. The physical transverse polarizations in the plane are orthogonal to their corresponding momenta and only approximately orthogonal to the other momenta. Altogether, this means for the three-vectors $\epsilon^{\|}$ and $\epsilon^{\perp}$

$$(\epsilon_i^{\|}\epsilon_j^{\|}) = -1 + \mathscr{O}(\theta) \quad (\epsilon_i^{\perp}\epsilon_j^{\perp}) = -1 \quad (\epsilon_i^{\perp}\epsilon_j^{\|}) = 0 \quad (\epsilon_i^{\perp}p_j) = 0 \quad (\epsilon_j^{\|}p_j) = 0\,, \tag{2.100}$$

with general $i \neq j$. For $i = j$ we find exactly one and zero. Using these kinematic relations we can tackle the splitting amplitude $g \to gg$. It is proportional to the vertex V_{ggg} which in terms of all incoming momenta reads

$$\begin{aligned}V_{ggg} &= ig_s f^{abc}\, \epsilon_a^\alpha \epsilon_b^\beta \epsilon_c^\gamma \left[g_{\alpha\beta}(p_a - p_b)_\gamma \right.\\
&\quad \left. + g_{\beta\gamma}(p_b - p_c)_\alpha + g_{\gamma\alpha}(p_c - p_a)_\beta \right] \\
&= ig_s f^{abc}\, \epsilon_a^\alpha \epsilon_b^\beta \epsilon_c^\gamma \left[g_{\alpha\beta}(-p_c - 2p_b)_\gamma \right.\\
&\quad \left. + g_{\beta\gamma}(p_b - p_c)_\alpha + g_{\gamma\alpha}(2p_c + p_b)_\beta \right] \qquad \text{with } p_a = -p_b - p_c \\
&= ig_s f^{abc}\, [-2(\epsilon_a \epsilon_b)(\epsilon_c p_b) + (\epsilon_b \epsilon_c)(\epsilon_a p_b) \\
&\quad - (\epsilon_b \epsilon_c)(\epsilon_a p_c) + 2(\epsilon_c \epsilon_a)(\epsilon_b p_c)] \qquad \text{with } (\epsilon_j p_j) = 0 \\
&= -2ig_s f^{abc}\, [(\epsilon_a \epsilon_b)(\epsilon_c p_b) - (\epsilon_b \epsilon_c)(\epsilon_a p_b) \\
&\quad - (\epsilon_c \epsilon_a)(\epsilon_b p_c)] \qquad \text{with } (\epsilon_a p_c) = -(\epsilon_a p_b) \\
&= -2ig_s f^{abc}\, \left[(\epsilon_a \epsilon_b)(\epsilon_c^\| p_b) - (\epsilon_b \epsilon_c)(\epsilon_a^\| p_b) \right.\\
&\quad \left. - (\epsilon_c \epsilon_a)(\epsilon_b^\| p_c)\right] \qquad \text{with } (\epsilon_i^\perp p_j) = 0\,.
\end{aligned}$$
(2.101)

Squaring the splitting matrix element to compute the $(n+1)$ and n particle matrix elements squared for the unpolarized case gives us

$$\begin{aligned}\overline{|\mathcal{M}_{n+1}|^2} &= \frac{1}{2}\left(\frac{1}{p_a^2}\right)^2 4g_s^2 \frac{1}{N_c^2 - 1}\frac{1}{N_a} \sum_{\text{color,pols}} \left[\sum_{\text{3 terms}} \pm f^{abc}(\epsilon\cdot\epsilon)(\epsilon\cdot p)\right]^2 \overline{|\mathcal{M}_n|^2} \\
&= \frac{2g_s^2}{p_a^4}\frac{1}{2(N_c^2 - 1)} \sum_{\text{color,pols}}\left[\sum_{\text{3 terms}} \pm f^{abc}(\epsilon\cdot\epsilon)(\epsilon\cdot p)\right]^2 \overline{|\mathcal{M}_n|^2}\,,
\end{aligned}$$
(2.102)

where the sums runs over all color and polarizations and over the three terms in the brackets of Eq. (2.101). The factor $1/2$ in the first line takes into account that for two final–state gluons the $(n+1)$-particle phase space is only half its usual size. Because we compute the color factor and spin sum for the decay of gluon a the formula includes averaging factors for the color $(N_c^2 - 1)$ and the polarization $N_a = 2$ of the mother particle.

Inside the color and polarization sum each term $(\epsilon\cdot\epsilon)(\epsilon\cdot p)$ is symmetric in two indices but gets multiplied with the anti–symmetric color factor. This means that the final result will only be finite if we square each term individually as a product of two symmetric and two anti–symmetric terms. In other words, the sum over the external gluons becomes an incoherent polarization sum,

$$\overline{|\mathcal{M}_{n+1}|^2} = \frac{2g_s^2}{p_a^2}\frac{N_c}{2}\sum_{\text{pols}}\left[\sum_{\text{3 terms}}\frac{(\epsilon\cdot\epsilon)^2(\epsilon\cdot p)^2}{p_a^2}\right]\overline{|\mathcal{M}_n|^2}\,, \qquad (2.103)$$

using $f^{abc} f^{abd} g^{cd} = N_c \delta^{cd}\delta^{cd} = N_c(N_c^2 - 1)$.

2.3 Infrared Divergences

Going through all possible combinations we know what can contribute inside the brackets of Eq. (2.101): $(\epsilon_a^\| \epsilon_b^\|)$ as well as $(\epsilon_a^\perp \epsilon_b^\perp)$ can be combined with $(\epsilon_c^\| p_b)$; $(\epsilon_b^\| \epsilon_c^\|)$ or $(\epsilon_b^\perp \epsilon_c^\perp)$ with $(\epsilon_a^\| p_b)$; and last but not least we can combine $(\epsilon_a^\| \epsilon_c^\|)$ and $(\epsilon_a^\perp \epsilon_c^\perp)$ with $(\epsilon_b^\| p_c)$. The finite combinations between polarization vectors and momenta which we appear in Eq. (2.103) are, in terms of z, E_a, and θ

$$(\epsilon_c^\| p_b) = -E_b \cos \angle(\vec{\epsilon}_c^{\,\|}, \vec{p}_b) = -E_b \cos\left(\frac{\pi}{2} - \theta\right) = -E_b \sin\theta \simeq -E_b \theta = -zE_a\theta$$

$$(\epsilon_a^\| p_b) = -E_b \cos \angle(\vec{\epsilon}_a^{\,\|}, \vec{p}_b) = -E_b \cos\left(\frac{\pi}{2} - \theta_b\right)$$
$$= -E_b \sin\theta_b \simeq -E_b \theta_b = -z(1-z)E_a\theta$$

$$(\epsilon_b^\| p_c) = -E_c \cos \angle(\vec{\epsilon}_b^{\,\|}, \vec{p}_c) = -E_c \cos\left(\frac{\pi}{2} - \theta\right)$$
$$= -E_c \sin\theta \simeq -E_c \theta = -(1-z)E_a\theta \ . \qquad (2.104)$$

For the four non–zero combinations of gluon polarizations, the splitting matrix elements ordered still the same way are

ϵ_a	ϵ_b	ϵ_c	$\pm(\epsilon \cdot \epsilon)(\epsilon \cdot p)$	$\dfrac{(\epsilon \cdot \epsilon)^2 (\epsilon \cdot p)^2}{p_a^2} = \dfrac{(\epsilon \cdot \epsilon)^2 (\epsilon \cdot p)^2}{z(1-z)E_a^2 \theta^2}$
$\|$	$\|$	$\|$		
$\perp$	$\perp$	$\|$	$(-1)(-z)E_a\theta$	$\dfrac{z}{1-z}$
$\|$	$\|$	$\|$		
$\|$	$\perp$	$\perp$	$-(-1)(-z)(1-z)E_a\theta$	$z(1-z)$
$\|$	$\|$	$\|$		
$\perp$	$\|$	$\perp$	$-(-1)(-1)(1-z)E_a\theta$	$\dfrac{1-z}{z}$

For the incoherent sum in Eq. (2.103) we find

$$\overline{|\mathcal{M}_{n+1}|^2} = \frac{2g_s^2}{p_a^2} \frac{N_c}{2} \, 2 \left[\frac{z}{1-z} + z(1-z) + \frac{1-z}{z} \right] \overline{|\mathcal{M}_n|^2}$$

$$\equiv \frac{2g_s^2}{p_a^2} \hat{P}_{g \leftarrow g}(z) \, \overline{|\mathcal{M}_n|^2}$$

$$\Leftrightarrow \quad \boxed{\hat{P}_{g \leftarrow g}(z) = C_A \left[\frac{z}{1-z} + \frac{1-z}{z} + z(1-z) \right]} , \qquad (2.105)$$

using $C_A = N_c$. The form of the splitting kernel is symmetric when we exchange the two gluons z and $(1-z)$. It diverges if either of the gluons become soft. The notation $\hat{P}_{i \leftarrow j} \sim \hat{P}_{ij}$ is inspired by a matrix notation which we can use to multiply the splitting matrix from the right with the incoming parton vector to get the final parton vector. Following the logic described above, with this calculation we prove that the factorized form of the $(n + 1)$-particle matrix element squared in Eq. (2.98) holds for gluons only.

The same kind of splitting kernel we can compute for the splitting of a *gluon into two quarks* and the splitting of a *quark into a quark and a gluon*

$$g(p_a) \to q(p_b) + \bar{q}(p_c) \qquad \text{and} \qquad q(p_a) \to q(p_b) + g(p_c) \, . \tag{2.106}$$

Both splittings include the quark–quark–gluon vertex, coupling the gluon current to the quark and antiquark spinors. For small angle scattering we can write the spinors of the massless quark $u(p_b)$ and the massless antiquark $v(p_c)$ in terms of two-component spinors

$$u(p) = \sqrt{E} \begin{pmatrix} \chi_\pm \\ \pm \chi_\pm \end{pmatrix} \qquad \text{with} \quad \chi_+ = \begin{pmatrix} 1 \\ \theta/2 \end{pmatrix} \text{ (spin up)}$$

$$\chi_- = \begin{pmatrix} -\theta/2 \\ 1 \end{pmatrix} \text{ (spin down)} \, . \tag{2.107}$$

For the massless antiquark we need to replace $\theta \to -\theta$ and take into account the different relative spin-momentum directions ($\sigma\hat{p}$, leading to the additional sign in the lower two spinor entries. The antiquark spinors then become

$$v(p) = -i\sqrt{E} \begin{pmatrix} \mp \epsilon \chi_\pm \\ \epsilon \chi_\pm \end{pmatrix} \qquad \text{with} \quad \chi_+ = \begin{pmatrix} 1 \\ -\theta/2 \end{pmatrix} \quad \epsilon\chi_+ = \begin{pmatrix} -\theta/2 \\ -1 \end{pmatrix} \text{ (spin up)}$$

$$\chi_- = \begin{pmatrix} \theta/2 \\ 1 \end{pmatrix} \quad \epsilon\chi_- = \begin{pmatrix} 1 \\ -\theta/2 \end{pmatrix} \text{ (spin down)} \, . \tag{2.108}$$

We again limit our calculations to the leading terms in the small scattering angle θ. In addition to the fermion spinors, for the coupling to a gluonic current we need the *Dirac matrices* which in the Dirac representation are conveniently expressed in terms of the Pauli matrices defined in Eq. (1.8)

$$\gamma^0 = \begin{pmatrix} \mathbb{1} & 0 \\ 0 & -\mathbb{1} \end{pmatrix} \quad \gamma^j = \begin{pmatrix} 0 & \tau^j \\ -\tau^j & 0 \end{pmatrix} \quad \Rightarrow \quad \gamma^0 \gamma^0 = \mathbb{1} \quad \gamma^0 \gamma^j = \begin{pmatrix} 0 & \tau^j \\ \tau^j & 0 \end{pmatrix} \tag{2.109}$$

We are particularly interested in the combination $\gamma^0 \gamma^j$ because of the definition of the conjugated spinor $\bar{u} = u^T \gamma^0$.

2.3 Infrared Divergences

In the notation introduced in Eq. (2.105) we first compute the splitting kernel $\hat{P}_{q \leftarrow g}$, sandwiching the qqg vertex between an outgoing quark $\bar{u}_\pm(p_b)$ and an outgoing antiquark $v_\pm(p_a)$ for all possible spin combinations. We start with all four gluon polarizations, i.e. all four gamma matrices, between two spin-up quarks and their spinors written out in Eqs. (2.107) and (2.108)

$$\frac{\bar{u}_+(p_b)\gamma^0 v_-(p_c)}{-i\sqrt{E_b}\sqrt{E_c}} = \left(1, \frac{\theta_b}{2}, 1, \frac{\theta_b}{2}\right) \begin{pmatrix} 1 \\ 1 \\ 1 \\ 1 \end{pmatrix} \begin{pmatrix} 1 \\ -\theta_c/2 \\ 1 \\ -\theta_c/2 \end{pmatrix}$$

$$= \left(1, \frac{\theta_b}{2}, 1, \frac{\theta_b}{2}\right) \begin{pmatrix} 1 \\ -\theta_c/2 \\ 1 \\ -\theta_c/2 \end{pmatrix} = 2$$

$$\frac{\bar{u}_+(p_b)\gamma^1 v_-(p_c)}{-i\sqrt{E_b}\sqrt{E_c}} = \left(1, \frac{\theta_b}{2}, 1, \frac{\theta_b}{2}\right) \begin{pmatrix} & & & 1 \\ & & 1 & \\ & 1 & & \\ 1 & & & \end{pmatrix} \begin{pmatrix} 1 \\ -\theta_c/2 \\ 1 \\ -\theta_c/2 \end{pmatrix}$$

$$= \left(1, \frac{\theta_b}{2}, 1, \frac{\theta_b}{2}\right) \begin{pmatrix} -\theta_c/2 \\ 1 \\ -\theta_c/2 \\ 1 \end{pmatrix} = \theta_b - \theta_c$$

$$\frac{\bar{u}_+(p_b)\gamma^2 v_-(p_c)}{-i\sqrt{E_b}\sqrt{E_c}} = \left(1, \frac{\theta_b}{2}, 1, \frac{\theta_b}{2}\right) \begin{pmatrix} & & & -i \\ & & i & \\ & -i & & \\ i & & & \end{pmatrix} \begin{pmatrix} 1 \\ -\theta_c/2 \\ 1 \\ -\theta_c/2 \end{pmatrix}$$

$$= i\left(1, \frac{\theta_b}{2}, 1, \frac{\theta_b}{2}\right) \begin{pmatrix} \theta_c/2 \\ 1 \\ \theta_c/2 \\ 1 \end{pmatrix} = i(\theta_b + \theta_c)$$

$$\frac{\bar{u}_+(p_b)\gamma^3 v_-(p_c)}{-i\sqrt{E_b}\sqrt{E_c}} = \left(1, \frac{\theta_b}{2}, 1, \frac{\theta_b}{2}\right) \begin{pmatrix} & & 1 & \\ & & & -1 \\ 1 & & & \\ & -1 & & \end{pmatrix} \begin{pmatrix} 1 \\ -\theta_c/2 \\ 1 \\ -\theta_c/2 \end{pmatrix}$$

$$= \left(1, \frac{\theta_b}{2}, 1, \frac{\theta_b}{2}\right) \begin{pmatrix} 1 \\ \theta_c/2 \\ 1 \\ \theta_c/2 \end{pmatrix} = 2 \ . \tag{2.110}$$

Somewhat surprisingly the unphysical scalar and longitudinal gluon polarizations seem to contribute to this vertex. However, after adding the two unphysical degrees of freedom they cancel because of the form of our metric. Assuming transverse gluons we compute this vertex factor also for the other diagonal spin combination

$$\frac{\bar{u}_-(p_b)\gamma^1 v_+(p_c)}{-i\sqrt{E_b}\sqrt{E_c}} = \left(-\frac{\theta_b}{2}, 1, \frac{\theta_b}{2}, -1\right) \begin{pmatrix} 1 \\ 1 \\ 1 \\ 1 \end{pmatrix} \begin{pmatrix} \theta_c/2 \\ 1 \\ -\theta_c/2 \\ -1 \end{pmatrix}$$

$$= \left(-\frac{\theta_b}{2}, 1, \frac{\theta_b}{2}, -1\right) \begin{pmatrix} -1 \\ -\theta_c/2 \\ 1 \\ \theta_c/2 \end{pmatrix} = \theta_b - \theta_c$$

$$\frac{\bar{u}_-(p_b)\gamma^2 v_+(p_c)}{-i\sqrt{E_b}\sqrt{E_c}} = \left(-\frac{\theta_b}{2}, 1, \frac{\theta_b}{2}, -1\right) \begin{pmatrix} -i \\ i \\ -i \\ i \end{pmatrix} \begin{pmatrix} \theta_c/2 \\ 1 \\ -\theta_c/2 \\ -1 \end{pmatrix}$$

$$= i\left(-\frac{\theta_b}{2}, 1, \frac{\theta_b}{2}, -1\right) \begin{pmatrix} 1 \\ -\theta_c/2 \\ -1 \\ \theta_c/2 \end{pmatrix} = -i(\theta_b + \theta_c) . \quad (2.111)$$

Before collecting the prefactors for this gluon–quark splitting, we also need the same–spin case

$$\frac{\bar{u}_+(p_b)\gamma^1 v_+(p_c)}{-i\sqrt{E_b}\sqrt{E_c}} = \left(1, \frac{\theta_b}{2}, 1, \frac{\theta_b}{2}\right) \begin{pmatrix} 1 \\ 1 \\ 1 \\ 1 \end{pmatrix} \begin{pmatrix} \theta_c/2 \\ 1 \\ -\theta_c/2 \\ -1 \end{pmatrix}$$

$$= \left(1, \frac{\theta_b}{2}, 1, \frac{\theta_b}{2}\right) \begin{pmatrix} -1 \\ -\theta_c/2 \\ 1 \\ \theta_c/2 \end{pmatrix} = 0$$

$$\frac{\bar{u}_+(p_b)\gamma^2 v_+(p_c)}{-i\sqrt{E_b}\sqrt{E_c}} = \left(1, \frac{\theta_b}{2}, 1, \frac{\theta_b}{2}\right) \begin{pmatrix} -i \\ i \\ -i \\ i \end{pmatrix} \begin{pmatrix} \theta_c/2 \\ 1 \\ -\theta_c/2 \\ -1 \end{pmatrix}$$

$$= i\left(1, \frac{\theta_b}{2}, 1, \frac{\theta_b}{2}\right) \begin{pmatrix} 1 \\ -\theta_c/2 \\ -1 \\ \theta_c/2 \end{pmatrix} = 0 , \quad (2.112)$$

2.3 Infrared Divergences

which vanishes. The gluon current can only couple to two fermions via a spin flip. For massless fermions this means that the gluon splitting into two quarks involves two quark spin cases, each of them coupling to two transverse gluon polarizations. Keeping track of all the relevant factors our vertex function for the splitting $g \to q\bar{q}$ becomes for each of the two quark spins

$$V_{qqg} = -ig_s T^a \bar{u}_\pm(p_b) \gamma_\mu \epsilon_a^\mu v_\mp(p_c) \equiv -ig_s T^a \epsilon_a^j F_\pm^{(j)} \quad \text{for} \quad j = 1, 2$$

$$\frac{|F_+^{(1)}|^2}{p_a^2} = \frac{|F_-^{(1)}|^2}{p_a^2} = \frac{E_b E_c (\theta_b - \theta_c)^2}{p_a^2} = \frac{E_a^2 z(1-z)(1-z-z)^2 \theta^2}{E_a^2 z(1-z)\theta^2} = (1-2z)^2$$

$$\frac{|F_+^{(2)}|^2}{p_a^2} = \frac{|F_-^{(2)}|^2}{p_a^2} = \frac{E_b E_c (\theta_b + \theta_c)^2}{p_a^2} = \frac{E_a^2 z(1-z)(1-z+z)^2 \theta^2}{E_a^2 z(1-z)\theta^2} = 1 .$$

(2.113)

We omit irrelevant factors i and (-1) which drop out once we compute the absolute value squared. In complete analogy to the gluon splitting case we can factorize the $(n + 1)$-particle matrix element into

$$\overline{|\mathcal{M}_{n+1}|^2}$$

$$= \left(\frac{1}{p_a^2}\right)^2 g_s^2 \frac{\mathrm{Tr}\, T^a T^a}{N_c^2 - 1} \frac{1}{N_a} \left[|F_+^{(1)}|^2 + |F_-^{(1)}|^2 + |F_+^{(2)}|^2 + |F_-^{(2)}|^2 \right] \overline{|\mathcal{M}_n|^2}$$

$$= \frac{g_s^2}{p_a^2} T_R \frac{N_c^2 - 1}{N_c^2 - 1} \left[(1-2z)^2 + 1 \right] \overline{|\mathcal{M}_n|^2} \qquad \text{with } \mathrm{Tr}\, T^a T^b = T_R \delta^{ab} \text{ and } N_a = 2$$

$$= \frac{2g_s^2}{p_a^2} T_R \left[z^2 + (1-z)^2 \right] \overline{|\mathcal{M}_n|^2}$$

$$\equiv \frac{2g_s^2}{p_a^2} \hat{P}_{q \leftarrow g}(z) \overline{|\mathcal{M}_n|^2}$$

$$\Leftrightarrow \quad \boxed{\hat{P}_{q \leftarrow g}(z) = T_R \left[z^2 + (1-z)^2 \right]} , \qquad (2.114)$$

with $T_R = 1/2$. In the first line we implicitly assume that the internal quark propagator can be written as something like $u\bar{u}/p_a^2$ and we only need to consider the denominator. This splitting kernel is again symmetric in z and $(1 - z)$ because QCD does not distinguish between the outgoing quark and the outgoing antiquark.

The third splitting we compute is gluon radiation off a quark, i.e. $q(p_a) \to q(p_b) + g(p_c)$, sandwiching the qqg vertex between an outgoing quark $\bar{u}_\pm(p_b)$ and an incoming quark $u_\pm(p_a)$. From the splitting of a gluon into a quark–antiquark pair we already know that we can limit our analysis to the physical gluon polarizations and a spin flip in the quarks. Inserting the spinors from Eq. (2.107) and the two relevant gamma matrices gives us

$$\frac{\bar{u}_+(p_b)\gamma^1 u_+(p_a)}{E} = \left(1, \frac{\theta_b^*}{2}, 1, \frac{\theta_b^*}{2}\right) \begin{pmatrix} & & & 1 \\ & & 1 & \\ & 1 & & \\ 1 & & & \end{pmatrix} \begin{pmatrix} 1 \\ \theta_a^*/2 \\ 1 \\ \theta_a^*/2 \end{pmatrix}$$

$$= \left(1, \frac{\theta_b^*}{2}, 1, \frac{\theta_b^*}{2}\right) \begin{pmatrix} \theta_a^*/2 \\ 1 \\ \theta_a^*/2 \\ 1 \end{pmatrix} = \theta_a^* + \theta_b^*$$

$$\frac{\bar{u}_+(p_b)\gamma^2 u_+(p_a)}{E} = \left(1, \frac{\theta_b^*}{2}, 1, \frac{\theta_b^*}{2}\right) \begin{pmatrix} & & & -i \\ & & i & \\ & -i & & \\ i & & & \end{pmatrix} \begin{pmatrix} 1 \\ \theta_a^*/2 \\ 1 \\ \theta_a^*/2 \end{pmatrix}$$

$$= i \left(1, \frac{\theta_b^*}{2}, 1, \frac{\theta_b^*}{2}\right) \begin{pmatrix} -\theta_a^*/2 \\ 1 \\ -\theta_a^*/2 \\ 1 \end{pmatrix} = i(\theta_b^* - \theta_a^*), \qquad (2.115)$$

with the angles θ_b^* and θ_a^* relative to the final state gluon direction $\vec{p}_c$. Comparing to the situation shown in Fig. 2.1 for the angle relative to the scattered gluon we now find $\theta_b^* = \theta$ while for the incoming quark $\theta_a^* = -\theta_c = -z\theta$. The spin–down case gives the same result, modulo a complex conjugation

$$\frac{\bar{u}_-(p_b)\gamma^1 u_-(p_a)}{\sqrt{E_b}\sqrt{E_a}} = \left(-\frac{\theta_b^*}{2}, 1, \frac{\theta_b^*}{2}, -1\right) \begin{pmatrix} & & & 1 \\ & & 1 & \\ & 1 & & \\ 1 & & & \end{pmatrix} \begin{pmatrix} -\theta_a^*/2 \\ 1 \\ \theta_a^*/2 \\ -1 \end{pmatrix}$$

$$= \left(-\frac{\theta_b^*}{2}, 1, \frac{\theta_b^*}{2}, -1\right) \begin{pmatrix} -1 \\ \theta_a^*/2 \\ 1 \\ -\theta_a^*/2 \end{pmatrix} = \theta_a^* + \theta_b^*$$

$$\frac{\bar{u}_-(p_b)\gamma^2 u_-(p_a)}{\sqrt{E_b}\sqrt{E_a}} = \left(-\frac{\theta_b^*}{2}, 1, \frac{\theta_b^*}{2}, -1\right) \begin{pmatrix} & & & -i \\ & & i & \\ & -i & & \\ i & & & \end{pmatrix} \begin{pmatrix} -\theta_a^*/2 \\ 1 \\ \theta_a^*/2 \\ -1 \end{pmatrix}$$

$$= i \left(-\frac{\theta_b^*}{2}, 1, \frac{\theta_b^*}{2}, -1\right) \begin{pmatrix} 1 \\ \theta_a^*/2 \\ -1 \\ -\theta_a^*/2 \end{pmatrix} = i(\theta_a^* - \theta_b^*). \qquad (2.116)$$

2.3 Infrared Divergences

In terms of θ the two combinations of angles become $\theta_a^* + \theta_b^* = \theta(1-z)$ and $\theta_a^* - \theta_b^* = \theta(-z-1)$. The vertex function for gluon radiation off a quark then reads

$$V_{qqg} = -ig_s T^a \, \bar{u}_\pm(p_b)\gamma_\mu \epsilon_a^\mu u_\pm(p_c) \equiv -ig_s T^a \, \epsilon_a^j \, F_\pm^{(j)} \qquad \text{for} \quad j = 1,2$$

$$\frac{|F_+^{(1)}|^2}{p_a^2} = \frac{|F_-^{(1)}|^2}{p_a^2} = \frac{E_a E_b (\theta_a^* + \theta_b^*)^2}{p_a^2} = \frac{E_a^2 z(z-1)^2 \theta^2}{E_a^2 z(1-z)\theta^2} = (1-z)$$

$$\frac{|F_+^{(2)}|^2}{p_a^2} = \frac{|F_-^{(2)}|^2}{p_a^2} = \frac{E_a E_b (\theta_b^* - \theta_a^*)^2}{p_a^2} = \frac{E_a^2 z(1+z)^2 \theta^2}{E_a^2 z(1-z)\theta^2} = \frac{(1+z)^2}{1-z}.$$
(2.117)

again dropping irrelevant prefactors. The factorized matrix element for this channel has the same form as Eq. (2.114), except for the color averaging factor of the now incoming quark,

$$\overline{|\mathcal{M}_{n+1}|^2} = \left(\frac{1}{p_a^2}\right)^2 g_s^2 \, \frac{\text{Tr}\, T^a T^a}{N_c} \, \frac{1}{N_a} \left[|F_+^{(1)}|^2 + |F_-^{(1)}|^2 + |F_+^{(2)}|^2 + |F_-^{(2)}|^2\right] \overline{|\mathcal{M}_n|^2}$$

$$= \frac{g_s^2}{p_a^2} \, \frac{N_c^2 - 1}{2N_c} \, \frac{(1+z)^2 + (1-z)^2}{1-z} \, \overline{|\mathcal{M}_n|^2}$$

$$= \frac{2g_s^2}{p_a^2} \, C_F \, \frac{1+z^2}{1-z} \, \overline{|\mathcal{M}_n|^2}$$

$$\equiv \frac{2g_s^2}{p_a^2} \, \hat{P}_{q \leftarrow g}(z) \, \overline{|\mathcal{M}_n|^2}$$

$$\Leftrightarrow \boxed{\hat{P}_{q \leftarrow q}(z) = C_F \frac{1+z^2}{1-z}}.$$
(2.118)

The color factor for gluon radiation off a quark is $C_F = (N^2 - 1)/(2N)$. The averaging factor $1/N_a = 2$ now is the number of quark spins in the intermediate state. Just switching $z \leftrightarrow (1-z)$ we can read off the kernel for a quark splitting written in terms of the final-state gluon

$$\boxed{\hat{P}_{g \leftarrow q}(z) = C_F \frac{1 + (1-z)^2}{z}}.$$
(2.119)

This result finalizes our calculation of *all QCD splitting kernels* $\hat{P}_{i \leftarrow j}(z)$ between quarks and gluons. As alluded to earlier, similar to ultraviolet divergences which get removed by counter terms these splitting kernels are universal. They do not depend on the hard n-particle matrix element which is part of the original $(n+1)$-particle process. We show all four results in Eqs. (2.105), (2.114), (2.118), and (2.119).

This means that by construction of the kernels $\hat{P}$ we have shown that the collinear factorization Eq. (2.99) holds at this level in perturbation theory.

Before using this splitting property to describe QCD effects at the LHC we need to look at the splitting of partons in the initial state, meaning $|p_a^2|, p_c^2 \ll |p_b^2|$ where p_b is the momentum entering the hard interaction. The difference to the final–state splitting is that now we can consider the split parton momentum $p_b = p_a - p_c$ as a t-channel diagram, so we already know $p_b^2 = t < 0$ from our usual Mandelstam variables argument. This *space–like splitting* version of Eq. (2.90) for p_b^2 gives us

$$\begin{aligned} t \equiv p_b^2 &= (-z p_a + \beta n + p_T)^2 \\ &= p_T^2 - 2z\beta(p_a n) \qquad &\text{with } p_a^2 = n^2 = (p_a p_T) = (n p_T) = 0 \\ &= p_T^2 + \frac{p_T^2 z}{1-z} \qquad &\text{using Eq. (2.93)} \\ &= \frac{p_T^2}{1-z} = -\frac{p_{T,1}^2 + p_{T,2}^2}{1-z} < 0 \,. \end{aligned} \qquad (2.120)$$

The calculation of the splitting kernels and matrix elements is the same as for the time–like case, with the one exception that for splitting in the initial state the flow factor has to be evaluated at the reduced partonic energy $E_b = z E_a$ and that the energy fraction entering the parton density needs to be replaced by $x_b \to z x_b$. The factorized matrix element for initial–state splitting then reads just like Eq. (2.99)

$$\sigma_{n+1} = \int \sigma_n \, \frac{dt}{t} \, dz \, \frac{\alpha_s}{2\pi} \, \hat{P}(z) \,. \qquad (2.121)$$

How to use this property to make statements about the quark and gluon content in the proton will be the focus of the next section.

2.3.3 DGLAP Equation

We can use everything we now know about collinear parton splitting to describe incoming partons at hadron colliders. For example in $pp \to Z$ production incoming partons inside the protons transform into each other via collinear splitting until they enter the Z production process as quarks. Taking Eq. (2.121) seriously, the parton density we insert into Eq. (2.28) depends on two parameters, the final energy fraction and the virtuality $f(x_n, -t_n)$. The second parameter t is new compared to the purely probabilistic picture in Eq. (2.28). However, it cannot be neglected unless we convince ourselves that it is unphysical. As we will see later it corresponds exactly to the artificial renormalization scale which appears when we resum the scaling logarithms which appear in counter terms.

2.3 Infrared Divergences

More quantitatively, we start with a quark inside the proton with an energy fraction x_0, as it enters the hadronic phase space integral shown in Sect. 2.1.4. Since this quark is confined inside the proton it can only have small transverse momentum, which means its four-momentum squared t_0 is negative and its absolute value $|t_0|$ is small. The variable t we call *virtuality*. For the incoming partons which if on–shell have $p^2 = 0$ it gives the distance to the mass shell. Let us simplify our kinematic argument by assuming that there exists only one splitting, namely successive gluon radiation off an incoming quark, where the outgoing gluons are not relevant

In that case each collinear gluon radiation will decrease the quark energy $x_{j+1} < x_j$ and increase its virtuality $|t_{j+1}| = -t_{j+1} > -t_j = |t_j|$ through its recoil.

From the last section we know what the successive splitting means in terms of splitting probabilities. We can describe how the parton density $f(x, -t)$ evolves in the $(x - t)$ plane as depicted in Fig. 2.2. The starting point (x_0, t_0) is at least probabilistically given by the energy and kind of the hadron, for example the proton. For a given small virtuality $|t_0|$ we start at some kind of fixed x_0 distribution. We then interpret each branching as a step strictly downward in $x_j \to x_{j+1}$ where the t value we assign to this step is the ever increasing virtuality $|t_{j+1}|$ after the branching. Each splitting means a synchronous shift in x and t, so the actual path in the $(x-t)$ plane really consists of discrete points. The probability of such a splitting to occur is given by $\hat{P}_{q\leftarrow q}(z) \equiv \hat{P}(z)$ as it appears in Eq. (2.121)

$$\frac{\alpha_s}{2\pi} \hat{P}(z) \frac{dt}{t} dz . \tag{2.122}$$

In this picture we consider this probability a smooth function in t and z. At the end of the path we will probe this evolved parton density, where x_n and t_n enter the hard scattering process and its energy momentum conservation.

When we convert a partonic into a hadronic cross section numerically we need to specify the probability of the parton density $f(x, -t)$ residing in an infinitesimal square $[x_j, x_j + \delta x]$ and, if this second parameter has anything to do with physics, $[|t_j|, |t_j| + \delta t]$. Using our (x, t) plane we compute the flows into this square and out of this square, which together define the net shift in f in the sense of a differential equation, similar to the derivation of Gauss' theorem for vector fields inside a surface

$$\delta f_{\text{in}} - \delta f_{\text{out}} = \delta f(x, -t) . \tag{2.123}$$

Fig. 2.2 Path of an incoming parton in the $(x - t)$ plane. Because we define t as a negative number its axis is labelled $|t|$

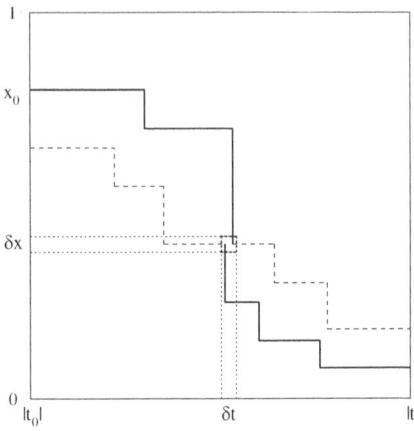

We compute the incoming and outgoing flows from the history of the (x, t) evolution. At this stage our picture becomes a little subtle; the way we define the path between two splittings in Fig. 2.2 it can enter and leave the square either vertically or horizontally. Because we do not consider the movement in the (x, t) plane continuous we can choose this direction as vertical or horizontal. Because we want to arrive at a differential equation in t we choose the vertical drop, such that the area the incoming and outgoing flows see is given by δt. If we define a splitting as such a vertical drop in x at the target value t_{j+1} an incoming path hitting the square at some value t can come from any x value above the square. Using this convention and following the fat solid lines in Fig. 2.2 the vertical flow into (and out of) the square (x, t) square is proportional to δt

$$\delta f_{\text{in}}(-t) = \delta t \; \left(\frac{\alpha_s \hat{P}}{2\pi t} \otimes f \right)(x, -t) = \frac{\delta t}{t} \int_x^1 \frac{dz}{z} \frac{\alpha_s}{2\pi} \hat{P}(z) f\left(\frac{x}{z}, -t\right)$$

$$= \frac{\delta t}{t} \int_0^1 \frac{dz}{z} \frac{\alpha_s}{2\pi} \hat{P}(z) f\left(\frac{x}{z}, -t\right) \quad \text{assuming } f(x', -t) = 0 \text{ for } x' > 1 ,$$

(2.124)

where δt is the size of the interval covered by the virtuality value t. We use the definition of a *convolution*

$$(f \otimes g)(x) = \int_0^1 dx_1 dx_2 \, f(x_1) g(x_2) \, \delta(x - x_1 x_2) \; = \int_0^1 \frac{dx_1}{x_1} f(x_1) g\left(\frac{x}{x_1}\right)$$

$$= \int_0^1 \frac{dx_2}{x_2} f\left(\frac{x}{x_2}\right) g(x_2) \; .$$

(2.125)

The outgoing flow we define in complete analogy, again leaving the infinitesimal square vertically. Following the fat solid line in Fig. 2.2 it is also proportional to δt

2.3 Infrared Divergences

$$\delta f_{\text{out}}(-t) = \delta t \int_0^1 dy \frac{\alpha_s \hat{P}(y)}{2\pi t} f(x,-t) = \frac{\delta t}{t} f(x,-t) \int_0^1 dy \frac{\alpha_s}{2\pi} \hat{P}(y). \tag{2.126}$$

The y integration, unlike the z integration for the incoming flow is not a convolution. This integration appears because we do not know the normalization of $\hat{P}(z)$ distribution which we interpret as a probability. The reason why it is not a convolution is that for the outgoing flow we know the starting condition and integrate over the final configurations; this aspect will become important later. Combining Eqs. (2.124) and (2.126) we can compute the change in the parton density of the quarks as

$$\delta f(x,-t) = \frac{\delta t}{t} \left[\int_0^1 \frac{dz}{z} \frac{\alpha_s}{2\pi} \hat{P}(z) f\left(\frac{x}{z},-t\right) - \int_0^1 dy \frac{\alpha_s}{2\pi} \hat{P}(y) f(x,-t) \right]$$

$$= \frac{\delta t}{t} \int_0^1 \frac{dz}{z} \frac{\alpha_s}{2\pi} \left[\hat{P}(z) - \delta(1-z) \int_0^1 dy \hat{P}(y) \right] f\left(\frac{x}{z},-t\right)$$

$$\equiv \frac{\delta t}{t} \int_x^1 \frac{dz}{z} \frac{\alpha_s}{2\pi} \hat{P}(z)_+ f\left(\frac{x}{z},-t\right)$$

$$\Leftrightarrow \quad \frac{\delta f(x,-t)}{\delta(-t)} = \frac{1}{(-t)} \int_x^1 \frac{dz}{z} \frac{\alpha_s}{2\pi} \hat{P}(z)_+ f\left(\frac{x}{z},-t\right), \tag{2.127}$$

again assuming $f(x) = 0$ for $x > 1$, strictly speaking requiring α_s to only depend on t but not on z, and using the specifically defined *plus subtraction* scheme

$$\boxed{F(z)_+ \equiv F(z) - \delta(1-z) \int_0^1 dy \, F(y)} \quad \text{or}$$

$$\int_0^1 dz \frac{f(z)}{(1-z)_+} = \int_0^1 dz \left(\frac{f(z)}{1-z} - \frac{f(1)}{1-z} \right). \tag{2.128}$$

For the second term we choose $F(z) = 1/(1-z)$, multiply it with an arbitrary test function $f(z)$ and integrate over z. In contrast to the original z integral the plus–subtracted integral is by definition finite in the limit $z \to 1$, where some of the splitting kernels diverge. For example, the quark splitting kernel including the plus prescription becomes $C_F((1+z^2)/(1-z))_+$. At this stage the plus prescription is simply a convenient way of writing a complicated combination of splitting kernels, but we will see that it also has a physics meaning.

Next, we check that the plus prescription indeed acts as a regularization technique for the parton densities. Obviously, the integral over $f(z)/(1-z)$ is divergent at the boundary $z \to 1$, which we know we can cure using *dimensional regularization*. The special case $f(z) = 1$ illustrates how dimensional regularization of infrared divergences in the phase space integration Eq. (2.88) works

$$\int_0^1 dz \, \frac{1}{(1-z)^{1-\epsilon}} = \int_0^1 dz \, \frac{1}{z^{1-\epsilon}} = \frac{z^\epsilon}{\epsilon}\bigg|_0^1 = \frac{1}{\epsilon} \qquad \text{with } \epsilon > 0, \quad (2.129)$$

for $4 + 2\epsilon$ dimensions. This change in sign avoids the analytic continuation of the usual value $n = 4 - 2\epsilon$ to $\epsilon < 0$. The dimensionally regularized integral we can write as

$$\int_0^1 dz \, \frac{f(z)}{(1-z)^{1-\epsilon}} = \int_0^1 dz \, \frac{f(z) - f(1)}{(1-z)^{1-\epsilon}} + f(1) \int_0^1 dz \, \frac{1}{(1-z)^{1-\epsilon}}$$

$$= \int_0^1 dz \, \frac{f(z) - f(1)}{1-z} (1 + \mathcal{O}(\epsilon)) + \frac{f(1)}{\epsilon}$$

$$= \int_0^1 dz \, \frac{f(z)}{(1-z)_+} (1 + \mathcal{O}(\epsilon)) + \frac{f(1)}{\epsilon}$$

$$\Leftrightarrow \int_0^1 dz \, \frac{f(z)}{(1-z)^{1-\epsilon}} - \frac{f(1)}{\epsilon} = \int_0^1 dz \, \frac{f(z)}{(1-z)_+} (1 + \mathcal{O}(\epsilon)). \qquad (2.130)$$

The dimensionally regularized integral minus the pole, i.e. the finite part of the dimensionally regularized integral, is the same as the plus–subtracted integral modulo terms of the order ϵ. The third line in Eq. (2.130) shows that the difference between a dimensionally regularized splitting kernel and a plus–subtracted splitting kernel manifests itself as terms proportional to $\delta(1-z)$. Physically, they represent contributions to a soft–radiation phase space integral.

Before we move on introducing a gluon density we can slightly reformulate the splitting kernel $\hat{P}_{q\leftarrow q}$ in Eq. (2.118). If the plus prescription regularizes the pole at $z \to 1$, what happens when we include the numerator of the regularized function, e.g. the quark splitting kernel? The finite difference between these results is

$$\left(\frac{1+z^2}{1-z}\right)_+ - (1+z^2)\left(\frac{1}{1-z}\right)_+ = \frac{1+z^2}{1-z} - \delta(1-z) \int_0^1 dy \, \frac{1+y^2}{1-y} - \frac{1+z^2}{1-z}$$

$$+ \delta(1-z) \int_0^1 dy \, \frac{1+z^2}{1-y}$$

$$= -\delta(1-z) \int_0^1 dy \, \left(\frac{1+y^2}{1-y} - \frac{2}{1-y}\right)$$

$$= \delta(1-z) \int_0^1 dy \, \frac{y^2 - 1}{y - 1}$$

$$= \delta(1-z) \int_0^1 dy \, (y+1) = \frac{3}{2}\delta(1-z).$$

$$(2.131)$$

2.3 Infrared Divergences

We can therefore write the quark's splitting kernel in two equivalent ways

$$P_{q \leftarrow q}(z) = C_F \left(\frac{1+z^2}{1-z} \right)_+ = C_F \left[\frac{1+z^2}{(1-z)_+} + \frac{3}{2}\delta(1-z) \right]. \qquad (2.132)$$

The infinitesimal version of Eq. (2.127) is the Dokshitzer–Gribov–Lipatov–Altarelli–Parisi or DGLAP *integro-differential equation* which describes the scale dependence of the quark parton density. As we already know quarks do not only appear in $q \to q$ splitting, but also in gluon splitting. Therefore, we generalize Eq. (2.127) to include the full set of QCD partons, i.e. quarks and gluons. This generalization involves a sum over all allowed splittings and the plus–subtracted splitting kernels. For the quark density on the left hand side it is

$$\frac{df_q(x,-t)}{d\log(-t)} = -t \frac{df_q(x,-t)}{d(-t)} = \sum_{j=q,g} \int_x^1 \frac{dz}{z} \frac{\alpha_s}{2\pi} P_{q \leftarrow j}(z) \, f_j\left(\frac{x}{z}, -t\right)$$

with $P_{q \leftarrow j}(z) \equiv \hat{P}_{q \leftarrow j}(z)_+$.

(2.133)

Going back to Eq. (2.127) we add all relevant parton indices and splittings and arrive at

$$\delta f_q(x,-t) = \frac{\delta t}{t} \left[\int_0^1 \frac{dz}{z} \frac{\alpha_s}{2\pi} \hat{P}_{q \leftarrow q}(z) \, f_q\left(\frac{x}{z}, -t\right) + \int_0^1 \frac{dz}{z} \frac{\alpha_s}{2\pi} \hat{P}_{q \leftarrow g}(z) \, f_g\left(\frac{x}{z}, -t\right) - \int_0^1 dy \, \frac{\alpha_s}{2\pi} \hat{P}_{q \leftarrow q}(y) \, f_q(x,-t) \right]. \qquad (2.134)$$

Of the three terms on the right hand side the first and the third together define the plus–subtracted splitting kernel $P_{q \leftarrow q}(z)$, just following the argument above. The second term is a proper convolution and the only term proportional to the gluon parton density. Quarks can be produced in gluon splitting but cannot vanish into it. Therefore, we have to identify the second term with $P_{q \leftarrow g}$ in Eq. (2.133) without adding a plus–regulator

$$P_{q \leftarrow g}(z) \equiv \hat{P}_{q \leftarrow g}(z) = T_R \left[z^2 + (1-z)^2 \right]. \qquad (2.135)$$

In principle, the splitting kernel $\hat{P}_{g \leftarrow q}$ also generates a quark, in addition to the final–state gluon. However, comparing this to the terms proportional to $\hat{P}_{q \leftarrow q}$ they both arise from the same splitting, namely a quark density leaving the infinitesimal square in the $(x-t)$ plane via the splitting $q \to qg$. Including the additional

$\hat{P}_{g \leftarrow q}(y)$ would be double counting and should not appear, as the notation $g \leftarrow q$ already suggests.

The second QCD parton density we have to study is the *gluon density*. The incoming contribution to the infinitesimal square is given by the sum of four splitting scenarios each leading to a gluon with virtuality $-t_{j+1}$

$$\begin{aligned}
\delta f_{\text{in}}(-t) &= \frac{\delta t}{t} \int_0^1 \frac{dz}{z} \frac{\alpha_s}{2\pi} \left[\hat{P}_{g \leftarrow g}(z) \left(f_g\left(\frac{x}{z}, -t\right) + f_g\left(\frac{x}{1-z}, -t\right) \right) \right. \\
&\qquad\qquad \left. + \hat{P}_{g \leftarrow q}(z) \left(f_q\left(\frac{x}{z}, -t\right) + f_{\bar{q}}\left(\frac{x}{z}, -t\right) \right) \right] \\
&= \frac{\delta t}{t} \int_0^1 \frac{dz}{z} \frac{\alpha_s}{2\pi} \left[2\hat{P}_{g \leftarrow g}(z) f_g\left(\frac{x}{z}, -t\right) + \hat{P}_{g \leftarrow q}(z) \left(f_q\left(\frac{x}{z}, -t\right) \right.\right. \\
&\qquad\qquad \left.\left. + f_{\bar{q}}\left(\frac{x}{z}, -t\right) \right) \right],
\end{aligned} \qquad (2.136)$$

using $P_{g \leftarrow \bar{q}} = P_{g \leftarrow q}$ in the first line and $P_{g \leftarrow g}(1-z) = P_{g \leftarrow g}(z)$ in the second. To leave the volume element in the (x, t) space a gluon can either split into two gluons or radiate one of n_f light-quark flavors. Combining the incoming and outgoing flows we find

$$\begin{aligned}
&\delta f_g(x, -t) \\
&= \frac{\delta t}{t} \int_0^1 \frac{dz}{z} \frac{\alpha_s}{2\pi} \left[2\hat{P}_{g \leftarrow g}(z) f_g\left(\frac{x}{z}, -t\right) + \hat{P}_{g \leftarrow q}(z) \left(f_q\left(\frac{x}{z}, -t\right) + f_{\bar{q}}\left(\frac{x}{z}, -t\right) \right) \right] \\
&\quad - \frac{\delta t}{t} \int_0^1 dy \frac{\alpha_s}{2\pi} \left[\hat{P}_{g \leftarrow g}(y) + n_f \hat{P}_{q \leftarrow g}(y) \right] f_g(x, -t)
\end{aligned} \qquad (2.137)$$

We have to evaluate the four terms in this expression one after the other. Unlike in the quark case they do not immediately correspond to regularizing the diagonal splitting kernel using the plus prescription.

First, there exists a contribution to δf_{in} proportional to f_q or $f_{\bar{q}}$ which is not matched by the outgoing flow. From the quark case we already know how to deal with it. For the corresponding splitting kernel there is no regularization through the plus prescription needed, so we define

$$\boxed{P_{g \leftarrow q}(z) \equiv \hat{P}_{g \leftarrow q}(z) = C_F \frac{1 + (1-z)^2}{z}.} \qquad (2.138)$$

This ensures that the off-diagonal contribution to the gluon density is taken into account when we extend Eq. (2.133) to a combined quark/antiquark and gluon form. Hence, the structure of the DGLAP equation implies that the two off-diagonal splitting kernels do not include any plus prescription $\hat{P}_{i \leftarrow j} = P_{i \leftarrow j}$. We could

2.3 Infrared Divergences

have expected this because off-diagonal kernels are finite in the soft limit, $z \to 1$. Applying a plus prescription would only have modified the splitting kernels at the isolated (zero-measure) point $y = 1$ which for a finite value of the integrand does not affect the integral on the right hand side of the DGLAP equation.

Second, the y integral describing the gluon splitting into a quark pair we can compute directly,

$$-\int_0^1 dy \, \frac{\alpha_s}{2\pi} \, n_f \, \hat{P}_{q \leftarrow g}(y) = -\frac{\alpha_s}{2\pi} \, n_f \, T_R \int_0^1 dy \, \left[1 - 2y + 2y^2\right] \qquad \text{using Eq. (2.135)}$$

$$= -\frac{\alpha_s}{2\pi} \, n_f \, T_R \left[y - y^2 + \frac{2y^3}{3}\right]_0^1$$

$$= -\frac{2}{3} \frac{\alpha_s}{2\pi} \, n_f \, T_R \, . \qquad (2.139)$$

Finally, the terms proportional to the purely gluonic splitting $P_{g \leftarrow g}$ appearing in Eq. (2.137) require some more work. The y integral coming from the outgoing flow has to consist of a finite term and a term we can use to define the plus prescription for $\hat{P}_{g \leftarrow g}$. We can compute the integral as

$$-\int_0^1 dy \, \frac{\alpha_s}{2\pi} \, \hat{P}_{g \leftarrow g}(y)$$

$$= -\frac{\alpha_s}{2\pi} \, C_A \int_0^1 dy \, \left[\frac{y}{1-y} + \frac{1-y}{y} + y(1-y)\right] \qquad \text{using Eq. (2.105)}$$

$$= -\frac{\alpha_s}{2\pi} \, C_A \int_0^1 dy \, \left[\frac{2y}{1-y} + y(1-y)\right]$$

$$= -\frac{\alpha_s}{2\pi} \, C_A \int_0^1 dy \, \left[\frac{2(y-1)}{1-y} + y(1-y)\right]$$

$$\quad - \frac{\alpha_s}{2\pi} \, C_A \int_0^1 dy \, \frac{2}{1-y}$$

$$= -\frac{\alpha_s}{2\pi} \, C_A \int_0^1 dy \, \left[-2 + y - y^2\right]$$

$$\quad - \frac{\alpha_s}{2\pi} \, 2C_A \int_0^1 dz \, \frac{1}{1-z}$$

$$= -\frac{\alpha_s}{2\pi} \, C_A \left[-2 + \frac{1}{2} - \frac{1}{3}\right] - \frac{\alpha_s}{2\pi} \, 2C_A \int_0^1 dz \, \frac{1}{1-z}$$

$$= \frac{\alpha_s}{2\pi} \, \frac{11}{6} \, C_A - \frac{\alpha_s}{2\pi} \, 2C_A \int_0^1 dz \, \frac{1}{1-z} \, . \qquad (2.140)$$

The second term in this result is what we need to replace the first term in the splitting kernel of Eq. (2.105) proportional to $1/(1-z)$ by $1/(1-z)_+$. We can see this using $f(z) = z$ and correspondingly $f(1) = 1$ in Eq. (2.128). The two finite terms in Eqs. (2.139) and (2.140) we have to include in the definition of $\hat{P}_{g \leftarrow g}$ ad hoc. Because the regularized splitting kernel appear inside a convolution the two finite terms require an additional term $\delta(1-z)$. Collecting all of them we arrive at

$$P_{g \leftarrow g}(z) = 2C_A \left(\frac{z}{(1-z)_+} + \frac{1-z}{z} + z(1-z) \right) + \frac{11}{6} C_A \, \delta(1-z) - \frac{2}{3} n_f T_R \, \delta(1-z). \tag{2.141}$$

This result concludes our computation of all four regularized splitting functions which appear in the DGLAP equation (2.133).

Before discussing and solving the DGLAP equation, let us briefly recapitulate: for the full quark and gluon particle content of QCD we have derived the DGLAP equation which describes a factorization scale dependence of the quark and gluon parton densities. The universality of the splitting kernels is obvious from the way we derive them—no information on the n-particle process ever enters the derivation.

The DGLAP equation is formulated in terms of four splitting kernels of gluons and quarks which are linked to the splitting probabilities, but which for the DGLAP equation have to be regularized. With the help of a plus–subtraction all kernels $P_{i \leftarrow j}(z)$ become finite, including in the soft limit $z \to 1$. However, splitting kernels are only regularized when needed, so the finite off-diagonal quark–gluon and gluon–quark splittings are unchanged. This means the plus prescription really acts as an infrared renormalization, moving universal infrared divergences into the definition of the parton densities. The original collinear divergence has vanished as well.

The only approximation we make in the computation of the splitting kernels is that in the y integrals we implicitly assume that the running coupling α_s does not depend on the momentum fraction. In its standard form and in terms of the factorization scale $\mu_F^2 \equiv -t$ the *DGLAP equation* reads

$$\frac{df_i(x, \mu_F)}{d \log \mu_F^2} = \sum_j \int_x^1 \frac{dz}{z} \frac{\alpha_s}{2\pi} P_{i \leftarrow j}(z) f_j \left(\frac{x}{z}, \mu_F \right) = \frac{\alpha_s}{2\pi} \sum_j \left(P_{i \leftarrow j} \otimes f_j \right)(x, \mu_F). \tag{2.142}$$

2.3.4 Parton Densities

Solving the integro-differential DGLAP equation (2.142) for the parton densities is clearly beyond the scope of this writeup. Nevertheless, we will sketch how we would approach this. This will give us some information on the structure of its solutions which we need to understand the physics of the DGLAP equation.

2.3 Infrared Divergences

One simplification we can make in this illustration is to postulate eigenvalues in parton space and solve the equation for them. This gets rid of the sum over partons on the right hand side. One such parton density is the *non–singlet parton density*, defined as the difference of two parton densities $f_q^{NS} = (f_u - f_{\bar{u}})$. Since gluons cannot distinguish between quarks and antiquarks, the gluon contribution to their evolution cancels, at least in the massless limit. This will be true at arbitrary loop order, since flavor $SU(3)$ commutes with the QCD gauge group. The corresponding DGLAP equation with leading order splitting kernels now reads

$$\frac{df_q^{NS}(x, \mu_F)}{d\log \mu_F^2} = \int_x^1 \frac{dz}{z} \frac{\alpha_s}{2\pi} P_{q \leftarrow q}(z) f_q^{NS}\left(\frac{x}{z}, \mu_F\right)$$

$$= \frac{\alpha_s}{2\pi} \left(P_{q \leftarrow q} \otimes f_q^{NS}\right)(x, \mu_F). \quad (2.143)$$

To solve it we need a transformation which simplifies a convolution, leading us to the *Mellin transform*. Starting from a function $f(x)$ of a real variable x we define the Mellin transform into moment space m

$$\boxed{\mathcal{M}[f](m) \equiv \int_0^1 dx \, x^{m-1} f(x)} \quad f(x) = \frac{1}{2\pi i} \int_{c-i\infty}^{c-i\infty} dm \, \frac{\mathcal{M}[f](m)}{x^m}.$$

$$(2.144)$$

The integration contour for the inverse transformation lies to the right of all singularities of the analytic continuation of $\mathcal{M}[f](m)$, which fixes the offset c. The Mellin transform of a convolution is the product of the two Mellin transforms, which gives us the transformed DGLAP equation

$$\mathcal{M}[P_{q \leftarrow q} \otimes f_q^{NS}](m) = \mathcal{M}\left[\int_0^1 \frac{dz}{z} P_{q \leftarrow q}\left(\frac{x}{z}\right) f_q^{NS}(z)\right](m)$$

$$= \mathcal{M}[P_{q \leftarrow q}](m) \, \mathcal{M}[f_q^{NS}](m, \mu_F)$$

$$\frac{d\mathcal{M}[f_q^{NS}](m, \mu_F)}{d\log \mu_F^2} = \frac{\alpha_s}{2\pi} \mathcal{M}[P_{q \leftarrow q}](m) \, \mathcal{M}[f_q^{NS}](m, \mu_F), \quad (2.145)$$

and its solution

$$\mathcal{M}[f_q^{NS}](m, \mu_F) = \mathcal{M}[f_q^{NS}](m, \mu_{F,0}) \exp\left(\frac{\alpha_s}{2\pi} \mathcal{M}[P_{q \leftarrow q}](m) \log \frac{\mu_F^2}{\mu_{F,0}^2}\right)$$

$$= \mathcal{M}[f_q^{NS}](m, \mu_{F,0}) \left(\frac{\mu_F^2}{\mu_{F,0}^2}\right)^{\frac{\alpha_s}{2\pi} \mathcal{M}[P_{q \leftarrow q}](m)}$$

$$\equiv \mathcal{M}[f_q^{\text{NS}}](m, \mu_{F,0}) \left(\frac{\mu_F^2}{\mu_{F,0}^2}\right)^{\frac{\alpha_s}{2\pi}\gamma(m)}, \qquad (2.146)$$

defining $\gamma(m) = \mathcal{M}[P](m)$.

The solution given by Eq. (2.146) still has the complication that it includes μ_F and α_s as two free parameters. To simplify this form we can include $\alpha_s(\mu_R^2)$ in the running of the DGLAP equation and identify the renormalization scale μ_R of the strong coupling with the factorization scale $\mu_F = \mu_R \equiv \mu$. This allows us to replace $\log \mu^2$ in the DGLAP equation by α_s, including the leading order Jacobian. This is clearly correct for all one-scale problems where we have no freedom to choose either of the two scales. We find

$$\frac{d}{d\log\mu^2} = \frac{d\log\alpha_s}{d\log\mu^2}\frac{d}{d\log\alpha_s} = \frac{1}{\alpha_s}\frac{d\alpha_s}{d\log\mu^2}\frac{d}{d\log\alpha_s} = -\alpha_s b_0 \frac{d}{d\log\alpha_s}. \qquad (2.147)$$

This additional factor of α_s on the left hand side will cancel the factor α_s on the right hand side of the DGLAP equation (2.145)

$$\frac{d\mathcal{M}[f_q^{\text{NS}}](m,\mu)}{d\log\alpha_s} = -\frac{1}{2\pi b_0}\gamma(m)\,\mathcal{M}[f_q^{\text{NS}}](m,\mu)$$

$$\mathcal{M}[f_q^{\text{NS}}](m,\mu) = \mathcal{M}[f_q^{\text{NS}}](m,\mu_0)\,\exp\left(-\frac{1}{2\pi b_0}\gamma(m)\log\frac{\alpha_s(\mu^2)}{\alpha_s(\mu_0^2)}\right)$$

$$= \mathcal{M}[f_q^{\text{NS}}](m,\mu_{F,0})\left(\frac{\alpha_s(\mu_0^2)}{\alpha_s(\mu^2)}\right)^{\frac{\gamma(m)}{2\pi b_0}}. \qquad (2.148)$$

Among other things, in this derivation we neglect that some splitting functions have singularities and therefore the Mellin transform is not obviously well defined. Our convolution is not really a convolution either, because we cut it off at Q_0^2 etc.; but the final structure in Eq. (2.148) really holds.

Because we will need it in the next section we emphasize that the same kind of solution appears in pure Yang–Mills theory, i.e. in QCD without quarks. Looking at the different color factors in QCD this limit can also be derived as the leading terms in N_c. In that case there also exists only one splitting kernel defining an anomalous dimension γ. We find in complete analogy to Eq. (2.148)

$$\boxed{\mathcal{M}[f_g](m,\mu) = \mathcal{M}[f_g](m,\mu_0)\left(\frac{\alpha_s(\mu_0^2)}{\alpha_s(\mu^2)}\right)^{\frac{\gamma(m)}{2\pi b_0}}.} \qquad (2.149)$$

To remind ourselves that in this derivation we unify the renormalization and factorization scales we denote them just as μ. This solution to the DGLAP equation

2.3 Infrared Divergences

is not completely determined: as a solution to a differential equation it also includes an integration constant which we express in terms of μ_0. The DGLAP equation therefore does not determine parton densities, it only describes their evolution from one scale μ_F to another, just like a renormalization group equation in the ultraviolet.

The structure of Eq. (2.149) already shows something we will in more detail discuss in the following Sect. 2.3.5: the splitting probability we find in the exponent. To make sense of such a structure we remind ourselves that such ratios of α_s values to some power can appear as a result of a resummed series. Such a series would need to include powers of $(\mathcal{M}[\hat{P}])^n$ summed over n which corresponds to a sum over splittings with a varying number of partons in the final state. Parton densities cannot be formulated in terms of a fixed final state because they include effects from any number of collinearly radiated partons summed over the number of such partons. For the processes we can evaluate using parton densities fulfilling the DGLAP equation this means that they always have the form

$$\boxed{pp \to \mu^+ \mu^- + X} \qquad \text{where } X \text{ includes any number of collinear jets.} \tag{2.150}$$

Why is γ is referred to as the *anomalous dimension* of the parton density? This is best illustrated using a running coupling with a finite mass dimension, like the gravitational coupling $G_{\text{Planck}} \sim 1/M_{\text{Planck}}^2$. When we attach a renormalization constant Z to this coupling we first define a dimensionless running bare coupling g. In n dimensions this gives us

$$g^{\text{bare}} = M^{n-2} G_{\text{Planck}} \to Zg(M^2) \,. \tag{2.151}$$

For the dimensionless gravitational coupling we can compute the running

$$\begin{aligned}\frac{dg(M^2)}{d \log M} &= \frac{d}{d \log M} \left(\frac{1}{Z} M^{n-2} G_{\text{Planck}} \right) \\ &= G_{\text{Planck}} \left(\frac{1}{Z} M \frac{dM^{n-2}}{dM} - \frac{1}{Z^2} \frac{dZ}{d \log M} M^{n-2} \right) \\ &= g(M)(n - 2 + \eta) \qquad \text{with} \quad \eta = -\frac{1}{Z} \frac{dZ}{d \log M}\end{aligned} \tag{2.152}$$

Hence, there are two sources of running for the renormalized coupling $g(M^2)$: first, there is the mass dimension of the bare coupling $n - 2$, and secondly there is η, a quantum effect from the coupling renormalization. For obvious reasons we call η the anomalous dimension of G_{Planck}.

This is similar to the running of the parton densities in Mellin space, as shown in Eq. (2.145), and with $\gamma(m)$ defined in Eq. (2.146), so we refer to γ as an anomalous dimension as well. The entire running of the transformed parton density arises from collinear splitting, parameterized by a finite γ. There is only a slight stumbling step in this analogy: usually, an anomalous dimension arises through renormalization involving a ultraviolet divergence and the renormalization scale. In our case we are discussing an infrared divergence and the factorization scale dependence.

2.3.5 Resumming Collinear Logarithms

Remembering how we arrive at the DGLAP equation we notice an analogy to the case of ultraviolet divergences and the running coupling. We start from universal infrared divergences. We describe them in terms of splitting functions which we regularize using the plus prescription. The DGLAP equation plays the role of a renormalization group equation for example for the running coupling. It links parton densities evaluated at different scales μ_F.

In analogy to the scaling logarithms considered in Sect. 2.2.3 we now test if we can point to a *type of logarithm* the DGLAP equation resums by reorganizing our perturbative series of parton splitting. To identify these resummed logarithms we build a physical model based on collinear splitting, but without using the DGLAP equation. We then solve it to see the resulting structure of the solutions and compare it to the structure of the DGLAP solutions in Eq. (2.149).

We start from the basic equation defining the physical picture of parton splitting in Eq. (2.99). Only taking into account gluons in pure Yang–Mills theory it precisely corresponds to the starting point of our discussion leading to the DGLAP equation, schematically written as

$$\sigma_{n+1} = \int \sigma_n \frac{dt}{t} dz \frac{\alpha_s}{2\pi} \hat{P}_{g \leftarrow g}(z) . \qquad (2.153)$$

This form of collinear factorization does not include parton densities and only applies to final state splittings. To include initial state splittings we need a definition of the virtuality variable t. If we remember that $t = p_b^2 < 0$ we can follow Eq. (2.120) and introduce a positive transverse momentum variable $\vec{p}_T^2$ in the usual Sudakov decomposition, such that

$$-t = \frac{p_T^2}{1-z} = \frac{\vec{p}_T^2}{1-z} > 0 \qquad \Rightarrow \qquad \frac{dt}{t} = \frac{dp_T^2}{p_T^2} = \frac{d\vec{p}_T^2}{\vec{p}_T^2} . \qquad (2.154)$$

From the definition of p_T in Eq. (2.90) we see that $\vec{p}_T^2$ is really the transverse three-momentum of the parton pair after splitting.

Beyond the single parton radiation discussed in Sect. 2.3.1 we consider a ladder of successive splittings of one gluon into two. For a moment, we forget about the actual parton densities and assume that they are part of the hadronic cross section σ_n. In the collinear limit the appropriate convolution gives us

2.3 Infrared Divergences

$$\sigma_{n+1}(x,\mu_F) = \int_{x_0}^{1} \frac{dx_n}{x_n} \hat{P}_{g\leftarrow g}\left(\frac{x}{x_n}\right) \sigma_n(x_n,\mu_0) \int_{\mu_0^2}^{\mu_F^2} \frac{d\vec{p}_{T,n}^{\,2}}{\vec{p}_{T,n}^{\,2}} \frac{\alpha_s(\mu_R^2)}{2\pi} \,. \quad (2.155)$$

The dz in Eq. (2.153) we replace by the proper convolution $\hat{P} \otimes \sigma_n$, evaluated at the momentum fraction x. Because the splitting kernel is infrared divergent we cut off the convolution integral at x_0. Similarly, the transverse momentum integral is bounded by an infrared cutoff μ_0 and the physical external scale μ_F. This is the range in which an additional collinear radiation is included in σ_{n+1}.

For splitting the two integrals in Eq. (2.155) it is crucial that μ_0 is the only scale the matrix element σ_n depends on. The other integration variable, the transverse momentum, does not feature in σ_n because collinear factorization is defined in the limit $\vec{p}_T^{\,2} \to 0$. For α_s we will see in the next step how μ_R can depend on the transverse momentum. All through the argument of this subsection we should keep in mind that we are looking for assumptions which allow us to solve Eq. (2.155) and compare the result to the solution of the DGLAP equation. In other words, these assumptions we will turn into a physics picture of the DGLAP equation and its solutions.

Making μ_F the *global upper boundary of the transverse momentum* integration for collinear splitting is our first assumption. We can then apply the recursion formula in Eq. (2.155) iteratively

$$\sigma_{n+1}(x,\mu_F) \sim \int_{x_0}^{1} \frac{dx_n}{x_n} \hat{P}_{g\leftarrow g}\left(\frac{x}{x_n}\right) \cdots \int_{x_0}^{1} \frac{dx_1}{x_1} \hat{P}_{g\leftarrow g}\left(\frac{x_2}{x_1}\right) \sigma_1(x_1,\mu_0)$$

$$\times \int_{\mu_0}^{\mu_F} \frac{d\vec{p}_{T,n}^{\,2}}{\vec{p}_{T,n}^{\,2}} \frac{\alpha_s(\mu_R^2)}{2\pi} \cdots \int_{\mu_0} \frac{d\vec{p}_{T,1}^{\,2}}{\vec{p}_{T,1}^{\,2}} \frac{\alpha_s(\mu_R^2)}{2\pi} \,. \quad (2.156)$$

The two sets of integrals in this equation we will solve one by one, starting with the $\vec{p}_T$ integrals.

To be able to make sense of the $\vec{p}_T^{\,2}$ integration in Eq. (2.156) and solve it we have to make two more assumptions in our multiple-splitting model. First, we identify the *scale of the strong coupling* α_s with the transverse momentum scale of the splitting $\mu_R^2 = \vec{p}_T^{\,2}$. This way we can fully integrate the integrand $\alpha_s/(2\pi)$ and link the final result to the global boundary μ_F.

In addition, we assume *strongly ordered splittings* in terms of the transverse momentum. If the ordering of the splitting is fixed externally by the chain of momentum fractions x_j, the first splitting, integrated over $\vec{p}_{T,1}^{\,2}$, is now bounded from above by the next external scale $\vec{p}_{T,2}^{\,2}$, which is then bounded by $\vec{p}_{T,3}^{\,2}$, etc. For the n-fold $\vec{p}_T^{\,2}$ integration this means

$$\mu_0^2 < \vec{p}_{T,1}^{\,2} < \vec{p}_{T,2}^{\,2} < \cdots < \mu_F^2 \quad (2.157)$$

We will study motivations for this ad hoc assumptions in Sect. 2.5.4.

Under these three assumptions the transverse momentum integrals in Eq. (2.156) become

$$\int_{\mu_0}^{\mu_F} \frac{d\vec{p}_{T,n}^2}{\vec{p}_{T,n}^2} \frac{\alpha_s(\vec{p}_{T,n}^2)}{2\pi} \cdots \int_{\mu_0}^{p_{T,3}} \frac{d\vec{p}_{T,2}^2}{\vec{p}_{T,2}^2} \frac{\alpha_s(\vec{p}_{T,2}^2)}{2\pi} \int_{\mu_0}^{p_{T,2}} \frac{d\vec{p}_{T,1}^2}{\vec{p}_{T,1}^2} \frac{\alpha_s(\vec{p}_{T,1}^2)}{2\pi} \cdots$$

$$= \int_{\mu_0}^{\mu_F} \frac{d\vec{p}_{T,n}^2}{\vec{p}_{T,n}^2} \frac{1}{2\pi b_0 \log \frac{\vec{p}_{T,n}^2}{\Lambda_{QCD}^2}} \cdots \int_{\mu_0}^{p_{T,3}} \frac{d\vec{p}_{T,2}^2}{\vec{p}_{T,2}^2} \frac{1}{2\pi b_0 \log \frac{\vec{p}_{T,2}^2}{\Lambda_{QCD}^2}}$$

$$\int_{\mu_0}^{p_{T,2}} \frac{d\vec{p}_{T,1}^2}{\vec{p}_{T,1}^2} \frac{1}{2\pi b_0 \log \frac{\vec{p}_{T,1}^2}{\Lambda_{QCD}^2}} \cdots$$

$$= \frac{1}{(2\pi b_0)^n} \int_{\mu_0}^{\mu_F} \frac{d\vec{p}_{T,n}^2}{\vec{p}_{T,n}^2} \frac{1}{\log \frac{\vec{p}_{T,n}^2}{\Lambda_{QCD}^2}} \cdots \int_{\mu_0}^{p_{T,3}} \frac{d\vec{p}_{T,2}^2}{\vec{p}_{T,2}^2} \frac{1}{\log \frac{\vec{p}_{T,2}^2}{\Lambda_{QCD}^2}}$$

$$\int_{\mu_0}^{p_{T,2}} \frac{d\vec{p}_{T,1}^2}{\vec{p}_{T,1}^2} \frac{1}{\log \frac{\vec{p}_{T,1}^2}{\Lambda_{QCD}^2}} \cdots . \tag{2.158}$$

We can solve the individual integrals by switching variables, for example in the last integral

$$\int_{\mu_0}^{p_{T,2}} \frac{d\vec{p}_{T,1}^2}{\vec{p}_{T,1}^2} \frac{1}{\log \frac{\vec{p}_{T,1}^2}{\Lambda_{QCD}^2}}$$

$$= \int_{\log\log \mu_0^2/\Lambda^2}^{\log\log p_{T,2}^2/\Lambda^2} d\log\log \frac{\vec{p}_{T,1}^2}{\Lambda_{QCD}^2} \quad \text{with} \quad \frac{d(ax)}{(ax)\log x} = d\log\log x$$

$$= \int_0^{\log\log p_{T,2}^2/\Lambda^2 - \log\log \mu_0^2/\Lambda^2} d\left(\log\log \frac{\vec{p}_{T,1}^2}{\Lambda_{QCD}^2} - \log\log \frac{\mu_0^2}{\Lambda_{QCD}^2}\right)$$

$$= \log \frac{\log \vec{p}_{T,1}^2/\Lambda_{QCD}^2}{\log \mu_0^2/\Lambda_{QCD}^2}\bigg|_0^{\vec{p}_{T,1}^2 \equiv \vec{p}_{T,2}^2}$$

$$= \log \frac{\log \vec{p}_{T,2}^2/\Lambda_{QCD}^2}{\log \mu_0^2/\Lambda_{QCD}^2} . \tag{2.159}$$

2.3 Infrared Divergences

This gives us for the chain of transverse momentum integrals

$$\int_0^{p_{T,n} \equiv \mu_F} d\log \frac{\log \vec{p}_{T,n}^2/\Lambda_{QCD}^2}{\log \mu_0^2/\Lambda_{QCD}^2} \cdots \int_0^{p_{T,2} \equiv p_{T,3}} d\log \frac{\log \vec{p}_{T,2}^2/\Lambda_{QCD}^2}{\log \mu_0^2/\Lambda_{QCD}^2} \int_0^{p_{T,1} \equiv p_{T,2}}$$
$$d\log \frac{\log \vec{p}_{T,1}^2/\Lambda_{QCD}^2}{\log \mu_0^2/\Lambda_{QCD}^2}$$

$$= \int_0^{p_{T,n} \equiv \mu_F} d\log \frac{\log \vec{p}_{T,n}^2/\Lambda_{QCD}^2}{\log \mu_0^2/\Lambda_{QCD}^2} \cdots$$
$$\int_0^{p_{T,2} \equiv p_{T,3}} d\log \frac{\log \vec{p}_{T,2}^2/\Lambda_{QCD}^2}{\log \mu_0^2/\Lambda_{QCD}^2} \left(\log \frac{\log \vec{p}_{T,2}^2/\Lambda_{QCD}^2}{\log \mu_0^2/\Lambda_{QCD}^2} \right)$$

$$= \int_0^{p_{T,n} \equiv \mu_F} d\log \frac{\log \vec{p}_{T,n}^2/\Lambda_{QCD}^2}{\log \mu_0^2/\Lambda_{QCD}^2} \cdots \frac{1}{2} \left(\log \frac{\log \vec{p}_{T,3}^2/\Lambda_{QCD}^2}{\log \mu_0^2/\Lambda_{QCD}^2} \right)^2$$

$$= \int_0^{p_{T,n} \equiv \mu_F} d\log \frac{\log \vec{p}_{T,n}^2/\Lambda_{QCD}^2}{\log \mu_0^2/\Lambda_{QCD}^2} \left(\frac{1}{2} \cdots \frac{1}{n-1} \right) \left(\log \frac{\log \vec{p}_{T,n}^2/\Lambda_{QCD}^2}{\log \mu_0^2/\Lambda_{QCD}^2} \right)^{n-1}$$

$$= \frac{1}{n!} \left(\log \frac{\log \mu_F^2/\Lambda_{QCD}^2}{\log \mu_0^2/\Lambda_{QCD}^2} \right)^n = \frac{1}{n!} \left(\log \frac{\alpha_s(\mu_0^2)}{\alpha_s(\mu_F^2)} \right)^n . \tag{2.160}$$

This is the final result for the chain of transverse momentum integrals in Eq. (2.156). By assumption, the strong coupling is evaluated at the factorization scale μ_F, which means we identify $\mu_R \equiv \mu_F$.

To compute the convolution integrals over the momentum fractions in Eq. (2.156),

$$\sigma_{n+1}(x,\mu) \sim \frac{1}{n!} \left(\frac{1}{2\pi b_0} \log \frac{\alpha_s(\mu_0^2)}{\alpha_s(\mu^2)} \right)^n \int_{x_0}^1 \frac{dx_n}{x_n} \hat{P}_{g \leftarrow g} \left(\frac{x}{x_n} \right) \cdots$$
$$\int_{x_0}^1 \frac{dx_1}{x_1} \hat{P}_{g \leftarrow g} \left(\frac{x_2}{x_1} \right) \sigma_1(x_1, \mu_0) , \tag{2.161}$$

we again Mellin transform the equation into moment space

$$\mathcal{M}[\sigma_{n+1}](m,\mu)$$
$$\times \sim \frac{1}{n!} \left(\frac{1}{2\pi b_0} \log \frac{\alpha_s(\mu_0^2)}{\alpha_s(\mu^2)} \right)^n \mathcal{M} \left[\int_{x_0}^1 \frac{dx_n}{x_n} \hat{P}_{g \leftarrow g} \left(\frac{x}{x_n} \right) \cdots \right.$$
$$\left. \int_{x_0}^1 \frac{dx_1}{x_1} \hat{P}_{g \leftarrow g} \left(\frac{x_2}{x_1} \right) \sigma_1(x_1, \mu_0) \right] (m)$$

$$= \frac{1}{n!} \left(\frac{1}{2\pi b_0} \log \frac{\alpha_s(\mu_0^2)}{\alpha_s(\mu^2)} \right)^n \gamma(m)^n \; \mathcal{M}[\sigma_1](m,\mu_0) \qquad \text{using } \gamma(m) \equiv \mathcal{M}[P](m)$$

$$= \frac{1}{n!} \left(\frac{1}{2\pi b_0} \log \frac{\alpha_s(\mu_0^2)}{\alpha_s(\mu^2)} \gamma(m) \right)^n \mathcal{M}[\sigma_1](m,\mu_0) \, . \tag{2.162}$$

We can now sum the production cross sections for n collinear jets and obtain

$$\sum_{n=0}^{\infty} \mathcal{M}[\sigma_{n+1}](m,\mu) = \mathcal{M}[\sigma_1](m,\mu_0) \sum_n \frac{1}{n!} \left(\frac{1}{2\pi b_0} \log \frac{\alpha_s(\mu_0^2)}{\alpha_s(\mu^2)} \gamma(m) \right)^n$$

$$= \mathcal{M}[\sigma_1](m,\mu_0) \, \exp\left(\frac{\gamma(m)}{2\pi b_0} \log \frac{\alpha_s(\mu_0^2)}{\alpha_s(\mu^2)} \right) \, . \tag{2.163}$$

This way we can write the Mellin transform of the $(n+1)$ particle production rate as the product of the n-particle rate times a ratio of the strong coupling at two scales

$$\boxed{\sum_{n=0}^{\infty} \mathcal{M}[\sigma_{n+1}](m,\mu) = \mathcal{M}[\sigma_1](m,\mu_0) \left(\frac{\alpha_s(\mu_0^2)}{\alpha_s(\mu^2)} \right)^{\frac{\gamma(m)}{2\pi b_0}} \, .} \tag{2.164}$$

This is the same structure as the DGLAP equation's solution in Eq. (2.149). It means that we should be able to understand the physics of the DGLAP equation using our model calculation of a gluon ladder emission, including the generically variable number of collinear jets in the form of $pp \to \mu^+\mu^- + X$, as shown in Eq. (2.150).

We should remind ourselves of the three assumptions we need to make to arrive at this form. There are two assumptions which concern the transverse momenta of the successive radiation: first, the global upper limit on all transverse momenta should be the factorization scale μ_F, with a strong ordering in the transverse momenta. This gives us a physical picture of the successive splittings as well as a physical interpretation of the factorization scale. Second, the strong coupling should be evaluated at the transverse momentum or factorization scale, so all scales are unified, in accordance with the derivation of the DGLAP equation.

Bending the rules of pure Yang–Mills QCD we can come back to the hard process σ_1 as the Drell–Yan process $q\bar{q} \to Z$. Each step in n means an additional parton in the final state, so σ_{n+1} is Z production with n collinear partons. On the left hand side of Eq. (2.164) we have the sum over any number of additional collinear partons; on the right hand side we see fixed order Drell–Yan production without any additional partons, but with an exponentiated correction factor. Comparing this to the running parton densities we can draw the analogy that any process computed with a scale dependent parton density where the scale dependence is governed by the DGLAP equation includes *any number* of collinear partons.

We can also identify the logarithms which are resummed by scale dependent parton densities. Going back to Eq. (2.87) reminds us that we start from the divergent collinear logarithms log $p_T^{\max}/p_T^{\min}$ arising from the collinear phase space integration. In our model for successive splitting we replace the upper boundary by μ_F. The collinear logarithm of successive initial–state parton splitting diverges for $\mu_0 \to 0$, but it gets absorbed into the parton densities and determines the structure of the DGLAP equation and its solutions. The upper boundary μ_F tells us to what extent we assume incoming quarks and gluons to be a coupled system of splitting partons and what the maximum momentum scale of these splittings is. Transverse momenta $p_T > \mu_F$ generated by hard parton splitting are not covered by the DGLAP equation and hence not a feature of the incoming partons anymore. They belong to the hard process and have to be consistently simulated, as we will see in Sects. 2.6.2 and 2.7. While this scale can be chosen freely we have to make sure that it does not become too large, because at some point the *collinear approximation* $C \simeq$ constant in Eq. (2.87) ceases to hold and with it our entire argument. Only if we do everything correctly, the DGLAP equation resums logarithms of the maximal *transverse momentum size* of the incoming gluon. They are universal and arise from simple kinematics.

The ordering of the splittings we have to assume is not relevant unless we simulate this splitting, as we will see in the next section. For the details of this we have to remember that our argument follows from the leading collinear approximation introduced in Sect. 2.3.1. Therefore, the strong p_T-ordering can in practice mean angular ordering or rapidity ordering, just applying a linear transformation.

2.4 Scales in LHC Processes

Looking back at Sects. 2.2 and 2.3 we introduce the factorization and renormalization scales step by step completely in parallel: first, computing perturbative higher order contributions to scattering amplitudes we encounter ultraviolet and infrared divergences. We regularize both of them using dimensional regularization with $n - 4 - 2\epsilon < 4$ for ultraviolet and $n > 4$ for infrared divergences, linked by analytic continuation. Both kinds of divergences are universal, which means that they are not process or observable dependent. This allows us to absorb ultraviolet and infrared divergences into a re-definition of the strong coupling and the parton density. This nominally infinite shift of parameters we refer to as renormalization for example of the strong coupling or as mass factorization absorbing infrared divergences into the parton distributions.

After renormalization as well as after mass factorization we are left with a *scale artifact*. Scales arise as part of a the pole subtraction: together with the pole $1/\epsilon$ we have a choice of finite contributions which we subtract with this pole. Logarithms of the renormalization and factorization scales will always be part of these finite terms. Moreover, in both cases the re-definition of parameters is not based on fixed order perturbation theory. Instead, it involves summing logarithms which otherwise can become large and spoil the convergence of our perturbative series in α_s. The only special feature of infrared divergences as compared to ultraviolet divergences is that to identify the resummed logarithms we have to unify both scales to one.

The hadronic production cross section for the Drell–Yan process or other LHC production channels, now including both scales, reads

$$\sigma_{\text{tot}}(\mu_F, \mu_R) = \int_0^1 dx_1 \int_0^1 dx_2 \sum_{ij} f_i(x_1, \mu_F) f_j(x_2, \mu_F)$$

$$\hat{\sigma}_{ij}(x_1 x_2 S, \alpha_s(\mu_R^2), \mu_F, \mu_R). \quad (2.165)$$

The Drell–Yan process has the particular feature that at leading order $\hat{\sigma}_{q\bar{q}}$ only involves weak couplings, it does not include α_s with its implicit renormalization scale dependence at leading order. Strictly speaking, in Eq. (2.165) the parton densities also depend on the renormalization scale because in their extraction we identify both scales. Carefully following their extraction we can separate the two scales if we need to. Lepton pair production and Higgs production in weak boson fusion are the two prominent electroweak production processes at the LHC.

The evolution of all running parameters from one renormalization/factorization scale to another is described either by renormalization group equation in terms of a beta function in the case of renormalization or by the DGLAP equation in the case of mass factorization. Our renormalization group equation for α_s is a single equation, but in general they are sets of coupled differential equations for all relevant parameters, which again makes them more similar to the DGLAP equation.

There is one formal difference between these two otherwise very similar approaches. The fact that we can absorb ultraviolet divergences into process–independent, universal counter terms is called renormalizability and has been proven to all orders for the kind of gauge theories we are dealing with. The universality of infrared splitting kernels has not (yet) in general been proven, but on the other hand we have never seen an example where is fails for sufficiently inclusive observables like production rates. For a while we thought there might be a problem with factorization in supersymmetric theories using the $\overline{\text{MS}}$ scheme, but this issue has been resolved. A summary of the properties of the two relevant scales for LHC physics we show in Table 2.2.

The way we introduce factorization and renormalization scales clearly labels them as an artifact of perturbation theories with divergences. What actually happens if we include *all orders* in perturbation theory? For example, the resummation of the self energy bubbles simply deals with one class of diagrams which have to be included, either order-by-order or rearranged into a resummation. Once we include all orders in perturbation theory it does not matter according to which combination of couplings and logarithms we order it. An LHC production rate will then not depend on arbitrarily chosen renormalization or factorization scales μ.

2.4 Scales in LHC Processes

Table 2.2 Comparison of renormalization and factorization scales appearing in LHC cross sections

	Renormalization scale μ_R	Factorization scale μ_F
Source	Ultraviolet divergence	Collinear (infrared) divergence
Poles cancelled	Counter terms (renormalization)	Parton densities (mass factorization)
Summation	Resum self energy bubbles	Resum parton splittings
Parameter	Running coupling $\alpha_s(\mu_R^2)$	Running parton density $f_j(x,\mu_F)$
Evolution	RGE for α_s	DGLAP equation
Large scales	Decrease of σ_{tot}	Increase of σ_{tot} for gluons/sea quarks
Theory background	Renormalizability Proven for gauge theories	Factorization Proven all orders for DIS Proven order-by-order DY...

Practically, in Eq. (2.165) we evaluate the renormalized parameters and the parton densities at some scale. This scale dependence will only cancel once we include all implicit and explicit appearances of the scales at all orders. Whatever scale we choose for the strong coupling or parton densities will eventually be compensated by explicit scale logarithms. In the ideal case, these logarithms are small and do not spoil perturbation theory. In a process with one distinct external scale, like the Z mass, we know that all scale logarithms should have the form $\log(\mu/m_Z)$. This logarithm vanishes if we evaluate everything at the 'correct' external energy scale, namely m_Z. In that sense we can think of the running coupling as a proper *running observable* which depends on the external energy of the process. This dependence on the external energy is not a perturbative artifact, because a cross section even to all orders does depend on the energy. The problem in particular for LHC analyses is that after analysis cuts every process will have more than one external energy scale.

We can turn around the argument of vanishing scale dependence to all orders in perturbation theory. This gives us an estimate of the minimum *theoretical error* on a rate prediction set by the scale dependence. The appropriate interval of what we consider reasonable scale choices depends on the process and the taste of the people doing this analysis. This error estimate is not at all conservative; for example the renormalization scale dependence of the Drell–Yan production rate or Higgs production in weak boson fusion is zero because α_s only enters are next–to–leading order. At the same time we know that the next–to–leading order correction to the Drell–Yan cross section is of the order of 30 %, which far exceeds the factorization scale dependence. Moreover, the different scaling behavior of a hadronic cross section shown in Table 2.2 implies that for example gluon–induced processes at typical x values around 10^{-2} show a cancellation of the factorization and renormalization scale variation. Estimating theoretical uncertainties from scale dependence therefore requires a good understanding of the individual process and the way it is affected by the two scales.

Guessing the right scale choice for a process is hard, often impossible. For example in Drell–Yan production at leading order there exists only one scale, m_Z. If we set $\mu = m_Z$ all scale logarithms vanish. In reality, LHC observables include several different scales. Some of them appear in the hard process, for example in the production of two or three particles with different masses. Others enter through the QCD environment where at the LHC we only consider final-state jets above a certain minimal transverse momentum. Even others appear though background rejection cuts in a specific analysis, for example when we only consider the Drell–Yan background for $m_{\mu\mu} > 1$ TeV to Kaluza–Klein graviton production. Using likelihood methods does not improve the situation because the phase space regions dominated by the signal will introduce specific energy scales which affect the perturbative prediction of the backgrounds. This is one of the reasons why an automatic comparison of LHC events with signal or background predictions is bound to fail once it requires an estimate of the theoretical uncertainty on the background simulation.

All that means that in practice there is no way to define a 'correct' scale. On the other hand, there are definitely *poor scale choices*. For example, using $1{,}000 \times m_Z$ as a typical scale in the Drell–Yan process will if nothing else lead to logarithms of the size log 1,000 whenever a scale logarithm appears. These logarithms eventually have to be cancelled to all orders in perturbation theory, inducing unreasonably large higher order corrections.

When describing jet radiation, we usually introduce a phase space dependent renormalization scale, evaluating the strong coupling at the transverse momentum of the radiated jet $\alpha_s(\vec{p}_{T,j}^{\,2})$. This choice gives the best kinematic distributions for the additional partons because in Sect. 2.3.5 we have shown that it resums large collinear logarithms.

The transverse momentum of a final–state particle is one of scale choices allowed by factorization; in addition to poor scale choices there also exist *wrong scale choices*, i.e. scale choices violating physical properties we need. Factorization or the Kinoshita–Lee–Nauenberg theorem which ensures that soft divergences cancel between real and virtual emission diagrams are such properties we should not violate—in QED the same property is called the Bloch–Nordsieck cancellation. Imagine picking a factorization scale defined by the partonic initial state, for example the partonic center–of–mass energy $s = x_1 x_2 S$. We know that this definition is not unique: for any final state it corresponds to the well defined sum of all momenta squared. However, virtual and real gluon emission generate different multiplicities in the final state, which means that the two sources of soft divergences only cancel until we multiply each of them with numerically different parton densities. Only scales which are uniquely defined in the final state can serve as factorization scales. For the Drell–Yan process such a scale could be m_Z, or the mass of heavy new-physics states in their production process. So while there is no such thing as a correct scale choice, there are more or less smart choices, and there are definitely very poor choices, which usually lead to an unstable perturbative behavior.

2.5 Parton Shower

In LHC phenomenology we are usually less interested in fixed-order perturbation theory than in logarithmically enhanced QCD effects. Therefore, we will not deepen our discussion of hadronic rates as shown in Eq. (2.165) based on fixed-order partonic cross sections convoluted with parton densities obeying the DGLAP equation. In Sect. 2.3.5 we have already seen that there exist more functions with the same structure as solutions to the DGLAP equation. In Sect. 2.5.1 we will derive one such object which we can use to describe jet radiation of incoming and outgoing partons in hard processes. These Sudakov factors will immediately lead us to a parton shower. They are based on universal patterns in jet radiation which we will in detail study in Sects. 2.5.2 and 2.5.3. In Sect. 2.5.4 we will introduce a key property of the parton shower, the ordered splitting of partons in several approximation.

2.5.1 Sudakov Form Factor

After introducing the kernels $\hat{P}_{i \leftarrow j}(z)$ as something like splitting probabilities we never applied a probabilistic approach to parton splitting. The basis of such an interpretation are *Sudakov form factors* describing the splitting of a parton i into any of the partons j based on the factorized form Eq. (2.99)

$$\Delta_i(t) \equiv \Delta_i(t, t_0) = \exp\left(-\sum_j \int_{t_0}^{t} \frac{dt'}{t'} \int_0^1 dy \, \frac{\alpha_s}{2\pi} \hat{P}_{j \leftarrow i}(y)\right). \quad (2.166)$$

Sudakov factors are an excellent example for technical terms hiding very simple concepts. If we instead referred to them as simple non–splitting probabilities everyone would immediately understand what we are talking about, taking away some of the mythical powers of theoretical physicists. The only parton splitting affecting a hard quark leg is $P_{q \leftarrow q}$ while a gluon leg can either radiate a gluon via $P_{g \leftarrow g}$ or $P_{q \leftarrow g}$. The fourth allowed splitting $P_{g \leftarrow q}$ also splits a quark into a quark-gluon pair, so we can decide to follow the quark direction instead of switching over to the gluon. We derive the form of the Sudakov factors for Δ_q,

$$\Delta_q(t) = \exp\left(-\int_{t_0}^{t} \frac{dt'}{t'} \int_0^1 dy \, \frac{\alpha_s}{2\pi} \hat{P}_{q \leftarrow q}(y)\right). \quad (2.167)$$

The unregularized splitting kernel is given by Eq. (2.118), so we can compute

$$\int_0^1 dz \, \frac{\alpha_s}{2\pi} \, \hat{P}_{q \leftarrow q}(y) = \frac{C_F}{2\pi} \int_0^1 dy \, \alpha_s \, \frac{1+y^2}{1-y}$$
$$= \frac{C_F}{2\pi} \int_0^1 dy \, \alpha_s \, \frac{-(1-y)(1+y) + 2}{1-y}$$
$$= \frac{C_F}{2\pi} \int_0^1 dy \, \alpha_s \left(\frac{2}{1-y} - 1 - y \right). \quad (2.168)$$

To compute the divergent first term we shift the y integration to $t'' = (1-y)^2 t$, which gives us

$$\frac{dt''}{dy} = \frac{d}{dy}(1-y)^2 t = 2(1-y)(-1)t = -2\frac{t''}{1-y} \quad \Leftrightarrow \quad \frac{dy}{1-y} = -\frac{1}{2}\frac{dt''}{t''}. \quad (2.169)$$

Without derivation we quote that the t'' integration, which naively has a lower boundary at zero is cut off at t'. In addition, we approximate $y \to 1$ wherever possible and find for the leading contributions to the splitting integral

$$\int_0^1 dz \, \frac{\alpha_s}{2\pi} \, \hat{P}_{q \leftarrow q}(y)$$
$$= \frac{C_F}{2\pi} \int_0^1 dy \, \alpha_s \left(\frac{2}{1-y} - 1 - y \right)$$
$$= \frac{C_F}{2\pi} \left(\int_{t'}^t dt'' \, \frac{\alpha_s}{t''} - \int_0^1 dy \, \alpha_s \, (1+y) \right)$$
$$= \frac{C_F}{2\pi} \alpha_s \left(\int_{t'}^t dt'' \, \frac{1}{t''} - \int_0^1 dy \, (1+y) \right) \qquad \text{leading power dependence on } y \text{ and } t''$$
$$= \frac{C_F}{2\pi} \alpha_s \left(\log \frac{t}{t'} - \frac{3}{2} \right) \equiv t' \, \Gamma_{q \leftarrow q}(t, t') . \quad (2.170)$$

The argument of the strong coupling was originally $y^2(1-y)^2 t'$, which turns into t' in the limit $y \to 1$. This way we can express all Sudakov factors in terms of splitting functions Γ_j,

$$\Delta_q(t) = \exp\left(-\int_{t_0}^t dt' \, \Gamma_{q \leftarrow q}(t, t') \right)$$
$$\Delta_g(t) = \exp\left(-\int_{t_0}^t dt' \, [\Gamma_{g \leftarrow g}(t, t') + \Gamma_{q \leftarrow g}(t')] \right), \quad (2.171)$$

which to leading logarithm in $\log t/t'$ read

2.5 Parton Shower

$$\Gamma_{q \leftarrow q}(t,t') = \frac{C_F}{2\pi} \frac{\alpha_s(t')}{t'} \left(\log \frac{t}{t'} - \frac{3}{2} \right)$$

$$\Gamma_{g \leftarrow g}(t,t') = \frac{C_A}{2\pi} \frac{\alpha_s(t')}{t'} \left(\log \frac{t}{t'} - \frac{11}{6} \right)$$

$$\Gamma_{q \leftarrow g}(t') = \frac{n_f}{6\pi} \frac{\alpha_s(t')}{t'} \,. \tag{2.172}$$

These formulas have a slight problem: terms arising from next–to–leading logarithms spoil the limit $t' \to t$, where a splitting probability should vanish. Technically, we can deal with the finite terms in the Sudakov factors by requiring them to be positive semi–definite, i.e. by replacing $\Gamma(t,t') < 0$ with zero. For the general argument this problem with the analytic expressions for the splitting functions is irrelevant.

Before we check that the Sudakov factors obey the DGLAP equation we confirm that such exponentials appear in probabilistic arguments, similar to our discussion of the central jet veto in Sect. 1.6.2. Using Poisson statistics for something expected to occur p times, the probability of observing it n times is given by

$$\mathscr{P}(n;p) = \frac{p^n e^{-p}}{n!} \qquad \mathscr{P}(0;p) = e^{-p} \,. \tag{2.173}$$

If the exponent in the Sudakov form factor in Eq. (2.166) describes the integrated splitting probability of a parton i this means that the Sudakov itself describes a *non–splitting probability* of the parton i into any final state j.

Based on such probabilistic Sudakov factors we can use a Monte Carlo, which is a *Markov process* without a memory of individual past steps, to compute a chain of parton splittings as depicted in Fig. 2.2. This will describe a quark or a gluon propagating forward in time. Starting from a point (x_1, t_1) in momentum–virtuality space we step by step move to the next splitting point (x_j, t_j). Following the original discussion t_2 is the target virtuality at x_2, and for time–like final–state branching the virtuality is positive $t_j > 0$ in all points j. The Sudakov factor is a function of t, so it gives us the probability of not seeing any branching between t_1 and t_2 as $\Delta(t_1)/\Delta(t_2) < 1$. The appropriate cutoff scale t_0 drops out of this ratio. Using a flat random number r_t the t_2 distribution is implicitly given by the solution to

$$\frac{\Delta(t_1)}{\Delta(t_2)} = r_t \in [0,1] \qquad \text{with} \quad t_1 > t_2 > t_0 > 0 \,. \tag{2.174}$$

Beyond the absolute cutoff scale t_0 we assume that no resolvable branching occurs.

In a second step we need to compute the matching energy fraction x_2 or the ratio x_2/x_1 describing the momentum fraction which is kept in the splitting at x_2. The y integral in the Sudakov factor in Eq. (2.166) gives us this probability distribution which we can again implicitly solve for x_2 using a flat random number r_x

$$\frac{\int_0^{x_2/x_1} dy \frac{\alpha_s}{2\pi} \hat{P}(y)}{\int_0^1 dy \frac{\alpha_s}{2\pi} \hat{P}(y)} = r_x \in [0, 1] \qquad \text{with } x_1 > x_2 > 0 \ . \qquad (2.175)$$

For splitting kernels with soft divergences at $y = 0$ or $y = 1$ we should include a numerical cutoff in the integration because the probabilistic Sudakov factor and the parton shower do not involve the regularized splitting kernels.

Of the four momentum entries of the radiated parton the two equations (2.174) and (2.175) give us two. The on–shell mass constraint fixes a third, so all we are left is the azimuthal angle distribution. We know from symmetry arguments that QCD splitting is insensitive to this angle, so we can generate it randomly between zero and 2π. For final–state radiation this describes probabilistic branching in a *Monte Carlo program*, just based on Sudakov form factors.

The same statement for initial–state radiation including parton densities we will put on a more solid or mathematical footing. The derivative of the Sudakov form factor Eq. (2.166)

$$\frac{1}{\Delta_i(t)} \frac{d\Delta_i(t)}{dt} = -\sum_j \frac{1}{t} \int_0^1 dy \frac{\alpha_s}{2\pi} \hat{P}_{j \leftarrow i}(y) \qquad (2.176)$$

is precisely the second term in $df(x,t)/dt$ for diagonal splitting, as shown in Eq. (2.127)

$$\frac{df_i(x,t)}{dt} = \frac{1}{t} \sum_j \left[\int_0^1 \frac{dz}{z} \frac{\alpha_s}{2\pi} \hat{P}_{i \leftarrow j}(z) f_j\left(\frac{x}{z}, t\right) - \int_0^1 dy \frac{\alpha_s}{2\pi} \hat{P}_{j \leftarrow i}(y) f_i(x,t) \right]$$

$$= \frac{1}{t} \sum_j \int_0^1 \frac{dz}{z} \frac{\alpha_s}{2\pi} \hat{P}_{i \leftarrow j}(z) f_j\left(\frac{x}{z}, t\right) + \frac{f_i(x,t)}{\Delta_i(t)} \frac{d\Delta_i(t)}{dt} \ . \qquad (2.177)$$

This relation suggests to consider the derivative of the f_i/Δ_i instead of the Sudakov factor alone to obtain something like the DGLAP equation

$$\frac{d}{dt} \frac{f_i(x,t)}{\Delta_i(t)} = \frac{1}{\Delta_i(t)} \frac{df_i(x,t)}{dt} - \frac{f_i(x,t)}{\Delta_i(t)^2} \frac{d\Delta_i(t)}{dt}$$

$$= \frac{1}{\Delta_i(t)} \left(\frac{1}{t} \sum_j \int_0^1 \frac{dz}{z} \frac{\alpha_s}{2\pi} \hat{P}_{i \leftarrow j}(z) f_j\left(\frac{x}{z}, t\right) + \frac{f_i(x,t)}{\Delta_i(t)} \frac{d\Delta_i(t)}{dt} \right)$$

$$- \frac{f_i(x,t)}{\Delta_i(t)^2} \frac{d\Delta_i(t)}{dt}$$

$$= \frac{1}{\Delta_i(t)} \frac{1}{t} \sum_j \int_0^{1-\epsilon} \frac{dz}{z} \frac{\alpha_s}{2\pi} \hat{P}_{i \leftarrow j}(z) f_j\left(\frac{x}{z}, t\right) \ . \qquad (2.178)$$

2.5 Parton Shower

In the last step we cancel what corresponds to the plus prescription for diagonal splitting, which means we remove the regularization of the splitting kernel at $z \to 1$. Therefore, we need to modify the upper integration boundary by a small parameter ϵ which can in principle depend on t. The resulting equation is the diagonal DGLAP equation with unsubtracted splitting kernels, solved by the ratio of parton densities and Sudakov factors

$$\boxed{t \frac{d}{dt} \frac{f_i(x,t)}{\Delta_i(t)} = \frac{d}{d \log t} \frac{f_i(x,t)}{\Delta_i(t)} = \sum_j \int_0^{1-\epsilon} \frac{dz}{z} \frac{\alpha_s}{2\pi} \hat{P}_{i \leftarrow j}(z) \frac{f_j\left(\frac{x}{z}, t\right)}{\Delta_i(t)}.}$$
(2.179)

We can study the structure of these solutions of the unsubtracted DGLAP equation by integrating f/Δ between appropriate points in t

$$\frac{f_i(x,t)}{\Delta_i(t)} - \frac{f_i(x,t_0)}{\Delta_i(t_0)} = \int_{t_0}^{t} \frac{dt'}{t'} \sum_j \int_0^{1-\epsilon} \frac{dz}{z} \frac{\alpha_s}{2\pi} \hat{P}_{i \leftarrow j}(z) \frac{f_j\left(\frac{x}{z}, t'\right)}{\Delta_i(t')}$$

$$f_i(x,t) = \frac{\Delta_i(t)}{\Delta_i(t_0)} f_i(x,t_0) + \int_{t_0}^{t} \frac{dt'}{t'} \frac{\Delta_i(t)}{\Delta_i(t')} \sum_j \int_0^{1-\epsilon} \frac{dz}{z} \frac{\alpha_s}{2\pi} \hat{P}_{i \leftarrow j}(z) f_j\left(\frac{x}{z}, t'\right)$$

$$= \Delta_i(t) f_i(x,t_0) + \int_{t_0}^{t} \frac{dt'}{t'} \frac{\Delta_i(t)}{\Delta_i(t')} \sum_j \int_0^{1-\epsilon} \frac{dz}{z} \frac{\alpha_s}{2\pi} \hat{P}_{i \leftarrow j}(z) f_j\left(\frac{x}{z}, t'\right)$$

$$= \Delta_i(t, t_0) f_i(x,t_0) + \int_{t_0}^{t} \frac{dt'}{t'} \Delta_i(t, t') \sum_j \int_0^{1-\epsilon} \frac{dz}{z} \frac{\alpha_s}{2\pi} \hat{P}_{i \leftarrow j}(z) f_j\left(\frac{x}{z}, t'\right),$$
(2.180)

where we choose t_0 such that $\Delta(t_0) = 1$ and introduce the notation $\Delta(t_1, t_2) = \Delta(t_1, t_0)/\Delta(t_2, t_0)$ for the ratio of two Sudakov factors in the last line. This equation is a *Bethe–Salpeter equation* describing the dependence of the parton density $f_i(x,t)$ on x and t. It has a suggestive interpretation: corresponding to Eq. (2.173) the first term can be interpreted as 'nothing happening to f between t_0 and t' because it is weighted by the Sudakov no-branching probability $\Delta_i(t, t_0)$. The second term includes the ratio of Sudakov factors which just like in Eq. (2.174) means no branching between t' and t. Integrating this factor times the splitting probability over $t' \epsilon [t_0, t]$ implies at least one branching between t_0 and t.

The key to using this probabilistic interpretation of the Sudakov form factor in conjunction with the parton densities is its numerical usability in a probabilistic approach: starting from a parton density somewhere in $(x-t)$ space we need to evolve it to a fixed point (x_n, t_n) given by the hard subprocess, e.g. $q\bar{q} \to Z$ with m_Z giving the scale and energy fraction of the two quarks. Numerically it would be much easier to simulate *backwards evolution* where we start from the known kinematics of the hard process and the corresponding point in the $(x-t)$ plane and evolve towards the partons in the proton, ideally to a point where the probabilistic picture of collinear, stable, non–radiating quarks and gluons in the proton holds. This means we need to define a probability that a parton evolved backwards from a space–like $t_2 < 0$ to $t_1 < 0$ with $|t_2| > |t_1|$ does not radiate or split.

For this final step we define a probability measure for the backwards evolution of partons $\Pi(t_1, t_2; x)$. Just like the two terms in Eq. (2.180) it links the splitting probability to a probability of an undisturbed evolution. For example, we can write the probability that a parton is generated by a splitting in the interval $[t, t+\delta t]$, evaluated at (t_2, x), as $dF(t; t_2)$. The measure corresponding to a Sudakov survival probability is then

$$\Pi(t_1, t_2; x) = 1 - \int_{t_1}^{t_2} dF(t; t_2) . \tag{2.181}$$

Comparing the definition of dF to the relevant terms in Eq. (2.180) and replacing $t \to t_2$ and $t' \to t$ we know what happens for the combination

$$\begin{aligned}
f_i(x, t_2) dF(t; t_2) &= \frac{dt}{t} \frac{\Delta_i(t_2)}{\Delta_i(t)} \sum_j \int_0^{1-\epsilon} \frac{dz}{z} \frac{\alpha_s}{2\pi} \hat{P}_{i \leftarrow j}(z) f_j\left(\frac{x}{z}, t\right) \\
&= dt\, \Delta_i(t_2) \frac{1}{t} \sum_j \int_0^{1-\epsilon} \frac{dz}{z} \frac{\alpha_s}{2\pi} \hat{P}_{i \leftarrow j}(z) \frac{f_j\left(\frac{x}{z}, t\right)}{\Delta_i(t)} \\
&= dt\, \Delta_i(t_2) \frac{d}{dt} \frac{f_i(x, t)}{\Delta_i(t)} \qquad \text{using Eq. (2.179)} .
\end{aligned} \tag{2.182}$$

This means

$$\Pi(t_1, t_2; x) = 1 - \left. \frac{f_i(x,t) \Delta_i(t_2)}{f_i(x, t_2) \Delta_i(t)} \right|_{t_1}^{t_2} = \frac{f_i(x, t_1) \Delta_i(t_2)}{f_i(x, t_2) \Delta_i(t_1)} , \tag{2.183}$$

2.5 Parton Shower

and gives us a probability measure for backwards evolution: the probability of evolving back from t_2 to t_1 is described by a Markov process with a flat random number as

$$\frac{f_i(x,t_1)\Delta_i(t_2)}{f_i(x,t_2)\Delta_i(t_1)} = r \in [0,1] \qquad \text{with } |t_2| > |t_1|. \qquad (2.184)$$

While we cannot write down this procedure in a closed form, it shows how we can algorithmically generate initial state as well as final state parton radiation patterns based on the unregularized DGLAP equation and the Sudakov factors solving this equation. One remaining issue is that in our derivation of the collinear resummation interpretation of the parton shower we assume some a strong ordering of the radiated partons which we will discuss in the next section.

2.5.2 Multiple Gluon Radiation

Following Eqs. (2.174) and (2.184) the parton shower is fundamentally a statistical approach. Sudakov form factors are nothing by no-emission probabilities. If we limit ourselves only to abelian splitting, i.e. radiating gluons off hard quark legs, the parton shower generates a statistical distribution of the number of radiated gluons. This guarantees finite results even in the presence of different infrared divergences. To understand the picture of parton splitting in terms of a Poisson process it is most instructive to consider soft gluon emission of a quark, ignoring the gluon self coupling. In other words, we study soft photon emission off an electron leg simply adding color factors C_F.

To this point we have built our parton shower on collinear parton splitting or radiation and its universal properties indicated by Eq. (2.99). Deriving the diagonal splitting kernels in Eqs. (2.105) and (2.118) we encounter an additional source of infrared divergences, namely *soft gluon emission* corresponding to energy fractions $z \to 0, 1$. Its radiation pattern is also universal, just like the collinear case. One way to study this soft divergence without an overlapping collinear pole is gluon radiation off a massless or massive hard quark leg

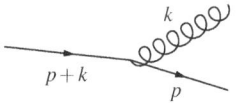

The original massive quark leg with momentum $p + k$ and mass m could be attached to some hard process as a splitting final state. It splits into a hard quark p and a soft gluon k. The general matrix element without any approximation reads

$$\mathcal{M}_{n+1} = g_s T^a \, \epsilon_\mu^*(k) \, \bar{u}(p) \gamma^\mu \frac{\slashed{p} + \slashed{k} + m}{(p+k)^2 - m^2} \, \mathcal{M}_n$$

$$= g_s T^a \, \epsilon_\mu^*(k) \, \bar{u}(p) \left[-\slashed{p}\gamma^\mu + 2p^\mu + m\gamma^\mu + \gamma^\mu \slashed{k} \right] \frac{1}{2(pk) + k^2} \, \mathcal{M}_n$$

$$= g_s T^a \, \epsilon_\mu^*(k) \, \bar{u}(p) \frac{2p^\mu + \gamma^\mu \slashed{k}}{2(pk) + k^2} \, \mathcal{M}_n \,, \tag{2.185}$$

using the Dirac equation $\bar{u}(p)(\slashed{p} - m) = 0$. At this level, a further simplification requires for example the soft gluon limit. In the presence of only hard momenta, except for the gluon, we can define it for example as $k^\mu = \lambda p^\mu$, where p^μ is an arbitrary four-vector combination of the surrounding hard momenta. The small parameter λ then characterizes the soft limit. For the invariant mass of the gluon we assume $k^2 = \mathcal{O}(\lambda^2)$, allowing for a slightly off–shell gluon. We find

$$\mathcal{M}_{n+1} = g_s T^a \, \epsilon_\mu^*(k) \, \bar{u}(p) \frac{p^\mu + \mathcal{O}(\lambda)}{(pk) + \mathcal{O}(\lambda^2)} \, \mathcal{M}_n$$

$$\sim g_s T^a \, \epsilon_\mu^*(k) \, \frac{p^\mu}{(pk)} \, \bar{u}(p) \, \mathcal{M}_n$$

$$\to g_s \, \epsilon_\mu^*(k) \left(\sum_j \hat{T}_j^a \frac{p_j^\mu}{(p_j k)} \right) \bar{u}(p) \, \mathcal{M}_n \tag{2.186}$$

The conventions are similar to Eq. (2.105), so $\mathcal{M}_n$ includes all additional terms except for the spinor of the outgoing quark with momentum $p + k$. Neglecting the gluon momentum altogether defines the leading term of the *eikonal approximation*.

In the last step of Eq. (2.186) we simply add all possible sources j of gluon radiation. This defines a color operator which we insert into the matrix element and which assumes values of $+T_{ij}^a$ for radiation off a quark, $-T_{ji}^a$ for radiation off an antiquark and $-if_{abc}$ for radiation off a gluon. For a color neutral process like our favorite Drell–Yan process adding an additional soft gluon $q\bar{q} \to Zg$ it returns $\sum_j \hat{T}_j = 0$. For a full QCD calculation, we would need to add single gluon radiation with a subsequent gluon splitting via the self interaction. This diagram does not appear in the QED case, it spoils our argument below, and it is not suppressed by any good arguments. In the following, we nevertheless strictly limit ourselves to the abelian part of QCD, i.e. gluon radiation off quarks and $f_{abc} \to 0$. This also means that all color factors are real. For the argument below we can think of gluon radiation off the process $e^+ e^- \to q\bar{q}$:

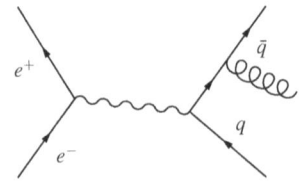

2.5 Parton Shower

The sum over the gluon radiation dipoles in Eq. (2.186) covers the two quarks in the final state. Next, we need to square this matrix element. It includes a polarization sum and will therefore depend on the gauge. We choose the general *axial gauge* for massless gauge bosons

$$\sum_{\text{pol}} \epsilon_\mu^*(k)\epsilon_\nu(k) = -g_{\mu\nu} + \frac{k_\mu n_\nu + n_\mu k_\nu}{(nk)} - n^2 \frac{k_\mu k_\nu}{(nk)^2} = -g_{\mu\nu} + \frac{k_\mu n_\nu + n_\mu k_\nu}{(nk)},$$
(2.187)

with a light-like reference vector n obeying $n^2 = 0$. The matrix element squared then reads

$$\overline{|\mathcal{M}_{n+1}|^2} = g_s^2 \left(-g_{\mu\nu} + \frac{k_\mu n_\nu + n_\mu k_\nu}{(nk)}\right) \left(\sum_j \hat{T}_j^a \frac{p_j^\mu}{(p_j k)}\right)^\dagger \left(\sum_j \hat{T}_j^a \frac{p_j^\nu}{(p_j k)}\right) \overline{|\mathcal{M}_n|^2}$$

$$= g_s^2 \left(-\left(\sum_j \hat{T}_j^a \frac{p_j^\mu}{(p_j k)}\right)^\dagger \left(\sum_j \hat{T}_j^a \frac{p_{j\mu}}{(p_j k)}\right)\right.$$

$$\left. + \frac{2}{(nk)} \left(\sum_j \hat{T}_j^a\right)^\dagger \left(\sum_j \hat{T}_j^a \frac{(p_j n)}{(p_j k)}\right)\right) \overline{|\mathcal{M}_n|^2}$$

$$= -g_s^2 \left(\sum_j \hat{T}_j^a \frac{p_j^\mu}{(p_j k)}\right)^\dagger \left(\sum_j \hat{T}_j^a \frac{p_{j\mu}}{(p_j k)}\right) \overline{|\mathcal{M}_n|^2}.$$
(2.188)

The insertion operator in the matrix element has the form of an insertion current multiplied by its hermitian conjugate. This current describes the universal form of soft gluon radiation off an n-particle process

$$\boxed{\overline{|\mathcal{M}_{n+1}|^2} \equiv -g_s^2 \, (J^\dagger \cdot J) \, \overline{|\mathcal{M}_n|^2}} \qquad \text{with} \quad J^{a\mu}(k,\{p_j\}) = \sum_j \hat{T}_j^a \frac{p_j^\mu}{(p_j k)}.$$
(2.189)

We can further simplify the squared current appearing in Eq. (2.188) to

$$(J^\dagger \cdot J) = \sum_j \hat{T}_j^a \hat{T}_j^a \frac{p_j^2}{(p_j k)^2} + 2 \sum_{i<j} \hat{T}_i^a \hat{T}_j^a \frac{(p_i p_j)}{(p_i k)(p_j k)}$$

$$= \sum_j \hat{T}_j^a \left(-\sum_{i \neq j} \hat{T}_i^a\right) \frac{p_j^2}{(p_j k)^2} + 2 \sum_{i<j} \hat{T}_i^a \hat{T}_j^a \frac{(p_i p_j)}{(p_i k)(p_j k)}$$

$$= -\left(\sum_{i<j}+\sum_{i>j}\right) \hat{T}_i^a \hat{T}_j^a \frac{p_j^2}{(p_jk)^2} + 2\sum_{i<j} \hat{T}_i^a \hat{T}_j^a \frac{(p_ip_j)}{(p_ik)(p_jk)}$$

$$= 2\sum_{i<j} \hat{T}_i^a \hat{T}_j^a \left(\frac{(p_ip_j)}{(p_ik)(p_jk)} - \frac{p_i^2}{2(p_ik)^2} - \frac{p_j^2}{2(p_jk)^2}\right)$$

$$= 2\sum_{i<j} \hat{T}_i^a \hat{T}_j^a \frac{(p_ip_j)}{(p_ik)(p_jk)} \qquad \text{for massless partons}$$

$$= 2\sum_{i<j} \hat{T}_i^a \hat{T}_j^a \frac{(p_ip_j)}{(p_ik)+(p_jk)} \left(\frac{1}{(p_ik)} + \frac{1}{(p_jk)}\right) . \tag{2.190}$$

In the last step we only bring the eikonal factor into a different form which sometimes comes in handy because it separates the two divergences associated with p_i and with p_j

At this point we return to massless QCD partons, keeping in mind that the ansatz Eq. (2.186) ensures that the insertion currents only model soft, not collinear radiation. Just as a side remark at this stage—our definition of the insertion current $J^{a\mu}$ in Eq. (2.189) can be generalized to colored processes, where the current becomes dependent on the gauge vector n to cancel the n dependence of the polarization sum

$$J^{a\mu}(k,\{p_j\}) = \sum_j \hat{T}_j^a \left(\frac{p_j}{(p_jk)} - \frac{n}{(nk)}\right) \tag{2.191}$$

We will study this *dipole radiation term* in Eqs. (2.189) and (2.191) in detail. Calling them dipoles is a little bit of a stretch if we compare it to a multipole series. To see the actual dipole structure we would need to look at the color structure.

Based on Eq. (2.188) for the eikonal limit we can write out the infinitesimal cross sections for soft gluon emission off a massless quark. The difference between the usual QED calculation and our QCD version of it are the color factors. If we take for example the Feynman diagram for the Drell–Yan process we see that the color factor for gluon ration off the outgoing quark diagram squared is $\hat{T}^a \hat{T}^a = -\text{Tr}(T^a T^a) = -(N_c^2 - 1)/2$. Similarly to the derivation of the splitting kernels we need to include a factor $1/(2N_c)$ to account for the averaging over the states of the intermediate quark. The phase space factor for any number of final state gluons we postpone at this stage. In terms of the momenta p_1 and p_2 of the outgoing quark and antiquark Eq. (2.190) gives us for the fully massive case

2.5 Parton Shower

$$d\sigma_{n+1} = g_s^2 \, \text{Tr}(T^a T^a) \frac{1}{2N_c} \, d\sigma_n \int \frac{d^3k}{(2\pi)^3 2k_0} \left(\frac{(p_1 p_2)}{(p_1 k)(p_2 k)} - \frac{p_1^2}{2(p_1 k)^2} - \frac{p_2^2}{2(p_2 k)^2} \right)$$

$$= g_s^2 d\sigma_n \frac{N_c^2 - 1}{4N_c} \int \frac{d^3k}{(2\pi)^3 2k_0} \left(\frac{(p_1 p_2)}{(p_1 k)(p_2 k)} - \frac{m^2}{2(p_1 k)^2} - \frac{m^2}{2(p_2 k)^2} \right). \tag{2.192}$$

Because we are interested in the dependence on the gluon energy it is not convenient to stick to kinematic invariants. Instead, we compute the phase space integral over the additional gluon momentum in a specific reference frame. We choose both, the quark and the antiquark energy to be the same $E_1 = E_2 \equiv E_q$, with the corresponding three-momenta $E_q \vec{v}_j$. The three-momentum of the gluon has the direction $\hat{k}$. In this frame we find

$$\frac{(p_1 p_2)}{(p_1 k)(p_2 k)} - \frac{m^2}{2(p_1 k)^2} - \frac{m^2}{2(p_2 k)^2}$$

$$= \frac{E_q^2(1 - \vec{v}_1 \vec{v}_2)}{E_q^2 k_0^2 (1 - \vec{v}_1 \hat{k})(1 - \vec{v}_2 \hat{k})} - \frac{m^2}{2 E_q^2 k_0^2 (1 - \vec{v}_1 \hat{k})^2} - \frac{m^2}{2 E_q^2 k_0^2 (1 - \vec{v}_2 \hat{k})^2}$$

$$= \frac{1}{k_0^2} \left(\frac{(1 - \vec{v}_1 \vec{v}_2)}{(1 - \vec{v}_1 \hat{k})(1 - \vec{v}_2 \hat{k})} - \frac{m^2}{2 E_q^2 (1 - \vec{v}_1 \hat{k})^2} - \frac{m^2}{2 E_q^2 (1 - \vec{v}_2 \hat{k})^2} \right). \tag{2.193}$$

The numerical most relevant axes in the angular integral over the gluon momentum direction $\hat{k}$ appear when the scalar products in three dimensions give $\vec{v}_j \hat{k} = 1$. We first deal with the second and third integrals in Eq. (2.193), using massless polar coordinates $d^3k = k_0^2 dk_0 d\cos\theta_k d\phi_k = 2\pi k_0^2 dk_0 d\cos\theta_k$,

$$\int \frac{d^3k}{(2\pi)^3 2k_0} \frac{1}{k_0^2} \frac{m^2}{2 E_q^2 (1 - \vec{v} \hat{k})^2} = \frac{1}{8\pi^2} \frac{m^2}{E_q^2} \int \frac{dk_0}{2k_0} \int_{-1}^{1} d\cos\theta_k \frac{1}{(1 - |\vec{v}| \cos\theta_k)^2}$$

$$= \frac{1}{8\pi^2} \frac{m^2}{E_q^2} \int \frac{dk_0}{2k_0} \left[\frac{(-1)}{|\vec{v}|} \frac{1}{|\vec{v}| \cos\theta_k - 1} \right]_{-1}^{1}$$

$$= -\frac{1}{8\pi^2} \frac{m^2}{E_q^2} \int \frac{dk_0}{2k_0} \frac{1}{|\vec{v}|} \left(\frac{1}{|\vec{v}| - 1} + \frac{1}{|\vec{v}| + 1} \right)$$

$$= -\frac{1}{8\pi^2} \frac{m^2}{E_q^2} \int \frac{dk_0}{2k_0} \frac{1}{|\vec{v}|} \frac{2|\vec{v}|}{|\vec{v}|^2 - 1}$$

$$= -\frac{m^2}{8\pi^2} \int \frac{dk_0}{k_0} \frac{1}{|\vec{p}_q|^2 - E_q^2}$$

$$= \frac{1}{4\pi^2} \int \frac{dk_0}{k_0} = \frac{1}{4\pi^2} \log \frac{k_0^{\max}}{k_0^{\min}}. \tag{2.194}$$

This result is logarithmically divergent in the limit of soft gluon radiation, but there are not issues with the angular integration over the gluon phase space and collinear configurations.

The first term in Eq. (2.193) has a more complex divergence structure. We can separate the two poles in the integrand and approximate the integrand by the respective residues,

$$\int \frac{d^3k}{(2\pi)^3 2k_0} \frac{1}{k_0^2} \frac{(1-\vec{v}_1\vec{v}_2)}{(1-\vec{v}_1\hat{k})(1-\vec{v}_2\hat{k})}$$

$$\simeq \int \frac{d^3k}{(2\pi)^3 2k_0} \frac{1}{k_0^2} \left(\frac{(1-\vec{v}_1\vec{v}_2)}{(1-\vec{v}_1\hat{k}_{\text{pole}})(1-\vec{v}_2\vec{v}_1)} \right.$$

$$\left. + \frac{(1-\vec{v}_1\vec{v}_2)}{(1-\vec{v}_1\vec{v}_2)(1-\vec{v}_2\hat{k}_{\text{pole}})} \right)$$

$$= \int \frac{d^3k}{(2\pi)^3 2k_0} \frac{1}{k_0^2} \left(\frac{1}{1-\vec{v}_1\hat{k}} + \frac{1}{1-\vec{v}_2\hat{k}} \right)_{\text{pole}}$$

$$= \frac{1}{8\pi^2} \int \frac{dk_0}{k_0} \int d\cos\theta_k \left(\frac{1}{1-\vec{v}_1\hat{k}} + \frac{1}{1-\vec{v}_2\hat{k}} \right)_{\text{pole}}$$

$$= \frac{1}{4\pi^2} \int \frac{dk_0}{k_0} \int d\cos\theta_k \frac{1}{1-\cos\theta_k} \qquad \text{with appropriate reference axes}$$

$$= \frac{1}{4\pi^2} \log \frac{k_0^{\max}}{k_0^{\min}} \log \frac{1-\cos\theta_k^{\max}}{1-\cos\theta_k^{\min}} . \qquad (2.195)$$

In this calculation we neglect effects of order m^2/E_q^2 when identifying $\vec{v}\hat{k} = \cos\theta_k$. We see that in the double integral of Eq. (2.195) both parts diverge logarithmically, usually referred to as the *Sudakov double logarithm*. The first integral develops an infrared divergence in the gluon energy k_0 when the gluon becomes soft, $k_0^{\min} \to 0$. The second integral diverges when the gluon is radiated collinearly with the hard quark or antiquark, $\cos\theta_k^{\max} \to 1$. The integrals in Eq. (2.194) are less divergent, so we can neglect them in the corresponding differential cross sections of Eq. (2.192)

$$d\sigma_{n+1} = d\sigma_n \frac{g_s^2 C_F}{2} \frac{1}{4\pi^2} \log \frac{k_0^{\max}}{k_0^{\min}} \log \frac{1-\cos\theta_k^{\max}}{1-\cos\theta_k^{\min}}$$

$$= d\sigma_n \frac{\alpha_s C_F}{2\pi} \log \frac{k_0^{\max}}{k_0^{\min}} \log \frac{1-\cos\theta_k^{\max}}{1-\cos\theta_k^{\min}} . \qquad (2.196)$$

To be able to continue with our calculation we resort to an obvious regularization scheme—the detector. Arbitrarily soft photons leave no trace in a calorimeter, so

2.5 Parton Shower

they are not observable. The detector threshold acts as a finite cutoff $k_0^{\min} > 0$. Similarly, a tracker cannot separate two tracks which are arbitrarily close to each other, which means that its resolution limits the $\cos\theta_k$ range.

From the general principles of field theory we know that soft divergences cancel once we combine virtual gluon exchange diagrams and real gluon emission at the same order in perturbation theory. The soft cutoff in the k_0 integration we assume to be linked between real and virtual diagrams using a Wick rotation. For QCD this is called the Kinoshita–Lee–Nauenberg theorem. The relevant Feynman diagrams are propagator corrections for the m^2 terms in Eq. (2.192) and vertex corrections for the double divergences. Certainly, the leading overlapping soft and collinear divergences in Eq. (2.195) should vanish after we combine real and virtual QCD corrections for the Drell–Yan process. This means that after adding virtual corrections we can assume $k_0^{\min}$ to be an experimental constraint without any issues in the limit $k_0^{\min} \to 0$. If everything is well defined we can exponentiate the successive dependence of Eq. (2.196). The only complication is that now we have to include the correction for the phase space integration of the many gluons in the final state. If we declare n the number of gluons radiated this factor is $1/n!$. The observable we are interested in is the cross section for any number of radiated gluons, for which we find

$$d\sigma_{n+1} = d\sigma_0 \frac{1}{(n+1)!} \left(-\frac{\alpha_s C_F}{2\pi} \log\frac{k_0^{\max}}{k_0^{\min}} \log\frac{1-\cos\theta_k^{\min}}{1-\cos\theta_k^{\max}}\right)^{n+1}$$

$$\Rightarrow \quad \sigma_{\text{tot}} = \sigma_0 \sum_n \frac{1}{n!} \left(-\frac{\alpha_s C_F}{2\pi} \log\frac{k_0^{\max}}{k_0^{\min}} \log\frac{1-\cos\theta_k^{\min}}{1-\cos\theta_k^{\max}}\right)^n$$

$$\boxed{\sigma_{\text{tot}} = \sigma_0 \exp\left(-\frac{\alpha_s C_F}{2\pi} \log\frac{k_0^{\max}}{k_0^{\min}} \log\frac{1-\cos\theta_k^{\min}}{1-\cos\theta_k^{\max}}\right)}. \quad (2.197)$$

This pattern implies that the number of radiated gluons in the Drell–Yan process, neglecting the triple gluon vertex and only taking into account the leading logarithms, follows a Poisson pattern. The total cross section as well as the distribution of the radiated gluons are both well defined even in the limit of $k_0^{\min} \to 0$.

2.5.3 Catani–Seymour Dipoles

From the previous discussions we know that parton or jet radiation is dominated by the collinear and soft limits and the double enhancement shown in Eq. (2.197). The universal collinear limit of the different parton splittings is described by the unregularized splitting kernels $\hat{P}_{i \leftarrow j}(z)$. They are the basis of the parton shower description of jet radiation. The problem with the unregularized splittings is that part of them are divergent in the soft limit $z \to 0$. The question is if we can find an approximate description of parton splitting including the soft divergence in addition to the collinear enhancement. Such a description is given by the Catani–Seymour dipoles and serves as the basis of the shower in the SHERPA event generator.

Radiating a soft gluon off a hard quark leg is kinematically easy: the eikonal limit shown in Eq. (2.186) leaves the quark momentum untouched, for example allowing us to define all three particles involved in the splitting to remain on their mass shells. For the collinear splitting the situation is less simple. To describe the splitting of a quark into a quark and a gluon (k^μ) we use the Sudakov decomposition of Eq. (2.90). During our computation of the splitting kernels it becomes obvious that this parameterization of the momenta has its shortcomings. The missing on–shell conditions for the partons involved in the splitting are a serious problem for the implementation of the splitting processes in a parton shower and its comparison to data. The question is if we can define the momenta involved in a parton splitting more appropriately.

Clearly, just moving around momentum definitions for example starting from the Sudakov decomposition will not be helpful. We are missing the necessary degrees of freedom to allow for all on–shell conditions. The trick is to include a third parton in the picture: let us assume an *emitter parton* splitting $\tilde{p}_{1k} \sim p_1 + k$ together with another, *spectator parton* $\tilde{p}_s$, where the splitting process and the spectator can exchange momentum

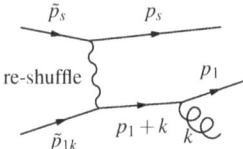

The momentum exchange between emitter and spectator respects momentum conservation,

$$\boxed{\tilde{p}_{1k}^\mu + \tilde{p}_s^\mu = p_1^\mu + k^\mu + p_s^\mu} \,. \tag{2.198}$$

In this picture the splitting process $\tilde{p}_{1k} \to p_1 + k$ does not conserve momentum. Instead, we prefer to require that also the emitter is on its mass shell, forgetting phase space factors in the splitting process for a moment. This means for massless partons that we simultaneously postulate

$$\tilde{p}_{1k}^2 = \tilde{p}_s^2 = p_1^2 = k^2 = p_s^2 = 0 \,. \tag{2.199}$$

The last three conditions we assume can be fulfilled. The second condition we can realize defining an appropriate momentum exchange with $\tilde{p}_s^2 \propto p_s^2$,

2.5 Parton Shower

$$\boxed{p_s^\mu = (1-y)\,\tilde{p}_s^\mu} \quad \Rightarrow \quad \tilde{p}_{1k}^\mu = p_1^\mu + k^\mu + p_s^\mu - \tilde{p}_s^\mu$$
$$= p_1^\mu + k^\mu - y\,\tilde{p}_s^\mu \qquad (2.200)$$

The exchanged momentum fraction y we can compute using the one remaining on–shell condition

$$0 \stackrel{!}{=} \tilde{p}_{1k}^2 = \left(p_1 + k + p_s - \frac{p_s}{1-y}\right)^2 = \left(p_1 + k - \frac{y}{1-y}\,p_s\right)^2$$
$$= \left(p_1 + k - \frac{1}{y^{-1}-1}\,p_s\right)^2$$

$$\Leftrightarrow \quad 0 = \left((y^{-1}-1)p_1 + (y^{-1}-1)k - p_s\right)^2$$
$$= 2(y^{-1}-1)\left[(y^{-1}-1)(p_1 k) - (p_1 p_s) - (p_s k)\right]$$

$$\Leftrightarrow \quad \frac{(p_1 k)}{y} = (p_1 k) + (p_1 p_s) + (p_s k)$$

$$\Leftrightarrow \quad y = \frac{(p_1 k)}{(p_1 k) + (p_1 p_s) + (p_s k)} \ . \qquad (2.201)$$

Next, we need to define the momentum fraction $\tilde{z}$ which the final–state particle p_1 carries away from the emitter. We define it in terms of the projection onto $\tilde{p}_s$ instead of the four-momentum itself,

$$(p_1 \tilde{p}_s) = \tilde{z}_1\,(\tilde{p}_{1k} \tilde{p}_s) \quad \Leftrightarrow \quad \tilde{z}_1 = \frac{(p_1 \tilde{p}_s)}{(\tilde{p}_{1k} \tilde{p}_s)} = \frac{(p_1 \tilde{p}_s)}{(p_1 \tilde{p}_s) + (k\,\tilde{p}_s) - y\,\tilde{p}_s^2}$$
$$= \frac{(p_1 p_s)}{(p_1 p_s) + (p_s k)} \ . \qquad (2.202)$$

The second momentum fraction then fulfills

$$(k\,\tilde{p}_s) = \tilde{z}_k\,(\tilde{p}_{1k} \tilde{p}_s) \quad \Leftrightarrow \quad \tilde{z}_1 + \tilde{z}_k = \frac{(p_1 p_s)}{(p_1 p_s) + (p_s k)} + \frac{(p_s k)}{(p_1 p_s) + (p_s k)} = 1 \ . \qquad (2.203)$$

In this parameterization we can look at the soft and collinear limits. According to Eq. (2.190) the relevant kinematic variable is $(p_1 k) + (p_s k)$, which we want to express in terms of $\tilde{z}_{1,k}$ and y. We find for the leading pole

$$1 - y = \frac{(p_1 p_s) + (p_s k)}{(p_1 k) + (p_1 p_s) + (p_s k)}$$

$$\tilde{z}_1(1 - y) = \frac{(p_1 p_s)}{(p_1 k) + (p_1 p_s) + (p_s k)}$$

$$\frac{1}{1 - \tilde{z}_1(1 - y)} = \frac{(p_1 k) + (p_1 p_s) + (p_s k)}{(p_1 k) + (p_s k)} , \qquad (2.204)$$

so the divergence of the intermediate propagator is described by the combination $1/(1 - \tilde{z}_1(1 - y))$. At this point we have convinced ourselves that the kinematic description including a spectator quark solves the problems with the on–shell partons consistently. The only prize we have to pay is the slightly more complicated form of the divergent kinematic variable in Eq. (2.204). The second question is if anything unexpected happens in the soft and collinear limits. As in many instances, we limit ourselves to final state gluon radiation, because the combination of initial-state and final-state partons leads to many different cases which are technically more involved.

In the *soft limit* the structure of the radiation matrix element is given by Eq. (2.189). In the presence of only hard momenta of the kind p^μ, except for the gluon, we can define the soft limit as $k^\mu = \lambda p^\mu$. The small parameter λ then characterizes the soft limit. From Eq. (2.203) we know that in this limit $\tilde{z}_k \to 0$ while $\tilde{z}_1 \to 1$. The parameter y computed in Eq. (2.201) scales like

$$y = \frac{(p_1 k)}{(p_1 k) + (p_1 p_s) + (p_s k)} = \lambda \frac{(p p_1)}{(p_s k) + \mathcal{O}(\lambda)} \to 0$$

$$\Leftrightarrow \quad \tilde{p}_s^\mu = p_s^\mu + \mathcal{O}(\lambda)$$

$$\Leftrightarrow \quad \tilde{p}_{1k}^\mu = p_1^\mu + \mathcal{O}(\lambda) . \qquad (2.205)$$

This is precisely the leading term in the eikonal approximation, assuming that the radiation of a soft gluon does not change the hard radiating quark leg. We can compute the form of the divergence following Eq. (2.204), namely

$$\frac{1}{1 - \tilde{z}_1(1 - y)} = \frac{(p_1 k) + (p_1 p_s) + (p_s k)}{(p_1 k) + (p_s k)} = \frac{1}{\lambda} \frac{(p_1 p_s) + \mathcal{O}(\lambda)}{(p_1 p) + (p_{sp})} , \qquad (2.206)$$

with an appropriate choice of a hard reference momentum p. This expression we can use to compute the soft splitting kernel for example for quark splitting into a hard quark and a soft gluon.

In the *collinear limit* we define a transverse momentum component, similar to Eq. (2.90). The only difference is that now we can require all participating partons to be on–shell, as seen in Eq. (2.199). We write the two momenta in the final state as

2.5 Parton Shower

$$p_1 = zp + p_T - \frac{p_T^2}{z}\frac{n}{2(pn)} \qquad k = (1-z)p - p_T - \frac{p_T^2}{1-z}\frac{n}{2(pn)},$$
(2.207)

where we postulate $p^2 = 0$, $n^2 = 0$, and $(pp_T) = 0 = (np_T)$. The momentum p is defined as the sum of p_1 and k, modulo contributions of order p_T. Aside from the on-shell conditions, this corresponds to the original Sudakov decomposition. We can confirm that the condition $p^2 = \mathcal{O}(p_T^2)$ in addition to the exact relations $p_1^2 = 0 = k^2$ is allowed,

$$\begin{aligned}
(p_1 + k)^2 &= 2(p_1 k) = 2\left(zp + p_T - \frac{p_T^2}{z}\frac{n}{2(pn)}\right) \\
&\quad \left((1-z)p - p_T - \frac{p_T^2}{1-z}\frac{n}{2(pn)}\right) \\
&= -2p_T^2 - \frac{2p_T^2}{z}\frac{(1-z)(pn)}{2(pn)} - \frac{2p_T^2}{1-z}\frac{z(pn)}{2(pn)} \\
&= -p_T^2\left(2 + \frac{1-z}{z} + \frac{z}{1-z}\right) \\
&= -p_T^2\frac{2z - 2z^2 + (1 - 2z + z^2) + z^2}{z(1-z)} = -\frac{1}{2z(1-z)}.
\end{aligned}$$
(2.208)

This is also the relevant small Mandelstam variable for the collinear splitting. The simple result also motivates the choice of pre-factors of n in the ansatz of Eq. (2.207). The additional factor $1/(z(1-z))$ is needed once we compute the proper divergence defined in Eq. (2.204). Again, we now describe the splitting kinematics in the collinear limit as

$$\begin{aligned}
y &= \frac{(p_1 k)}{(p_1 k) + (p_1 p_s) + (p_s k)} = -\frac{p_T^2}{2z(1-z)}\frac{1}{(p_s(p_1+k)) + \mathcal{O}(p_T^2)} \to 0 \\
\tilde{p}_s^\mu &= p_s^\mu + \mathcal{O}(p_T^2) \\
\tilde{p}_{1k}^\mu &= p_1^\mu + k^\mu + \mathcal{O}(p_T^2) \\
\tilde{z}_1 &= z + \mathcal{O}(p_T^2).
\end{aligned}$$
(2.209)

As for the soft case, the momentum re-shuffling does not affect the leading terms in the collinear limit. Our divergent Mandelstam variable becomes

$$\frac{1}{1 - \tilde{z}_1(1-y)} = \frac{1}{1-z}\left(1 + \mathcal{O}(p_T^2)\right),$$
(2.210)

which is exactly the z behavior of the unregularized spitting kernel $\hat{P}_{q \leftarrow q}$ in Eq. (2.118). We see that the Catani–Seymour description of parton splitting with its spectator parton not only allows us to keep all participating particles on their mass shell, it also correctly describes the soft as well as the collinear splitting point–by–point in phase space. Without going into the reasons we should mention that this soft–collinear description of jet matrix elements turns out to be much more successful than one would expect. The momentum regime in which the Catani–Seymour dipoles describe LHC results extends far beyond $p_T \lesssim m_Z$.

We will see in Sect. 2.7.1 that the correct modelling of parton splittings in the soft and collinear limits is a key ingredient to higher order calculations of LHC cross sections. These calculations are the main application of Catani–Seymour dipoles. However, for this calculation we need to also integrate the expressions for soft and collinear splitting amplitudes over phase space. The specific parameterization which allows us to assume that all particles in the splitting process are on their mass shells. When we include the mother and the spectator momenta $\tilde{p}_{1k}$ and $\tilde{p}_s$ in the factorized form of the n-particle and $(n+1)$-particle phase space the re-mapping in Eq. (2.198) leads to an additional Jacobian $(1-y)/(1-\tilde{z}_1)$.

2.5.4 Ordered Emission

From the derivation of the Catani–Seymour dipoles we know that for example the emission of a gluon off a hard quark line is governed by distinctive soft and collinear phase space regimes. In our argument for the exponentiation of gluon radiation matrix elements in Eq. (2.197) there is one piece missing: multiple gluon emission has to be ordered by some parameter, such that in squaring the multiple emission matrix element we can neglect interference terms. These interference diagrams contributing to the full amplitude squared are called non–planar diagrams. The question is if we can justify to neglect them from first principles field theory and QCD. There are three reasons to do this, even though none of them gives exactly zero for soft and collinear splittings. On the other hand, in combination they make for a very good reason.

First, an arguments for a strongly ordered gluon emission comes from the *divergence structure* of soft and collinear gluon emission. Two successively radiated gluons look like

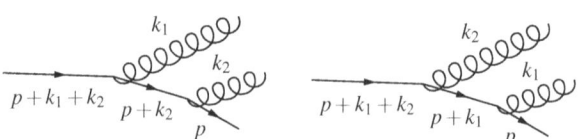

2.5 Parton Shower

According to Eq. (2.186) single gluon radiation with momentum k off a hard quark with momentum p is described by a kinematic term $(\epsilon^* p)(pk)$. For successive radiation the two Feynman diagrams give us the combined kinetic terms

$$\frac{(\epsilon_1 p)}{(p+k_1+k_2)^2-m^2}\frac{(\epsilon_2 p)}{(p+k_2)^2-m^2} + \frac{(\epsilon_2 p)}{(p+k_1+k_2)^2-m^2}\frac{(\epsilon_1 p)}{(p+k_1)^2-m^2}$$

$$= \frac{(\epsilon_1 p)}{2(pk_1)+2(pk_2)+(k_1+k_2)^2}\frac{(\epsilon_2 p)}{2(pk_2)}$$

$$+ \frac{(\epsilon_2 p)}{2(pk_1)+2(pk_2)+(k_1+k_2)^2}\frac{(\epsilon_1 p)}{2(pk_1)} \qquad k_1^2 = 0 = k_2^2$$

$$\simeq \frac{(\epsilon_1 p)}{2\max_j(pk_j)}\frac{(\epsilon_2 p)}{2(pk_2)} + \frac{(\epsilon_2 p)}{2\max_j(pk_j)}\frac{(\epsilon_1 p)}{2(pk_1)} \qquad (pk_j) \text{ strongly ordered}$$

$$\simeq \begin{cases} \dfrac{(\epsilon_1 p)(\epsilon_2 p)}{2\max_j(pk_j)}\dfrac{1}{2(pk_2)} & (pk_2) \ll (pk_1)\ k_2 \text{ softer} \\ \dfrac{(\epsilon_1 p)(\epsilon_2 p)}{2\max_j(pk_j)}\dfrac{1}{2(pk_1)} & (pk_1) \ll (pk_2)\ k_1 \text{ softer} . \end{cases} \qquad (2.211)$$

Going back to the two Feynman diagrams this means that once one of the gluons is significantly softer than the other the Feynman diagrams with the later soft emission dominates. After squaring the amplitude there will be no phase space regime where interference terms between the two diagrams are numerically relevant. The coherent sum over gluon radiation channels reduces to a incoherent sum, ordered by the softness of the gluon.

This argument can be generalized to multiple gluon emission by recognizing that the kinematics will always be dominated by the more divergent propagators towards the final state quark with momentum p. Note, however, that it is based on an ordering of the scalar products (pk_j) interpreted as the softness of the gluons. We already know that a small value of (pk_j) can as well point to a collinear divergence; every step in the argument of Eq. (2.211) still applies.

Second, we can derive ordered multiple gluon emission from the phase space integration in the *soft or eikonal approximation*. There, gluon radiation is governed by the so-called radiation dipoles given in Eq. (2.190). Because each dipole includes a sum over all radiating legs in the amplitude, the square includes a double sum over the hard legs. Diagonal terms vanish at least for over–all color–neutral processes. Because the following argument is purely based on kinematics we will ignore all color charges and other factors.

For successive gluon radiation off a quark leg the question we are interested in is where the soft gluon k is radiated, for example in relation to the hard quark p_1 and the harder gluon p_2. The kinematics of this process is the same as soft gluon radiation of a quark–antiquark pair produced in an electroweak process. For the dipoles we let the indices i, j run over the harder quark, antiquark, and possibly gluon legs. A well–defined process with all momenta defined as outgoing is

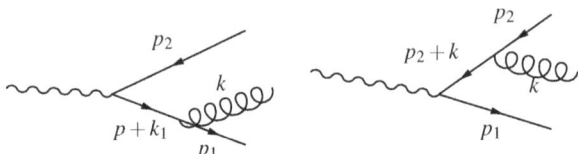

in the approximation of abelian QCD, i.e. no triple gluon vertices. We start by symmetrizing the leading soft radiation dipole with respect to the two hard momenta in a particular way,

$$\begin{aligned}(J^\dagger \cdot J)_{12} &= \frac{(p_1 p_2)}{(p_1 k)(p_2 k)} \\ &= \frac{1}{k_0^2} \frac{1 - \cos\theta_{12}}{(1 - \cos\theta_{1k})(1 - \cos\theta_{2k})} \qquad \text{in terms of opening angles } \theta \\ &= \frac{1}{2k_0^2} \left(\frac{1 - \cos\theta_{12}}{(1 - \cos\theta_{1k})(1 - \cos\theta_{2k})} + \frac{1}{1 - \cos\theta_{1k}} - \frac{1}{1 - \cos\theta_{2k}} \right) \\ &\quad + (1 \leftrightarrow 2) \\ &\equiv \frac{W_{12}^{[1]} + W_{12}^{[2]}}{k_0^2}.\end{aligned} \qquad (2.212)$$

The last term is an implicit definition of the two terms $W_{12}^{[1]}$. The pre-factor $1/k_0^2$ is given by the leading soft divergence. The original form of $(J^\dagger J)$ is symmetric in the two indices, which means that both hard partons can take the role of the hard parton and the interference partner. In the new form the symmetry in each of the two terms is broken. Each of the two terms we need to integrate over the gluon's phase space, including the azimuthal angle ϕ_{1k}. Note, however, that this splitting into two contributions is not the standard separation into the two diagrams. It is a specific ansatz to show the ordering patterns we will see below.

To compute the actual integral we express the three parton vectors in polar coordinates where the initial parton p_1 propagates into the x direction, the interference partner p_2 in the $(x-y)$ plane, and the soft gluon in the full three-dimensional space described by polar coordinates,

$$\hat{p}_1 = (1, 0, 0) \qquad \text{hard parton}$$

2.5 Parton Shower

$$\hat{p}_2 = (\cos\theta_{12}, \sin\theta_{12}, 0) \qquad \text{interference partner}$$

$$\hat{k} = (\cos\theta_{1k}, \sin\theta_{1k}\cos\phi_{1k}, \sin\theta_{1k}\sin\phi_{1k}) \qquad \text{soft gluon}$$

$$\Rightarrow \quad \cos\theta_{2k} \equiv (\hat{p}_2\hat{k}) = \cos\theta_{12}\cos\theta_{1k} + \sin\theta_{12}\sin\theta_{1k}\cos\phi_{1k} \,. \tag{2.213}$$

From the scalar product between these four-vectors we see that of the terms appearing in Eq. (2.212) only the opening angle θ_{2k} includes ϕ_{1k}, which for the azimuthal angle integration means

$$\int_0^{2\pi} d\phi_{1k}\, W_{12}^{[1]} = \frac{1}{2}\int_0^{2\pi} d\phi_{1k}\left(\frac{1-\cos\theta_{12}}{(1-\cos\theta_{1k})(1-\cos\theta_{2k})} + \frac{1}{1-\cos\theta_{1k}}\right.$$

$$\left. - \frac{1}{1-\cos\theta_{2k}}\right).$$

$$= \frac{1}{2}\frac{1}{1-\cos\theta_{1k}}\int_0^{2\pi} d\phi_{1k}\left(\frac{1-\cos\theta_{12}}{1-\cos\theta_{2k}} + 1 - \frac{1-\cos\theta_{1k}}{1-\cos\theta_{2k}}\right)$$

$$= \frac{1}{2}\frac{1}{1-\cos\theta_{1k}}\left(2\pi + (\cos\theta_{1k} - \cos\theta_{12})\int_0^{2\pi} d\phi_{1k}\,\frac{1}{1-\cos\theta_{2k}}\right). \tag{2.214}$$

The azimuthal angle integral in this expression for $W_{12}^{[i]}$ we can solve

$$\int_0^{2\pi} d\phi_{1k}\,\frac{1}{1-\cos\theta_{2k}}$$

$$= \int_0^{2\pi} d\phi_{1k}\,\frac{1}{1-\cos\theta_{12}\cos\theta_{1k} - \sin\theta_{12}\sin\theta_{1k}\cos\phi_{1k}}$$

$$= \int_0^{2\pi} d\phi_{1k}\,\frac{1}{a - b\cos\phi_{1k}}$$

$$= \oint_{\text{unit circle}} dz\,\frac{1}{iz}\,\frac{1}{a - b\frac{z+1/z}{2}} \qquad \text{with}\quad z = e^{i\phi_{1k}},\ \cos\phi_{1k} = \frac{z+1/z}{2}$$

$$= \frac{2}{i}\oint dz\,\frac{1}{2az - b - bz^2}$$

$$= \frac{2i}{b}\oint \frac{dz}{(z-z_-)(z-z_+)} \qquad \text{with}\quad z_\pm = \frac{a}{b} \pm \sqrt{\frac{a^2}{b^2} - 1}\,. \tag{2.215}$$

This integral is related to the sum of all residues of poles inside the closed integration contour. Of the two poles z_- is the one which typically lies within the unit circle, so we find

$$\int_0^{2\pi} d\phi_{1k} \frac{1}{1-\cos\theta_{2k}} = \frac{2i}{b} 2\pi i \frac{1}{z_- - z_+} = \frac{2\pi}{\sqrt{a^2-b^2}}$$

$$= \frac{2\pi}{\sqrt{(\cos\theta_{1k}-\cos\theta_{12})^2}} = \frac{2\pi}{|\cos\theta_{1k}-\cos\theta_{12}|} . \quad (2.216)$$

The entire integral in Eq. (2.214) then becomes

$$\int_0^{2\pi} d\phi_{1k} W_{12}^{[1]} = \frac{1}{2} \frac{1}{1-\cos\theta_{1k}} \left(2\pi + (\cos\theta_{1k}-\cos\theta_{12}) \frac{2\pi}{|\cos\theta_{1k}-\cos\theta_{12}|} \right)$$

$$= \frac{\pi}{1-\cos\theta_{1k}} (1+\text{sign}(\cos\theta_{1k}-\cos\theta_{12}))$$

$$= \begin{cases} \dfrac{2\pi}{1-\cos\theta_{1k}} & \text{if } \theta_{1k} < \theta_{12} \\ 0 & \text{else.} \end{cases} \quad (2.217)$$

The soft gluon is only radiated at angles between zero and the opening angle of the initial parton p_1 and its hard interference partner or spectator p_2. The same integral over $W_{12}^{[2]}$ gives the same result, with switched roles of p_1 and p_2. Combining the two permutations this means that the soft gluon is always radiated within a cone centered around one of the hard partons and with a radius given by the distance between the two hard partons. Again, the coherent sum of diagrams reduces to an incoherent sum. This derivation *angular ordering* is exact in the soft limit.

There is a simple physical argument for this suppressed radiation outside a cone defined by the radiating legs. Part of the deviation is that the over-all process is color-neutral. This means that once the gluon is far enough from the two quark legs it will not resolve their individual charges but only feel the combined charge. This screening leads to an additional suppression factor of the kind $\theta_{12}^2/\theta_{1k}^2$. This effect is called coherence.

The third argument for ordered emission comes from *color factors*. Crossed successive splittings or interference terms between different orderings are color suppressed. For example in the squared diagram for three jet production in e^+e^- collisions the additional gluon contributes a color factor

$$\text{Tr}(T^a T^a) = \frac{N_c^2 - 1}{2} = N_c C_F \quad (2.218)$$

When we consider the successive radiation of two gluons the ordering matters. As long as the gluon legs do not cross each other we find the color factor

2.5 Parton Shower

$$\text{Tr}(T^a T^a T^b T^b)$$
$$= (T^a T^a)_{il}(T^b T^b)_{li}$$
$$= \frac{1}{4}\left(\delta_{il}\delta_{jj} - \frac{\delta_{ij}\delta_{jl}}{N_c}\right)\left(\delta_{il}\delta_{jj} - \frac{\delta_{ij}\delta_{jl}}{N_c}\right) \quad \text{using} \quad T^a_{ij} T^a_{kl} = \frac{1}{2}\left(\delta_{il}\delta_{jk} - \frac{\delta_{ij}\delta_{kl}}{N_c}\right)$$
$$= \frac{1}{4}\left(\delta_{il} N_c - \frac{\delta_{il}}{N_c}\right)\left(\delta_{il} N_c - \frac{\delta_{il}}{N_c}\right)$$
$$= N_c \left(\frac{N_c^2 - 1}{2N_c}\right)^2 = N_c C_F^2 = \frac{16}{3} \tag{2.219}$$

Similarly, we can compute the color factor when the two gluon lines cross. We find

$$\text{Tr}(T^a T^b T^a T^b) = -\frac{N_c^2 - 1}{4N_c} = -\frac{C_F}{2} = -\frac{2}{3}. \tag{2.220}$$

Numerically, this color factor is suppressed compared to 16/3. This kind of behavior is usually quoted in powers of N_c where we assume N_c to be large. In those terms non–planar diagrams are suppressed by a factor $1/N_c^2$ compared to the planar diagrams.

Once we also include the triple gluon vertex we can radiate two gluons off a quark leg with the color factor

$$\text{Tr}(T^a T^b) \, f^{acd} f^{bcd} = \frac{\delta^{ab}}{2} N_c \delta^{ab} = \frac{N_c(N_c^2 - 1)}{2} = N_c^2 C_F = \frac{36}{3}. \tag{2.221}$$

This is not suppressed compared to successive planar gluon emission, neither in actual numbers not in the large-N_c limit.

We can try the same argument for a purely gluonic theory, i.e. radiating gluons off two hard gluons in the final state. The color factor for single gluon emission after squaring is

$$f^{abc} f^{abc} = N_c \delta^{aa} = N_c(N_c^2 - 1) \sim N_c^3, \tag{2.222}$$

using the large-N_c limit in the last step. For planar double gluon emission with the exchanged gluon indices b and f we find

$$f^{abd} f^{abe} f^{dfg} f^{efg} = N_c \delta^{de} N_c \delta^{de} = N_c^4. \tag{2.223}$$

Splitting one radiated gluon into two gives

$$f^{abc} f^{cef} f^{def} f^{abd} = N_c \delta^{cd} N_c \delta^{cd} = N_c^4. \tag{2.224}$$

This means that planar emission and successive splittings cannot be separated based on the color factor for either hard radiating quarks or gluons. We can use the color factor argument only for abelian splittings to justify ordered gluon emission.

2.6 Multi–Jet Events

Up to now we have derived and established the parton shower as a probabilistic tool to simulate the successive emission of jets in hard processes. This includes a careful look at the collinear and soft structure of parton splitting as well as the crucial assumption of ordered emission. The starting point of this whole argument was that the Sudakov factors obey the DGLAP equation, as shown in Eq. (2.179).

In the following we will introduce an alternative object which allows us to the compute rates and patterns of jet radiation. In Sect. 2.6.1 we will introduce generating functionals and their evolution equations. Their underlying approximations are related to the parton shower, but their main results hold more generally. We have already used some of the key features which will derive here in Higgs physics applications in Sect. 1.6.2 Per se it is not clear how jet radiation described by the parton shower and jet radiation described by fixed-order QCD processes are linked. In Sect. 2.6.2 we will discuss ways to combine the two approaches in realistic LHC simulations, bringing us very close to contemporary research topics.

2.6.1 Jet Radiation Patterns

From Sect. 2.5.2 we know that unlike other observables related to multi–jet events the number of radiated jets is well defined after a simple resummation. Generating functionals for the jet multiplicity allow us to calculate resummed jet quantities from first principles in QCD. We construct a generating functional in an arbitrary parameter u by demanding that repeated differentiation at $u = 0$ gives *exclusive multiplicity* distributions $P_n \equiv \sigma_n / \sigma_{\text{tot}}$,

$$\boxed{\Phi = \sum_{n=1}^{\infty} u^n P_{n-1}} \qquad \text{with} \quad P_{n-1} = \frac{\sigma_{n-1}}{\sigma_{\text{tot}}} = \frac{1}{n!} \frac{d^n}{du^n} \Phi \bigg|_{u=0} . \qquad (2.225)$$

For the generating functional Φ we will suppress the argument u. In the application to gluon emission the explicit factor $1/n!$ corresponds to the phase space factor for identical bosons. Because in P_n we only count radiated jets, our definition uses P_{n-1} where other conventions use P_n. A second observable we can extract from Φ is the *average jet multiplicity*,

2.6 Multi-Jet Events

$$\left.\frac{d\Phi}{du}\right|_{u=1} = \sum_{n=1}^{\infty} n\, u^{n-1}\, \left.\frac{\sigma_{n-1}}{\sigma_{\text{tot}}}\right|_{u=1} = 1 + \frac{1}{\sigma_{\text{tot}}} \sum_{n=1}^{\infty} (n-1)\, \sigma_{n-1}\,. \tag{2.226}$$

Note again that P_{n-1} describes $n-1$ radiated jets, in the simplest case corresponding to n observed jets in the final state.

The question is what we can say about such generating functionals. In analogy to the DGLAP equation we can derive an evolution equation for Φ. We start by reminding ourselves that for the parton densities and the Sudakov factors the integrated version of the evolution equation given in Eq. (2.180) reads

$$f_i(x,t) = \Delta_i(t,t_0) f_i(x,t_0) + \int_{t_0}^{t} \frac{dt'}{t'}\, \Delta_i(t,t')$$
$$\sum_j \int_0^{1-\epsilon} \frac{dz}{z}\, \frac{\alpha_s}{2\pi}\, \hat{P}_{i \leftarrow j}(z)\, f_j\left(\frac{x}{z}, t'\right)\,. \tag{2.227}$$

The sum over the splittings is organized by initial states j which turn into the relevant parton i in the collinear approximation. The third particle involved in the splitting $j \to i$ follows automatically.

Instead of deriving the corresponding equation for the generating functional Φ we motivate it by analogy. In the Sudakov picture we can apply our probabilistic picture to parton splittings $i \to jk$ in the final state. This should correspond to an evolution equation for the generating functionals for the number of jets. All three external particles are then described by generating functionals Φ instead of parton densities, giving us

$$\boxed{\Phi_i(t) = \Delta_i(t,t_0)\Phi_i(t_0) + \int_{t_0}^{t} \frac{dt'}{t'} \Delta_i(t,t') \sum_{i \to j,k} \int_0^1 dz\, \frac{\alpha_s}{2\pi}\, \hat{P}_{i \to jk}(z)\, \Phi_j(z^2 t')\Phi_k((1-z)^2 t')}\,. \tag{2.228}$$

This evolution equation for general functionals is the same DGLAP equation we use for parton densities in the initial state. The difference is that the generating functionals count jets in the final state. The precise link between the generating functionals Φ and a parton–density–inspired partition function we skip at this stage. Similarly, we introduce the argument of the strong coupling without any further motivation as $\alpha_s(z^2(1-z)^2 t')$. It will become clear during our computation that this scale choice is appropriate.

The argument in this section will go two ways: first, we write down a proper *differential evolution equation* for $\Phi_q(t)$. Then, we solve this equation for quarks, only including the abelian splitting $q \to qg$. This solution will give us the known *jet scaling patterns*. To start with, we insert the unregularized splitting kernel from Eq. (2.118) into the evolution equation,

$$\Phi_q(t) = \Delta_q(t,t_0)\Phi_q(t_0) + \int_{t_0}^{t}\frac{dt'}{t'}\Delta_q(t,t')$$
$$\int_0^1 dz\,\frac{\alpha_s}{2\pi}\,C_F\,\frac{1+z^2}{1-z}\,\Phi_q(z^2 t')\Phi_g((1-z)^2 t')$$
$$= \Delta_q(t,t_0)\Phi_q(t_0) + \int_{t_0}^{t}\frac{dt'}{t'}\Delta_q(t,t')$$
$$\int_0^1 dz\,\frac{\alpha_s C_F}{2\pi}\,\frac{-(1-z)(1+z)+2}{1-z}\,\Phi_q(z^2 t')\Phi_g((1-z)^2 t')$$
$$= \Delta_q(t,t_0)\Phi_q(t_0) + \int_{t_0}^{t}\frac{dt'}{t'}\Delta_q(t,t')$$
$$\int_0^1 dz\,\frac{\alpha_s C_F}{2\pi}\,\left(\frac{2}{1-z}-1-z\right)\Phi_q(z^2 t')\Phi_g((1-z)^2 t')\,. \quad (2.229)$$

First, we simplify the divergent part of Eq. (2.229), using the new integration parameter $t'' = (1-z)^2 t'$. This gives us the same Jacobian as in Eq. (2.169),

$$\frac{dt''}{dz} = \frac{d}{dz}(1-z)^2 t' = 2(1-z)(-1)t' = -2\frac{t''}{1-z} \quad \Leftrightarrow \quad \frac{dz}{1-z} = -\frac{1}{2}\frac{dt''}{t''}\,. \quad (2.230)$$

In addition, we approximate $z \to 1$ wherever possible and cut off all t integrations at the infrared resolution scale t_0,

$$\int_0^1 dz\,\frac{\alpha_s(z^2(1-z)^2 t')C_F}{2\pi}\,\frac{2}{1-z}\,\Phi_q(z^2 t')\Phi_g((1-z)^2 t')$$
$$= \Phi_q(t')\int_{t_0}^{t'}dt''\,\frac{\alpha_s(t'')C_F}{2\pi}\,\frac{1}{t''}\Phi_g(t'')\,. \quad (2.231)$$

For the finite part in Eq. (2.229) we neglect the logarithmic z dependence of all functions and integrate the leading power dependence $1+z$ to $3/2$,

$$-\int_0^1 dz\,\frac{\alpha_s(z^2(1-z)^2 t')C_F}{2\pi}\,(1+z)\,\Phi_q(z^2 t')\Phi_g((1-z)^2 t')$$
$$\simeq -\frac{\alpha_s(t')C_F}{2\pi}\,\frac{3}{2}\,\Phi_q(t')\Phi_g(t')\,. \quad (2.232)$$

2.6 Multi–Jet Events

After these two simplifying steps Eq. (2.229) reads

$$\Phi_q(t) = \Delta_q(t,t_0)\Phi_q(t_0) + \frac{C_F}{2\pi}\int_{t_0}^{t}\frac{dt'}{t'}\,\Delta_q(t,t')$$

$$\left(\int_{t_0}^{t'}dt''\frac{\alpha_s(t'')}{t''}\Phi_q(t')\Phi_g(t'') - \frac{3}{2}\alpha_s(t')\Phi_q(t')\Phi_g(t')\right)$$

$$= \Delta_q(t,t_0)\Phi_q(t_0) + \frac{C_F}{2\pi}\Delta_q(t,t_0)\int_{t_0}^{t}\frac{dt'}{t'}\,\frac{1}{\Delta_q(t',t_0)}\Phi_q(t')$$

$$\left(\int_{t_0}^{t'}dt''\frac{\alpha_s(t'')}{t''}\Phi_g(t'') - \frac{3}{2}\alpha_s(t')\Phi_g(t')\right). \qquad (2.233)$$

The original Sudakov factor $\Delta_q(t,t')$ is split into a ratio of two Sudakov factors. This allows us to differentiate both sides with respect to t,

$$\frac{d}{dt}\Phi_q(t) = \frac{d\Delta_q(t,t_0)}{dt}\Phi_q(t_0) + \frac{C_F}{2\pi}\frac{d\Delta_q(t,t_0)}{dt}\int_{t_0}^{t}\frac{dt'}{t'}\,\frac{1}{\Delta_q(t',t_0)}\Phi_q(t')$$

$$\left(\int_{t_0}^{t'}dt''\frac{\alpha_s(t'')}{t''}\Phi_g(t'') - \frac{3}{2}\alpha_s(t')\Phi_g(t')\right) + \frac{C_F}{2\pi}\Delta_q(t,t_0)\frac{1}{t}\frac{1}{\Delta_q(t,t_0)}\Phi_q(t)$$

$$\left(\int_{t_0}^{t}dt''\frac{\alpha_s(t'')}{t''}\Phi_g(t'') - \frac{3}{2}\alpha_s(t)\Phi_g(t)\right) = \frac{d\Delta_q(t,t_0)}{dt}$$

$$\left[\Phi_q(t_0) + \frac{C_F}{2\pi}\int_{t_0}^{t}\frac{dt'}{t'}\,\frac{1}{\Delta_q(t',t_0)}\Phi_q(t')\right.$$

$$\left.\left(\int_{t_0}^{t'}dt''\frac{\alpha_s(t'')}{t''}\Phi_g(t'') - \frac{3}{2}\alpha_s(t')\Phi_g(t')\right)\right]$$

$$+\frac{C_F}{2\pi}\frac{1}{t}\Phi_q(t)\left(\int_{t_0}^{t}dt''\frac{\alpha_s(t'')}{t''}\Phi_g(t'') - \frac{3}{2}\alpha_s(t)\Phi_g(t)\right) = \frac{1}{\Delta_q(t,t_0)}$$

$$\frac{d\Delta_q(t,t_0)}{dt}\Phi_q(t) + \Phi_q(t)\frac{C_F}{2\pi}\frac{1}{t}\left(\int_{t_0}^{t}dt''\frac{\alpha_s(t'')}{t''}\Phi_g(t') - \frac{3}{2}\alpha_s(t)\Phi_g(t)\right).$$
$$(2.234)$$

In the last step we use the definition in Eq. (2.233). This simplified equation has a *solution* which we can write in a closed form, namely

$$\Phi_q(t) = \Phi_q(t_0) \, \Delta_q(t,t_0) \, \exp\left[\frac{C_F}{2\pi} \int_{t_0}^{t} dt' \, \frac{\alpha_s(t')}{t'} \left(\log \frac{t}{t'} - \frac{3}{2}\right) \Phi_g(t')\right]$$

$$= \Phi_q(t_0) \, \exp\left[-\int_{t_0}^{t} dt' \, \Gamma_{q\leftarrow q}(t,t')\right] \exp\left[\int_{t_0}^{t} dt' \, \Gamma_{q\leftarrow q}(t,t')\Phi_g(t')\right]$$

$$= \Phi_q(t_0) \, \exp\left[\int_{t_0}^{t} dt' \, \Gamma_{q\leftarrow q}(t,t')\left(\Phi_g(t') - 1\right)\right]. \tag{2.235}$$

We can prove this by straightforward differentiation of the first line in Eq. (2.235),

$$\frac{d\Phi_q(t)}{dt} = \Phi_q(t_0) \frac{d\Delta_q(t,t_0)}{dt} \exp\left[\frac{C_F}{2\pi} \int_{t_0}^{t} dt' \, \frac{\alpha_s(t')}{t'} \left(\log \frac{t}{t'} - \frac{3}{2}\right) \Phi_g(t')\right]$$

$$+ \Phi_q(t) \frac{d}{dt}\left[\frac{C_F}{2\pi} \int_{t_0}^{t} dt' \, \frac{\alpha_s(t')}{t'} \left(\log t - \log t' - \frac{3}{2}\right) \Phi_g(t')\right]$$

$$= \frac{1}{\Delta_q(t,t_0)} \frac{d\Delta_q(t,t_0)}{dt} \Phi_q(t) + \Phi_q(t) \frac{C_F}{2\pi} \frac{\alpha_s(t)}{t} \left(-\log t - \frac{3}{2}\right) \Phi_g(t)$$

$$+ \Phi_q(t) \frac{C_F}{2\pi} \frac{1}{t} \int_{t_0}^{t} dt' \, \frac{\alpha_s(t')}{t'} \Phi_g(t') + \Phi_q(t) \frac{C_F}{2\pi} \log t \, \frac{\alpha_s(t)}{t} \Phi_g(t)$$

$$= \frac{1}{\Delta_q(t,t_0)} \frac{d\Delta_q(t,t_0)}{dt} \Phi_q(t) + \Phi_q(t) \frac{C_F}{2\pi} \frac{1}{t}$$

$$\left(\int_{t_0}^{t} dt' \, \frac{\alpha_s(t')}{t'} \Phi_g(t') - \alpha_s(t)\frac{3}{2}\Phi_g(t)\right). \tag{2.236}$$

The expression given in Eq. (2.235) indeed solves the evolution equation in Eq. (2.234). The corresponding computation for $\Phi_g(t)$ follows the same path.

By definition, the generating functional evaluated at the resolution scale t_0 describes an ensemble of jets which have had no opportunity to split. This means $\Phi_{q,g}(t_0) = u$. The quark and gluon generating functionals to next-to-leading logarithmic accuracy are

$$\Phi_q(t) = u \, \exp\left[\int_{t_0}^{t} dt' \, \Gamma_{q\leftarrow q}(t,t')\left(\Phi_g(t') - 1\right)\right]$$

$$\Phi_g(t) = u \, \exp\left[\int_{t_0}^{t} dt' \, \left(\Gamma_{g\leftarrow g}(t,t')\left(\Phi_g(t') - 1\right) + \Gamma_{q\leftarrow g}(t')\left(\frac{\Phi_q^2(t')}{\Phi_g(t')} - 1\right)\right)\right]. \tag{2.237}$$

2.6 Multi–Jet Events

The splitting kernels are defined in Eq. (2.172); gluon splitting to quarks described by $\Gamma_{q \leftarrow g}$ is suppressed by a power of the logarithm $\log t/t'$.

The logarithm $\log t/t'$ combined with the coupling constant α_s included in the splitting kernels is the small parameter which we will use for the following argument. If this logarithmically enhanced term dominates the physics, the evolution equations for quark and gluons are structurally identical. In both cases, the Φ dependence of the exponent spoils an effective solution of Eq. (2.237). However, the general form of $\Gamma(t, t')$ ensures that the main contribution to the t' integral comes from the region where $t' \sim t_0$. Unless something drastic happens with the integrands in Eq. (2.237) this means that under the integral we can approximate $\Phi_{q,g}(t_0) = u$ and, if necessary, iteratively insert the solution for $\Phi(t)$ into the differential equation. The leading terms for both, quark and gluon evolution equations turn into the closed form

$$\Phi_{q,g}(t) = u \, \exp\left[\int_{t_0}^{t} dt' \, \Gamma_{q,g}(t,t') \, (u-1)\right] = u \, \exp\left[-(1-u) \int_{t_0}^{t} dt' \, \Gamma_{q,g}(t,t')\right]. \tag{2.238}$$

Using the Sudakov factor defined in Eq. (2.166) the generating functional in the approximation of large logarithmically enhanced parton splitting is

$$\boxed{\Phi_{q,g}(t) = u \, \Delta_{q,g}(t)^{1-u}}. \tag{2.239}$$

For the jet rates this corresponds to a *Poisson distribution*

$$P_{n-1} = \Delta_{q,g}(t) \, \frac{|\log \Delta_{q,g}(t)|^{n-1}}{(n-1)!} \quad \text{or} \quad \boxed{R_{(n+1)/n} = \frac{|\log \Delta_{q,g}(t)|}{n+1}}. \tag{2.240}$$

We can prove this result by induction. For general values of u we will show that the n-th derivative of the generating functional $\Phi_{q,g}$ reads

$$\frac{1}{n!} \frac{d^n}{du^n} \Phi_{q,g}(t) = \frac{(-\log \Delta_{q,g})^{n-1}}{n!} \Delta_{q,g} \left(n - u \log \Delta_{q,g}\right) e^{-u \log \Delta_{q,g}}$$

$$\stackrel{u=0}{=} \frac{(-\log \Delta_{q,g})^{n-1}}{(n-1)!} \Delta_{q,g} = P_{n-1} . \tag{2.241}$$

By construction, the Sudakov factor is smaller than unity, so $|\log \Delta_{q,g}| = -\log \Delta_{q,g}$. For the case $n = 1$ and general values of u Eq. (2.239) indeed gives us

$$\frac{d}{du}\Phi_{q,g}(t) = \Delta_{q,g}\frac{d}{du}ue^{-u\log\Delta_{q,g}}$$
$$= \Delta_{q,g}\left[e^{-u\log\Delta_{q,g}} + u\left(-\log\Delta_{q,g}\right)e^{-u\log\Delta_{q,g}}\right]$$
$$= \Delta_{q,g}\left(1 - u\log\Delta_{q,g}\right)e^{-u\log\Delta_{q,g}}, \qquad (2.242)$$

in agreement with our aim in Eq. (2.241). Next, we compute the step

$$\frac{1}{n!}\frac{d^n}{du^n}\Phi_{q,g}(t)$$
$$= \frac{1}{n}\frac{d}{du}\left(\frac{1}{(n-1)!}\frac{d^{n-1}}{du^{n-1}}\Phi_{q,g}(t)\right)$$
$$= \frac{1}{n}\frac{d}{du}\left[\frac{(-\log\Delta_{q,g})^{n-2}}{(n-1)!}\Delta_{q,g}\left(n-1-u\log\Delta_{q,g}\right)e^{-u\log\Delta_{q,g}}\right] \qquad \text{using Eq. (2.241)}$$
$$= \frac{(-\log\Delta_{q,g})^{n-2}}{n!}\Delta_{q,g}\left[(-\log\Delta_{q,g})e^{-u\log\Delta_{q,g}}\right.$$
$$\left.+\left(n-1-u\log\Delta_{q,g}\right)(-\log\Delta_{q,g}e^{-u\log\Delta_{q,g}}\right]$$
$$= \frac{(-\log\Delta_{q,g})^{n-1}}{n!}\Delta_{q,g}\left[1+n-1-u\log\Delta_{q,g}\right]e^{-u\log\Delta_{q,g}}. \qquad (2.243)$$

This is indeed Eq. (2.241), completing our proof of this general solution and the special case $u=0$ in Eq. (2.240).

In addition to this Poisson case we can find a second, recursive solution for the generating functionals. It holds in the limit of small emission probabilities. The emission probability is governed by $\Gamma_{i\leftarrow j}(t,t')$, as defined in Eq. (2.172). We can make it small by avoiding a logarithmic enhancement, corresponding to no large scale ratios t/t_0. In addition, we would like to get rid of $\Gamma_{q\leftarrow g}$ while keeping $\Gamma_{g\leftarrow g}$. Theoretically, this means removing the gluon splitting into two quarks and limiting ourselves to pure Yang-Mills theory. In that case the scale derivative of Eq. (2.237) reads

$$\frac{d\Phi_g(t)}{dt}$$
$$= u\frac{d}{dt}\exp\left[\int_{t_0}^t dt'\,\Gamma_{g\leftarrow g}(t,t')\left(\Phi_g(t')-1\right)\right]$$
$$= \Phi_g(t)\frac{C_A}{2\pi}\frac{d}{dt}\int_{t_0}^t dt'\,\frac{\alpha_s(t')}{t'}\left(\log t - \log t' - \frac{11}{6}\right)\left(\Phi_g(t')-1\right) \qquad \text{inserting Eq. (2.172)}$$
$$= \Phi_g(t)\frac{C_A}{2\pi}\left[\frac{\alpha_s(t)}{t}\left(-\log t - \frac{11}{6}\right)\left(\Phi_g(t)-1\right)\right.$$
$$\left.+\frac{1}{t}\int_{t_0}^t dt'\,\frac{\alpha_s(t')}{t'}\left(\Phi_g(t')-1\right) + \log t\,\frac{\alpha_s(t)}{t}\left(\Phi_g(t)-1\right)\right]$$
$$= \Phi_g(t)\frac{C_A}{2\pi t}\left[-\frac{11}{6}\alpha_s(t)\left(\Phi_g(t)-1\right) + \int_{t_0}^t dt'\,\frac{\alpha_s(t')}{t'}\left(\Phi_g(t')-1\right)\right]. \qquad (2.244)$$

2.6 Multi–Jet Events

This form is already greatly simplified, but in the combination of the integral and the running strong coupling it is not clear what the limit of small but finite $\log t/t_0$ would be. Integrating by parts we find a form which we can estimate systematically,

$$\begin{aligned}\frac{d\Phi_g(t)}{dt} &= \Phi_g(t)\frac{C_A}{2\pi t}\left[-\frac{11}{6}\alpha_s(t)\left(\Phi_g(t)-1\right)\right.\\ &\quad -\int_{t_0}^t dt'\log\frac{t'}{t_0}\frac{d}{dt'}\left(\alpha_s(t')\left(\Phi_g(t')-1\right)\right)+\log\frac{t'}{t_0}\alpha_s(t')\left(\Phi_g(t')-1\right)\Big|_{t_0}^t\Big]\\ &= \Phi_g(t)\frac{C_A}{2\pi t}\left[\alpha_s(t)\left(\log\frac{t}{t_0}-\frac{11}{6}\right)\left(\Phi_g(t)-1\right)\right.\\ &\quad -\int_{t_0}^t dt'\,\log\frac{t'}{t_0}\frac{d}{dt'}\left(\alpha_s(t')\left(\Phi_g(t')-1\right)\right)\Big]. \end{aligned} \qquad (2.245)$$

We can evaluate this expression in the limit of $t = t_0 + \delta$ or $t_0/t = 1 - \delta/t$. The two leading terms read

$$\begin{aligned}\frac{d\Phi_g(t)}{dt} &= \Phi_g(t)\frac{C_A}{2\pi t}\left[\alpha_s(t)\left(\frac{\delta}{t}-\frac{11}{6}\right)\left(\Phi_g(t)-1\right)\right.\\ &\quad -(t-t_0)\frac{\delta}{t}\frac{d}{dt}\left(\alpha_s(t)\left(\Phi_g(t)-1\right)\right)\Big]\\ &= \Phi_g(t)\frac{C_A}{2\pi}\frac{\alpha_s(t)}{t}\left(\frac{\delta}{t}-\frac{11}{6}\right)\left(\Phi_g(t)-1\right)+\mathcal{O}\left(\frac{\delta^2}{t^2}\right). \end{aligned} \qquad (2.246)$$

To next–to–leading order in δ/t the equation for the generating functional becomes

$$\frac{d\Phi_g(t)}{dt} = \Phi_g(t)\,\tilde{\Gamma}_{g\leftarrow g}(t,t_0)\left(\Phi_g(t)-1\right)$$

$$\text{with}\qquad \tilde{\Gamma}_{g\leftarrow g}(t,t_0) = \frac{C_A}{2\pi}\frac{\alpha_s(t)}{t}\left(\log\frac{t}{t_0}-\frac{11}{6}\right). \qquad (2.247)$$

With $\tilde{\Gamma}$ we define a slightly modified splitting kernel, where the prefactor α_s/t is evaluated at the first argument t instead of the second argument t_0. Including the boundary condition $\Phi_g(t_0) = u$ we can solve this equation for the generating functional, again using the method of the known solution,

$$\boxed{\Phi_g(t) = \frac{1}{1 + \dfrac{1-u}{u\tilde{\Delta}_g(t)}}} \quad \text{with} \quad \tilde{\Delta}_g(t) = \exp\left(-\int_{t_0}^t dt'\, \tilde{\Gamma}_{g\leftarrow g}(t',t_0)\right). \quad (2.248)$$

The derivative of this solution is

$$\frac{d\Phi_g(t)}{t} = \frac{d}{dt}\left(1 + \frac{1-u}{u\tilde{\Delta}_g(t)}\right)^{-1}$$

$$= -\Phi_g(t)^2\, \frac{1-u}{u}\, \frac{d}{dt} \exp\left(+\int_{t_0}^t dt'\, \tilde{\Gamma}_{g\leftarrow g}(t',t_0)\right)$$

$$= -\Phi_g(t)^2\, \frac{1-u}{u\tilde{\Delta}_g(t)}\, \frac{d}{dt} \int_{t_0}^t dt'\, \tilde{\Gamma}_{g\leftarrow g}(t',t_0)$$

$$= -\Phi_g(t)^2 \left(\frac{1}{\Phi_g(t)} - 1\right)$$

$$\frac{d}{dt}\int_{t_0}^t dt'\, \tilde{\Gamma}_{g\leftarrow g}(t',t_0) = \Phi_g(t)\, (\Phi_g(t)-1)\, \tilde{\Gamma}_{g\leftarrow g}(t,t_0), \quad (2.249)$$

which is precisely the evolution equation in Eq. (2.247).

While we have suggestively defined a modified splitting kernel $\tilde{\Gamma}$ in Eq. (2.247) and even extended this analogy to a Sudakov-like factor in Eq. (2.248) it is not entirely clear what this object represents. In the limit of large $\log t/t_0 \gg 1$ or $t \gg t_0$, which is not the limit we rely on for the pure Yang–Mills case, we find

$$\int_{t_0}^t dt'\, \tilde{\Gamma}_{g\leftarrow g}(t',t_0) - \int_{t_0}^t dt'\, \Gamma_{g\leftarrow g}(t',t_0) = -\frac{C_A}{2\pi}\alpha_s(t_0) \int_{t_0}^t d\frac{t'}{t_0}\, \log\frac{t'}{t_0}$$

$$= -\frac{C_A}{2\pi}\alpha_s(t_0) \left[\frac{t'}{t_0}\log\frac{t'}{t_0} - \frac{t'}{t_0}\right]_1^{t/t_0}$$

$$= -\frac{C_A}{2\pi}\alpha_s(t_0)\, \frac{t}{t_0}\log\frac{t}{t_0}. \quad (2.250)$$

In the staircase limit $t \sim t_0$ and consistently neglecting $\log t/t_0$ the two kernels $\Gamma_{g\leftarrow g}$ and $\tilde{\Gamma}_{g\leftarrow g}$ become identical. In the same limit we find $\Delta_g \sim \tilde{\Delta}_g \sim 1$. Again using $t' = t_0 + \delta$ and only keeping the leading terms in δ we can compute the leading difference

2.6 Multi–Jet Events

$$\tilde{\Gamma}_{g \leftarrow g}(t', t_0) - \Gamma_{g \leftarrow g}(t', t_0)$$

$$= \frac{C_A}{2\pi} \left(\frac{\alpha_s(t')}{t'} - \frac{\alpha_s(t_0)}{t_0} \right) \left(\frac{\delta}{t'} - \frac{11}{6} \right)$$

$$= -\frac{C_A}{2\pi} \frac{11}{6} (t' - t_0) \left. \frac{d\, \alpha_s(t)}{dt\ t} \right|_{t_0}$$

$$= -\frac{C_A}{2\pi} \frac{11}{6} \delta \left[\frac{1}{t} \frac{d\alpha_s(t)}{dt} - \frac{\alpha_s(t)}{t^2} \right]_{t_0}$$

$$= -\frac{C_A}{2\pi} \frac{11}{6} \delta \left[-\frac{1}{t} \frac{\alpha_s^2(t) b_0}{t} - \frac{\alpha_s(t)}{t^2} \right]_{t_0} \quad \text{using Eq. (2.73)}$$

$$= \frac{C_A \alpha_s(t_0)}{2\pi} \frac{11}{6} \frac{\delta}{t_0^2} (1 + b_0 \alpha_s(t_0))$$

$$\int_{t_0}^{t} dt'\ \tilde{\Gamma}_{g \leftarrow g}(t', t_0) - \int_{t_0}^{t} dt'\ \Gamma_{g \leftarrow g}(t', t_0) = \frac{C_A \alpha_s(t_0)}{2\pi} \frac{11}{6} \frac{\delta^2}{t_0^2} (1 + b_0 \alpha_s(t_0)) \ . \tag{2.251}$$

In the pure Yang–Mills theory the running of the strong coupling is described by $b_0 = 1/(4\pi) 11 N_c/3$. In both limits the true and the modified splitting kernels differ by the respective small parameter.

The closed form for the generating functional in Eq. (2.248) allows us to compute the number of jets in purely gluonic events. The first derivative is

$$\frac{d}{du} \Phi_g(t) = \frac{d}{du} u \left(u + \frac{1-u}{\tilde{\Delta}_g(t)} \right)^{-1}$$

$$= \left(u + \frac{1-u}{\tilde{\Delta}_g} \right)^{-1} + u(-1) \left(u + \frac{1-u}{\tilde{\Delta}_g^2} \right)^{-2} \left(1 - \frac{1}{\tilde{\Delta}_g} \right) \ . \tag{2.252}$$

The form of the n-th derivative we can again prove by induction. Clearly, for $n = 1$ the above result is identical with the general solution

$$\frac{d^n}{du^n} \Phi_g(t) = n! \left(\frac{1}{\tilde{\Delta}_g} - 1 \right)^{n-1} \left(u + \frac{1-u}{\tilde{\Delta}_g} \right)^{-n} \left[1 + u \left(u + \frac{1-u}{\tilde{\Delta}_g} \right)^{-1} \left(\frac{1}{\tilde{\Delta}_g} - 1 \right) \right] \ . \tag{2.253}$$

The induction step from n to $n+1$ is

$$\frac{d^{n+1}}{du^{n+1}}\Phi_g(t)$$

$$= \frac{d}{du} n! \left(\frac{1}{\tilde{\Delta}_g} - 1\right)^{n-1} \left(u + \frac{1-u}{\tilde{\Delta}_g}\right)^{-n} \left[1 + u\left(u + \frac{1-u}{\tilde{\Delta}_g}\right)^{-1}\left(\frac{1}{\tilde{\Delta}_g} - 1\right)\right]$$

$$= n! \left(\frac{1}{\tilde{\Delta}_g} - 1\right)^{n-1} \left\{-n\left(u + \frac{1-u}{\tilde{\Delta}_g}\right)^{-n-1}\left(1 - \frac{1}{\tilde{\Delta}_g}\right)\left[1 + u\left(u + \frac{1-u}{\tilde{\Delta}_g}\right)^{-1}\left(\frac{1}{\tilde{\Delta}_g} - 1\right)\right]\right.$$

$$\left. + \left(u + \frac{1-u}{\tilde{\Delta}_g}\right)^{-n}\left[\left(u + \frac{1-u}{\tilde{\Delta}_g}\right)^{-1}\left(\frac{1}{\tilde{\Delta}_g} - 1\right) + u\left(u + \frac{1-u}{\tilde{\Delta}_g}\right)^{-2}\left(\frac{1}{\tilde{\Delta}_g} - 1\right)^2\right]\right\}$$

$$= n! \left(\frac{1}{\tilde{\Delta}_g} - 1\right)^{n-1} \left\{n\left(u + \frac{1-u}{\tilde{\Delta}_g}\right)^{-n-1}\left(\frac{1}{\tilde{\Delta}_g} - 1\right)\left[1 + u\left(u + \frac{1-u}{\tilde{\Delta}_g}\right)^{-1}\left(\frac{1}{\tilde{\Delta}_g} - 1\right)\right]\right.$$

$$\left. + \left(u + \frac{1-u}{\tilde{\Delta}_g}\right)^{-n-1}\left(\frac{1}{\tilde{\Delta}_g} - 1\right)\left[1 + u\left(u + \frac{1-u}{\tilde{\Delta}_g}\right)^{-1}\left(\frac{1}{\tilde{\Delta}_g} - 1\right)\right]\right\}$$

$$= (n+1)! \left(\frac{1}{\tilde{\Delta}_g} - 1\right)^{n} \left(u + \frac{1-u}{\tilde{\Delta}_g}\right)^{-n-1}\left[1 + u\left(u + \frac{1-u}{\tilde{\Delta}_g}\right)^{-1}\left(\frac{1}{\tilde{\Delta}_g} - 1\right)\right]. \quad (2.254)$$

Evaluating the solution given by Eq. (2.253) for $u = 0$ gives us the jet rates

$$P_{n-1} = \frac{1}{n!}\frac{d^n}{du^n}\Phi_g(t)\bigg|_{u=0} = \left(\frac{1}{\tilde{\Delta}_g} - 1\right)^{n-1} \tilde{\Delta}_g^n = \tilde{\Delta}_g\left(1 - \tilde{\Delta}_g\right)^{n-1}, \quad (2.255)$$

which predicts constant ratios

$$\boxed{R_{(n+1)/n} = 1 - \tilde{\Delta}_g(t)}. \quad (2.256)$$

Such constant ratios define a *staircase pattern*. The name describes the form of the n-distribution on a logarithmic scale. This pattern was first seen in W+jets production at UA1 in 1985. It has for a long time been considered an accidental sweet spot where many QCD effects cancel each other to produce constant ratios of successive exclusive n-jet rates. Our derivation from the generating functionals suggest that staircase scaling is one of two *pure jet scaling patterns*:

1. In the presence of large scale differences abelian splittings generate a Poisson pattern with $R_{(n+1)/n} \propto 1/(n+1)$, as seen in Eq. (2.240).
2. For democratic scales non–abelian splittings generate a staircase pattern with constant $R_{(n+1)/n}$ shown in Eq. (2.256).

2.6 Multi–Jet Events

We have shown them for final state radiation only, so they should be observable in $e^+e^- \to$ jets events. Our derivation of the scaling patterns is exclusively based on the parton shower. However, it turns out that corrections from hard matrix element corrections, described in the next section, do not change the staircase scaling patterns.

To generalize the final–state jet scaling patterns to initial state radiation we need to include parton densities. For simplicity, we again resort to the Drell–Yan process. We can approximate the parton densities based on two assumptions: threshold kinematics and $x_1 \approx x_2$. In the absence of additional jets the hadronic and partonic energy scales in on–shell Z production are linked by $x^{(0)} \approx m_Z/\sqrt{s}$. For other jet configurations we denote the threshold value as $x^{(n)}$. Putting everything together we find for the generating function

$$\Phi_{\text{Drell-Yan}} = \sum_{i,j} f_i\left(\frac{x^{(n)}}{2}\right) \tilde{\Phi}_i \times f_j\left(\frac{x^{(n)}}{2}\right) \tilde{\Phi}_j . \qquad (2.257)$$

This means we can use the jet rates from the e^+e^- case multiplied by a parton density factor. The structural new feature in $\Phi_{\text{Drell-Yan}}$ is the n-dependence from the parton densities. This is different from the resummed and n-independent definition in Eq. (2.225). However, the generating function is not a physical object. Using our usual formalism and notation we find

$$\begin{aligned} P_{n-1} &= \frac{1}{n!}\frac{d^n}{du^n}\Phi_{\text{Drell-Yan}}\bigg|_{u=0} \\ &= \frac{1}{n!}\sum_{i,j} f_i\left(\frac{x^{(n)}}{2}\right) f_j\left(\frac{x^{(n)}}{2}\right) \times \frac{d^n}{du^n}\tilde{\Phi}_i\tilde{\Phi}_j\bigg|_{u=0} . \end{aligned} \qquad (2.258)$$

2.6.2 CKKW and MLM Schemes

The main problem with QCD at the LHC is the range of energy scales of the jets we encounter. *Collinear jets* with their small transverse momenta are well described by a parton shower. From Sect. 2.3.5 we know that strictly speaking the parton shower only fills the phase space region up to a maximum transverse momentum $p_T < \mu_F$. In contrast, *hard jets* with large transverse momentum are described by matrix elements which we compute using the QCD Feynman rules. They fill the non–collinear part of phase space which is not covered by the parton shower. Because of the collinear logarithmic enhancement we discussed in Sect. 2.3.5 we expect many more collinear and soft jets than hard jets at the LHC.

The natural question then becomes: what is the range of 'soft' or 'collinear' and what is 'hard'? Applying a consistency condition we can define collinear jet radiation by the validity of the collinear approximation in Eq. (2.87). The maximum

p_T of a collinear jet is the upper end of the region for which the jet radiation cross section behaves like $1/p_T$, or the point where the distribution $p_T d\sigma/dp_T$ leaves its plateau. For harder and harder jets we will at some point become limited by the partonic energy available at the LHC, which means the p_T distribution of additional jets will start dropping faster than $1/p_T$. Collinear logarithms will become numerically irrelevant and jets will be described by the regular matrix element squared without any resummation.

Quarks and gluons produced in association with gauge bosons at the Tevatron behave like collinear jets for $p_T \lesssim 20\,\text{GeV}$, because quarks at the Tevatron are limited in energy. At the LHC, jets produced in association with tops behave like collinear jets to $p_T \sim 150\,\text{GeV}$, jets produced with new particles of mass 500 GeV behave like collinear jets to p_T scales larger than 300 GeV. This is not good news, because collinear jets means many jets, and many jets produce *combinatorial backgrounds* and ruin the missing momentum resolution of the detector: if we are looking for example for two jets to reconstruct an invariant mass you can simply plot all events as a function of this invariant mass and remove the backgrounds by requiring all event to sit around a peak in m_{jj}. If we have for example three jets in the event we have to decide which of the three jet–jet combinations should go into this distribution. If this is not possible we have to consider two of the three combinations as uncorrelated 'background' events. In other words, we make three histogram entries out of each signal or background event and consider all three background events plus two of the three signal combinations as background. This way the signal-to-background ratio decreases from N_S/N_B to $N_S/(3N_B + 2N_S)$. A famous victim of such combinatorics is the (former) Higgs discovery channel $pp \to t\bar{t}H$ with $H \to b\bar{b}$.

For theorists this means that at the LHC we have to reliably model collinear and hard jets. For simplicity, in this section we will first limit our discussion to final-state radiation, for example off the R-ratio process $e^+e^- \to q\bar{q}$ from Sect. 2.1.1. Combining collinear and hard jets in the final state has to proceed in two steps. The first of them has nothing to do with the actual jet simulation. If we categorize the generated events by counting the number of jets in the final state we can refer to an *exclusive rate*, which requires a process to have exactly a given number of jets, or an *inclusive rate*, where we for example identify n jets and ignore everything else appearing in the event. Additional collinear jets which we usually denote as '$+X$' will be included. We already know that a total rate for any hard process we compute as $e^+e^- \to q\bar{q} + X$, with any additional number of collinear jets in the final state. Predictions involving parton densities and the DGLAP equation are jet-inclusive. Any scheme combining the parton shower and hard matrix elements for events with arbitrary jet multiplicity has to follow the path

1. Define jet-exclusive events from the hard matrix elements and the parton shower
2. Combine final states with *different numbers of final-state particles*
3. Reproduce matrix element results in high-p_T and well separated phase space region
4. Reproduce parton shower results for collinear and soft radiation
5. Interpolate smoothly and avoid double counting of events

2.6 Multi–Jet Events

For specific processes at the Tevatron the third and fourth point on this list have actually been tackled by so-called matrix element corrections in the parton shower Monte Carlos PYTHIA and HERWIG. At the LHC this structure of event generation has become standard.

The final state of the process $e^+e^- \to q\bar{q} + X$ often involves more than two jets due to final state splitting. Even for the first step of defining jet–exclusive predictions from the matrix element we have to briefly consider the geometry of different jets. To separate jet–inclusive event samples into jet–exclusive event samples we have to define some kind of jet separation parameter. If we radiate a gluon off one of the quark legs, it gives us a $q\bar{q}g$ final state. This additional gluon can be collinear with and hence geometrically close to one of the quarks or not. Jet algorithms which decide if we count such a splitting as one or two jets we describe in detail in Sect. 3.1.1. They are based on a choice of collinearity measure y_{ij} which we can for example construct as a function of the distance in R space, introduced in Eq. (2.35), and the transverse momenta. We define two jets as collinear and hence as one jet if $y_{ij} < y_{\text{resol}}$ where y_{resol} is a free parameter in the algorithm. As a result, the number of jets in an event will depend on this *resolution parameter* y_{resol}.

For the second step of combining hard and collinear jet simulation the same resolution parameter appears in a form where it becomes a collinear vs *hard matching parameter* y_{match}. It allows us to clearly assign each hadron collider event a number of collinear jets and a number of hard jets. Such an event with its given number of more or less hard jets we can then describe either using matrix elements or using a parton shower, where 'describe' means computing the relative probability of different phase space configurations. The parton shower will do well for jets with $y_{ij} < y_{\text{match}}$. In contrast, if for our closest jets we find $y_{ij} > y_{\text{match}}$, we know that collinear logarithms did not play a major role, so we should use the hard matrix element. If we assign the hard process a typical energy or virtuality scale t_{hard} we can translate the matching parameter y_{match} into a virtuality scale $t_{\text{match}} = y_{\text{match}}^2 t_{\text{hard}}$, below which we do not trust the hard matrix element. For example for the Drell–Yan process the hard scale would be something like the Z mass.

The CKKW jet combination scheme first tackles the problem of defining and combining jet–exclusive final states with different numbers of jets. The main ingredient to translating one into the other are non–splitting probabilities called Sudakov factors. They can transform inclusive n-particle rates into exact n-particle rates, with no additional final–state jet outside a given resolution scale. We can compute integrated splitting probabilities $\Gamma_j(t_{\text{hard}}, t)$ which for quarks and gluons are implicitly defined through the Sudakov factors introduced in Eq. (2.166)

$$\Delta_q(t_{\text{hard}}, t_{\text{match}}) = \exp\left(-\int_{t_{\text{match}}}^{t_{\text{hard}}} dt\, \Gamma_{q \leftarrow q}(t_{\text{hard}}, t)\right)$$

$$\Delta_g(t_{\text{hard}}, t_{\text{match}}) = \exp\left(-\int_{t_{\text{match}}}^{t_{\text{hard}}} dt\, \left[\Gamma_{g \leftarrow g}(t_{\text{hard}}, t) + \Gamma_{q \leftarrow g}(t)\right]\right). \qquad (2.259)$$

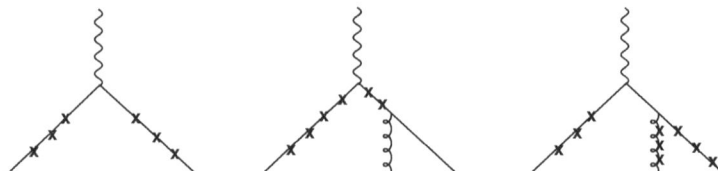

Fig. 2.3 Vetoed showers on two-jet and three-jet contributions. The scale at the gauge boson vertex is t_{hard}. The two-jet (three-jet) diagram implies exactly two (three) jets at the resolution scale t_{match}, below which we rely on the parton shower (Figure from Ref. [5])

For final–state radiation t corresponds to the original $\sqrt{p_a^2}$. Moving forward in time it is ordered according to $t_{\text{hard}} > t > t_{\text{match}}$. The resolution of individual jets we identify with the matrix element–shower matching scale t_{match}. To leading logarithm the explicit form of the splitting kernels is given in Eq. (2.172). The virtualities $t_{\text{hard}} > t$ correspond to the incoming (mother) and outgoing (daughter) parton. Note that the wrong limit for $\Gamma_j(t_{\text{hard}}, t_{\text{hard}}) \neq 0$ can be circumvented technically. To avoid unnecessary approximations in the y integration more recent CKKW implementations integrate the splitting kernels numerically.

To get a first idea how to transform inclusive into exact n-jet rates we compute the probability of seeing exactly *two jets* in the process $e^+e^- \to q\bar{q}$. Looking at Fig. 2.3 this means that none of the two quarks in the final state radiate a resolved gluon between the virtualities t_{hard} (given by the qqZ vertex) and $t_{\text{match}} < t_{\text{hard}}$. As will become important later, we specify that this no-radiation statement assumes a jet resolution as given by end point of the external quark and gluon legs. The probability we have to multiply the inclusive two-jet rate with is then

$$\left[\Delta_q(t_{\text{hard}}, t_{\text{match}})\right]^2 , \qquad (2.260)$$

once for each quark. Whatever happens at virtualities below t_{match} will be governed by the parton shower and does not matter anymore. Technically, this requires us to define a so-called vetoed parton shower which we will describe in Sect. 2.7.3.

What is the probability that the initially two-jet final state evolves exactly into *three jets*, again following Fig. 2.3? We know that it contains a factor $\Delta_q(t_{\text{hard}}, t_{\text{match}})$ for one untouched quark.

After splitting at t_q with the probability $\Gamma_{q \leftarrow q}(t_q, t_{\text{hard}})$ the second quark survives to t_{match}, giving us a factor $\Delta_q(t_q, t_{\text{match}})$. If we assign the virtuality t_g to the radiated gluon at the splitting point we find the gluon's survival probability as $\Delta_g(t_g, t_{\text{match}})$. Together with the quark Sudakovs this gives us

$$\Delta_q(t_{\text{hard}}, t_{\text{match}}) \; \Gamma_{q \leftarrow q}(t_{\text{hard}}, t_q) \; \Delta_q(t_q, t_{\text{match}}) \; \Delta_g(t_g, t_{\text{match}}) \cdots \qquad (2.261)$$

That's all there is, with the exception of the intermediate quark. There has to appear another factor describing that the quark, starting from t_{hard}, gets to the splitting point t_q untouched. Naively we would guess that this probability is given by $\Delta_q(t_{\text{hard}}, t_q)$.

2.6 Multi–Jet Events

However, this Sudakov factor describes no splittings resolved at the lower scale t_q. What we really mean is no splitting between t_{hard} and t_q resolved at a third scale $t_{\text{match}} < t_q$ given by the quark leg hitting the parton shower regime. We get this information by computing the probability of no splitting between t_{hard} and t_q, namely $\Delta_q(t_{\text{hard}}, t_{\text{match}})$, but under the condition that splittings from t_q down to t_{match} are explicitly allowed.

If zero splittings give us a probability factor $\Delta_q(t_{\text{hard}}, t_{\text{match}})$, to describe exactly one splitting from t_q on, we add a factor $\Gamma(t_q, t)$ with an unknown splitting point t. This point t we integrate over between the resolution point t_{match} and the endpoint of the no-splitting range, t_q. This is the same argument as in our physical interpretation of the Sudakov factors solving the DGLAP equation (2.180). For an arbitrary number of possible splittings between t_q and t_{match} we find the sum

$$\Delta_q(t_{\text{hard}}, t_{\text{match}}) \left[1 + \int_{t_{\text{match}}}^{t_q} dt \, \Gamma_{q \leftarrow q}(t_q, t) + \text{more splittings} \right] =$$
$$= \Delta_q(t_{\text{hard}}, t_{\text{match}}) \, \exp \left[\int_{t_{\text{match}}}^{t_q} dt \, \Gamma_{q \leftarrow q}(t_q, t) \right] = \frac{\Delta_q(t_{\text{hard}}, t_{\text{match}})}{\Delta_q(t_q, t_{\text{match}})}.$$
(2.262)

The factors $1/n!$ in the Taylor series appear because for example radiating two ordered jets in the same t interval can proceed two ways, both of which lead to the same final state. Once again: we compute the probability of nothing happening between t_{hard} and t_q from the probability of nothing happening between t_{hard} and t_{match} times any number of possible splittings between t_q and t_{match}.

Collecting all factors from Eqs. (2.261) and (2.262) gives us the probability to find exactly three partons resolved at t_{match} as part of the inclusive sample

$$\Delta_q(t_{\text{hard}}, t_{\text{match}}) \, \Gamma_{q \leftarrow q}(t_{\text{hard}}, t_q) \, \Delta_q(t_q, t_{\text{match}}) \, \Delta_g(t_g, t_{\text{match}}) \frac{\Delta_q(t_{\text{hard}}, t_{\text{match}})}{\Delta_q(t_q, t_{\text{match}})}$$
$$= \Gamma_{q \leftarrow q}(t_{\text{hard}}, t_q) \, [\Delta_q(t_{\text{hard}}, t_{\text{match}})]^2 \, \Delta_g(t_g, t_{\text{match}}).$$
(2.263)

This result is what we expect: both quarks go through untouched, just like in the two-parton case. In addition, we need exactly one splitting producing a gluon, and this gluon cannot split further. This example illustrates how we can compute these probabilities using Sudakov factors: adding a gluon corresponds to adding a splitting probability times the survival probability for this gluon, everything else magically drops out. At the end, we integrate over the splitting point t_q.

This discussion allows us to write down the first step of the *CKKW algorithm*, combining different hard n-jet channels into one consistent set of events. One by one we turn inclusive n-jet events into exact n-jet events. We can write down the slightly simplified algorithm for final–state radiation. As a starting point, we generate events and compute leading order cross sections for all n-jet processes. A universal lower jet radiation cutoff t_{match} ensures that all jets are hard and that all corresponding cross sections $\sigma_{n,i}$ are finite. The second index i describes different non–interfering parton configurations for a given number of final–state jets, like $q\bar{q}gg$ and $q\bar{q}q\bar{q}$ for $n = 4$. The purpose of the algorithm is to assign a weight (probability, matrix element squared,...) to a given phase space point, statistically picking the correct process and combining them properly. It proceeds event by event:

1. For each jet final state (n,i) compute the relative probability $P_{n,i} = \sigma_{n,i} / \sum_{k,j} \sigma_{k,j}$
   and select a final state (n,i) with this probability $P_{n,i}$
2. Assign the momenta from the phase space generator to, assumed, hard external particles
   and compute the transition matrix element $|\mathcal{M}|^2$ including parton shower below t_{match}
3. Use a jet algorithm to compute the shower history, i.e. all splitting points t_j in each event
   and check that this history corresponds to possible Feynman diagrams and does not violate any symmetries
4. For each internal and external line compute the Sudakov non–splitting probability down to t_{match}
5. Re-weight the α_s values of each splitting using the k_T scale from the shower history
6. Combine matrix element, Sudakovs, and α_s into a final weight

We can use this final event weight to compute distributions from weighted events or to decide if to keep or discard an event when producing unweighted events. The construction ensures that the relative weight of the different n–jet rates is identical to the probabilities we initially computed. In step 2 the CKKW event generation first chooses the appropriate hard scale in the event; in step 3 we compute the individual starting scale for the parton shower applied to each of the legs. Following our example, this might be t_{hard} for partons leaving the hard process itself or t_g for a parton appearing via later splitting.

In a second step of the CKKW scheme we match this combined hard matrix element with the parton shower, given the matching point t_{match}. From the final experimental resolution scale t_{resol} up to a matching scale t_{match} we rely on the parton shower to describe jet radiation while above the matching scale jet radiation is explicitly forbidden by the Sudakov non–splitting probabilities. Individually, both regimes consistently combine different n–jet processes. All we need to make sure is that there is no double counting.

From the discussion of Eq. (2.262) we know that Sudakovs describing the evolution between two scales and using a third scale as the resolution are going to be the problem. Carefully distinguishing the scale of the actual splitting from the scale of jet resolution is the key. The CKKW scheme starts each parton shower at the point where the parton first appears, and it turns out that we can use this argument to keep the regimes $y > y_{\text{match}}$ and $y < y_{\text{match}}$ separate. There is a simple way to check this, namely if the y_{match} *dependence* drops out of the final combined probabilities. The

2.6 Multi–Jet Events

answer for final–state radiation is yes, as proven in the original paper, including a hypothetical next–to–leading logarithm parton shower. A modified CKKW scheme is implemented in the publicly available SHERPA event generator.

An alternative to the CKKW scheme which has been developed independently but incorporates essentially the same physics is the *MLM scheme*, for example implemented in ALPGEN or Madgraph. Its main difference to the CKKW scheme is that it avoids computing the survival properties using Sudakov form factors. Instead, it vetos those events which CKKW removes by applying the Sudakov non–splitting probabilities. This way MLM avoids problems with splitting probabilities beyond the leading logarithms, for example the finite terms appearing in Eq. (2.172), which can otherwise lead to a mismatch between the actual shower evolution and the analytic expressions of the Sudakov factors. In addition, the veto approach allows the MLM scheme to combine a set of independently generated n–parton events, which can be convenient.

In the MLM scheme we again start by independently simulating n-jet events including hard jet radiation as well as the parton shower. In this set of complete events we then veto events which are simulated the wrong way. This avoids double counting of events which on the one hand are generated with n hard jets from the matrix element and on the other hand appear for example as $(n-1)$ hard jets with an additional jet from the parton shower.

After applying a jet algorithm, which in the case of ALPGEN is a cone algorithm and in case of Madgraph is a k_T algorithm, we compare the showered event with the un-showered hard event by identifying each reconstructed showered jet with the partons we started from. If all jet–parton combinations match and there exist no additional resolved jets we know that the showering has not altered the hard structure of the event. If there is a significant change between the original hard parton event and the showered event this event has to go. This choice corresponds to an event weight including the Sudakov non–splitting probabilities in the CKKW scheme. The only exception to this rule is the set of events with the highest jet multiplicity for which additional jets can only come from the parton shower. After defining the proper exclusive n-jet event sets we can again use the parton shower to describe more collinear jet radiation between t_match and t_resol.

After combining the samples we still need a backwards evolution of a generated event to know the virtuality scales which fix $\alpha_s(Q^2)$. As a side effect, if we also know the Feynman diagrams describing an event we can check that a certain splitting with its color structure is actually possible. For the parton shower or splitting simulation we need to know the interval of virtualities over which for example the additional gluon in the previous two-jet example can split. The lower end of this interval is universally given by t_match, but the upper end we cannot extract from the record event by event. Therefore, to compute the α_s values at each splitting point we start the parton shower at an universal hard scale, chosen as the hard(est) scale of the process.

Aside from such technical details all merging schemes are conceptually similar enough that we should expect them to reproduce each others' results, and they largely do. But the devil is in the details, so experiment tells us which scheme as

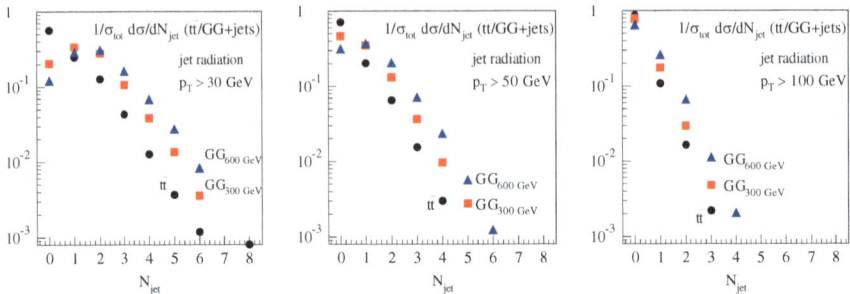

Fig. 2.4 Number of additional jets with a transverse momentum of at least 30, 50 or 100 GeV radiated off top pair production and the production of heavy states at the LHC. An example for such heavy states are scalar gluons with a mass of 300 or 600 GeV, pair-produced in gluon fusion (Figures from Ref. [17])

part of which event generator produces the most usable results for a given LHC measurement.

To summarize, we can use the CKKW and MLM schemes to first combine n-jet events with variable n and then consistently add the parton shower. In other words, we can for example simulate $Z + n$ jets production at the LHC to arbitrarily large numbers of jets, limited only by computational resources and the physical argument that at some point any additional jet radiation will be described by the parton shower. This combination will describe all jets correctly over the entire collinear and hard phase space. In Fig. 2.4 we show the number of jets expected to be produced in association with a pair of top quarks and a pair of heavy new states at the LHC. The details of these heavy scalar gluons are secondary for the basic features of these distributions. The only parameter which matters is their mass serving as the hard scale of the process, setting the factorization scale, and defining the upper limit of collinearly enhanced initial–state radiation. We see that heavy states come with many jets radiated at $p_T \lesssim 30$ GeV, where most of these jets vanish once we require transverse momenta of at least 100 GeV. This figure tells us that an analysis which asks for a reconstruction of two W-decay jets may well be swamped by combinatorics.

Looking at the individual columns in Fig. 2.4 there is one thing we have to keep in mind: each of the merged matrix elements combined into this sample is computed at leading order. The emission of real particles is included, virtual corrections are not. In other words, the CKKW and MLM schemes give us all *jet distributions*, but only to leading order in the strong coupling. When we combine the different jet multiplicities to evaluate total rates, jet merging improves the rate prediction because it includes contributions from all orders in α_s, provided they come with a potentially large logarithm from jet emission. From all we know, these leading logarithms dominate the higher order QCD corrections for most LHC processes, but it is not obvious how general this feature is and how we can quantify it. This is certainly true for all cases where higher order effects appear unexpectedly large and can be traced back to new partonic processes or phase space configurations opening

2.6 Multi–Jet Events

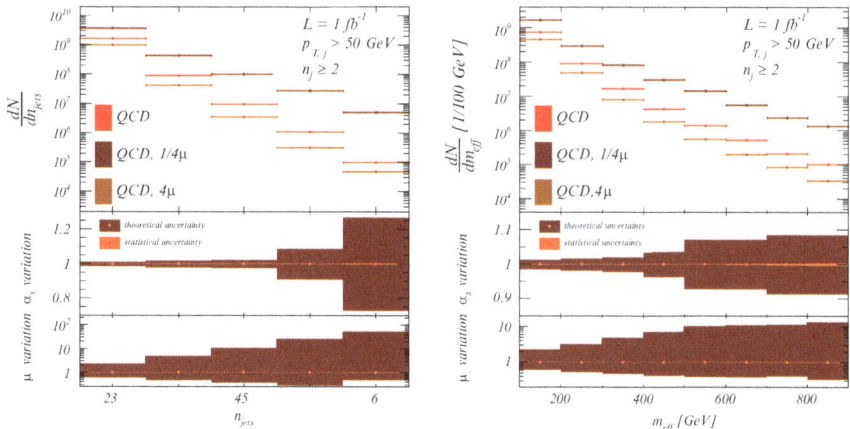

Fig. 2.5 Exclusive number of jets and effective mass distributions for pure QCD jet events at the LHC with a center–of–mass energy of 7 TeV and $p_{T,j} > 50$ GeV. The curves including the α_s uncertainty and a scale variation (tuning parameter) are computed with SHERPA and a fully merged sample including up to six hard jets. These distributions describe typical backgrounds for searches for jets plus missing energy with fake missing energy, which could originate from supersymmetric squark and gluino production (Figures from Ref. [9])

up at higher jet multiplicities. Systematically curing some of this shortcoming (but at a prize) will be the topic of the next section.

Before moving on to an alternative scheme we will illustrate why Higgs or exotics searches at the LHC really care about progress in QCD simulations: one way to look for heavy particles decaying into jets, leptons and missing energy is the variable

$$\begin{aligned} m_{\text{eff}} &= \not{E}_T + \sum_j E_{T,j} + \sum_\ell E_{T,\ell} \\ &= \not{p}_T + \sum_j p_{T,j} + \sum_\ell p_{T,\ell} \qquad \text{(for massless quarks, leptons)} \end{aligned} \qquad (2.264)$$

This variable and its relatives we will discuss in detail in Sect. 3.3.2. For gluon–induced QCD processes the effective mass should be small while the new physics signal's effective mass scale will be determined by the heavy masses.

For QCD jets as well as for W and Z plus jets backgrounds we can study the production of many jets using the CKKW scheme. Figure 2.5 shows the two critical distributions. First, in the number of hard jets we see the so-called staircase scaling behavior, namely constant ratios of exclusive $(n+1)$-jet and n-jet rates σ_{n+1}/σ_n. Such a scaling is closely related to the pattern we discuss in Eq. (1.225), in the context of the central jet veto of Sect. 1.6.2. The particularly interesting aspect of staircase scaling is that the constant ratio is the same for jet–inclusive and jet–exclusive cross sections $R^{\text{incl}}_{(n+1)/n} = R_{(n+1)/n}$, as shown in Eq. (1.226).

The consistent variation of α_s gives a small parametric uncertainty on these rates. A common scaling factor μ/μ_0 for all factorization, renormalization and shower scales in the process following our argument of Sect. 2.4 is strictly speaking not fixed by our physical interpretation in terms of resummation; such a factor as part of the leading logarithm can be factored out as a subleading finite term, so it should really be considered a tuning parameters for each simulation tool. Using the same simulation we also show the effective mass and observe a drop towards large values of m_{eff}. However, this drop is nowhere as pronounced as in some parton shower predictions. This analysis shows that the naive parton shower is not a good description of QCD background processes to the production of heavy particles. Using a very pragmatic approach and tune the parton shower to correctly describe LHC data even in this parameter region will most likely violate basic concepts like factorization, so we would be well advised to use merging schemes like CKKW or MLM for such predictions.

2.7 Next–to–Leading Orders and Parton Shower

As we know for example for the R ratio from Sect. 2.1.1 the precision of a leading order QCD calculation in terms of the strong coupling constant α_s is not always sufficient to match the experimental accuracy. In such a case we need to compute observables to higher order in QCD. On the other hand, in Sect. 2.3.5 we have seen that the parton shower does not fit into fixed order perturbation theory. With its collinear logarithm it sums particular terms to all orders in α_s. So how can we on the one hand compute higher order corrections to for example the Drell–Yan cross section and distributions and in addition consistently combine them with the parton shower?

Such a combination would remove one of the historic shortcomings of parton shower Monte Carlos. Apart from the collinear approximation for jet radiation they were always limited by the fact that in the words of one of the authors they 'only do shapes'. In other words, the normalization of the simulated event sample will always be leading order in perturbative QCD and hence subject to large theoretical uncertainties. The reason for this shortcoming is that collinear jet radiation relies on a hard process and the corresponding production cross section and works with splitting probabilities, but never touches the total cross section it started from.

As a solution we compute higher order cross sections to normalize the total cross section entering the respective Monte Carlo simulation. This is what we call a K factor: $K = \sigma^{\text{improved}}/\sigma^{\text{MC}} = \sigma^{\text{improved}}/\sigma^{\text{LO}}$. It is crucial to remember that higher order cross sections integrate over unobserved additional jets in the final state. So when we normalize the Monte Carlo we assume that we can first integrate over additional jets and obtain σ^{improved} and then just normalize the Monte Carlo which puts back these jets in the collinear approximation. Obviously, we should try to do better than that, and there are two ways to improve this traditional Monte Carlo approach, the MC@NLO scheme and the POWHEG scheme.

2.7.1 Next–to–Leading Order in QCD

When we compute the next–to–leading order correction to a cross section, for example to Drell–Yan production, we consider all contributions of the order $G_F^2 \alpha_s$. There are three obvious sets of Feynman diagrams we have to add and then square: the Born contribution $q\bar{q} \to Z$, the virtual gluon exchange for example between the incoming quarks, and the real gluon emission $q\bar{q} \to Zg$. An additional set of diagrams we should not forget are the crossed channels $qg \to Zq$ and $\bar{q}g \to Z\bar{q}$. Only amplitudes with the same external particles can be squared, so we find the matrix-element-squared contributions

$$|\mathscr{M}_B|^2 \propto G_F^2$$
$$2\mathrm{Re}\,\mathscr{M}_V^* \mathscr{M}_B \propto G_F^2 \alpha_s \qquad |\mathscr{M}_{Zg}|^2 \propto G_F^2 \alpha_s \qquad |\mathscr{M}_{Zq}|^2, |\mathscr{M}_{Z\bar{q}}|^2 \propto G_F^2 \alpha_s \;.$$
(2.265)

Strictly speaking, we have to include counter terms, which following Eq. (2.54) are a modification of $|\mathscr{M}_B|^2$. We add these counter terms to the interference of Born and virtual gluon diagrams to remove the ultraviolet divergences. However, this is not the issue we want to discuss.

Infrared poles arise from two sources, soft and collinear divergences. To avoid the complication of overlapping collinear and soft divergences we will follow a toy model by Bryan Webber. It describes simplified particle radiation off a hard process: the energy of the system before radiation is x_s and the energy of the outgoing particle, call it photon or gluon, is x, so $x < x_s < 1$. When we compute *next–to–leading order corrections* to a hard process, the different contributions, neglecting crossed channels, are

$$\left.\frac{d\sigma}{dx}\right|_B = B\,\delta(x) \qquad \left.\frac{d\sigma}{dx}\right|_V = \alpha_s \left(-\frac{B}{2\epsilon} + V\right)\delta(x) \qquad \left.\frac{d\sigma}{dx}\right|_R = \alpha_s \frac{R(x)}{x} \;.$$
(2.266)

The constant B describes the Born process and the factorizing poles in the virtual contribution. The coupling constant α_s ignores factors 2 and π or color factors. We immediately see that the integral over x in the real emission rate is logarithmically divergent in the soft limit, similar to the collinear divergences we know. Because we are interested in infrared divergences we choose $n = 4 + 2\epsilon$ dimensions with $\epsilon > 0$, just like in Sect. 2.3.1, which will regularize the real emission and compensate the resulting pole $1/\epsilon$ with the virtual corrections. This means that the Kinoshita–Lee–Nauenberg theorem is built into our toy model.

From factorization which we have illustrated based on the universality of the leading splitting kernels we know that in the collinear and soft limits the real emission has to follow the Born matrix element

$$\boxed{\lim_{x \to 0} R(x) = B} \ . \tag{2.267}$$

An *observable* computed beyond leading order includes contributions from real gluon emission and virtual gluon exchange. If the observable is infrared safe it will have a smooth limit towards vanishing gluon energy $O(x) \to O(0)$. The virtual corrections alone diverge, but the expectation value including virtual and real gluon contributions after dimensional regularization is finite. In dimensional regularization this cancellation schematically reads

$$\langle O \rangle \sim \int_0^1 dx \, \frac{O(x)}{x^{1-2\epsilon}} - \frac{O(0)}{2\epsilon} \ . \tag{2.268}$$

This kind of combination has a finite limit for $\epsilon \to 0$. However, for numerical applications and event simulation we need to implement this cancellation differently.

The expectation value of any *infrared safe observable* over the entire phase space, including Born terms, virtual corrections and real emission, is given by

$$\langle O \rangle \equiv \langle O \rangle_B + \langle O \rangle_V + \langle O \rangle_R = \int_0^1 dx \, O(x) \left[\frac{d\sigma}{dx}\bigg|_B + \frac{d\sigma}{dx}\bigg|_V + \frac{1}{x^{-2\epsilon}} \frac{d\sigma}{dx}\bigg|_R \right] . \tag{2.269}$$

The same way in which the renormalization and factorization scales appear, dimensional regularization now yields an additional factor $1/x^{-2\epsilon}$. Because we know its structure, we will omit the factorization scale factor in the following.

When we compute for example a distribution of the energy of one of the heavy particles in the process, we can extract a histogram from of the integral for $\langle O \rangle$ in Eq. (2.269) and obtain a normalized distribution. The problem is that we have to numerically integrate over x, and the individual parts of the integrand in Eq. (2.269) are not integrable.

There exist two methods to combine the virtual and real contributions to an observable and produce a finite physics result. The first way historically introduced by the Dutch loop school for example to compute QCD corrections to top pair production is *phase space slicing*: we divide the divergent phase space integral into a finite part and a pole, by introducing a small parameter Δ, which acts like

$$\langle O \rangle_R + \langle O \rangle_V$$

$$= \int_0^1 dx \, \frac{O(x)}{x^{-2\epsilon}} \frac{d\sigma}{dx}\bigg|_R + \langle O \rangle_V$$

$$= \left(\int_0^\Delta + \int_\Delta^1 \right) dx \, \alpha_s \frac{R(x) O(x)}{x^{1-2\epsilon}} + \langle O \rangle_V$$

$$= \alpha_s R(0) \, O(0) \int_0^\Delta dx \, \frac{1}{x^{1-2\epsilon}} + \alpha_s \int_\Delta^1 dx \, \frac{R(x) O(x)}{x} + \langle O \rangle_V \qquad \text{with } \Delta \ll 1$$

2.7 Next–to–Leading Orders and Parton Shower

$$= \alpha_s B \ O(0) \ \frac{\Delta^{2\epsilon}}{2\epsilon} + \alpha_s \int_\Delta^1 dx \ \frac{R(x)O(x)}{x} + \langle O \rangle_V \qquad \text{using Eq. (2.267)}$$

$$= \alpha_s \frac{B \ O(0)}{2} \ \frac{2\epsilon \log \Delta + \mathcal{O}(\epsilon^2)}{\epsilon}$$

$$+ \alpha_s \int_\Delta^1 dx \ \frac{R(x)O(x)}{x} + \alpha_s V O(0) \qquad \text{using Eq. (2.266)}$$

$$= \alpha_s B O(0) \ \log \Delta + \alpha_s \int_\Delta^1 dx \ \frac{R(x)O(x)}{x} + \alpha_s V O(0) + \mathcal{O}(\epsilon) \ . \tag{2.270}$$

The two sources of $\log \Delta$ dependence have to cancel in the final expression, so we can evaluate the integral at finite but small values of Δ. An amusing numerical trick is to re-write the explicit $\log \Delta$ contribution into a real–emission–type phase space integral. If the eikonal approximation is given in terms of a Mandelstam variable $\delta(s_4)$ and the cutoff has mass dimension two we can write

$$\log \frac{\Delta}{\mu^2} = \int_0^{s_4^{\max}} ds_4 \ \log \frac{\Delta}{\mu^2} \ \delta(s_4) = \int_\Delta^{s_4^{\max}} ds_4 \ \left[\frac{\log \frac{s_4^{\max}}{\mu^2}}{s_4^{\max} - \Delta} - \frac{1}{s_4} \right] \tag{2.271}$$

and similarly for $\log^2 \Delta$. We can conveniently integrate this representation along with the real emission phase space. The result will be a finite value for the next–to–leading order rate in the limit $\Delta \to 0$ and exactly $\epsilon = 0$. This means that using phase space slicing we have exchanged dimensional regularization for an energy cutoff. The advantage is that we can compute cross section more easily, the disadvantage is that numerically the large cancellation between the real and virtual correction appears in the single $x = 0$ bin.

To avoid such cancellations between integrals and replace them by cancellations among integrands we use a *subtraction method* to define integrable functions under the x integral in Eq. (2.269). While our toy model appears more similar to the Frixione–Kunszt–Signer subtraction scheme than to the *Catani Seymour* scheme, both of them really are equivalent at the level of the soft–collinear toy model. The special features of the Catani–Seymour dipoles only feature when we include the full modelling of the soft and collinear divergences described in Sect. 2.5.3.

Starting from the individually divergent virtual and real contributions we first subtract and then add again a smartly chosen term, in this toy model identical to a plus–subtraction following Eq. (2.128)

$$\langle O \rangle_R + \langle O \rangle_V = \int_0^1 dx\, \alpha_s \frac{R(x)O(x)}{x^{1-2\epsilon}} + \langle O \rangle_V$$

$$= \int_0^1 dx\, \left(\frac{\alpha_s R(x)O(x)}{x^{1-2\epsilon}} - \frac{\alpha_s R(0)O(0)}{x^{1-2\epsilon}} \right)$$

$$+ \alpha_s\, B\, O(0) \int_0^1 dx \frac{1}{x^{1-2\epsilon}} + \langle O \rangle_V$$

$$= \alpha_s \int_0^1 dx\, \frac{R(x)O(x) - BO(0)}{x} + \alpha_s \frac{B\, O(0)}{2\epsilon} + \langle O \rangle_V$$

$$= \alpha_s \int_0^1 dx\, \frac{R(x)O(x) - BO(0)}{x} + \alpha_s V O(0) \quad \text{using Eq. (2.266)}.$$

$$\tag{2.272}$$

In the subtracted real emission integral we take the limit $\epsilon \to 0$ because the asymptotic behavior of $R(x \to 0)$ regularizes this integral without any dimensional regularization required. In our toy model we omit finite contributions from the integrated subtraction term which will have to be added to the finite virtual corrections. In proper QCD exactly the same happens with the Catani–Seymour dipoles and their integrated form. We end up with a perfectly finite x integral for the sum of all three contributions, so even in the limit $\epsilon = 0$ there is no numerically small parameter in the expression

$$\langle O \rangle = \langle O \rangle_B + \langle O \rangle_V + \langle O \rangle_R$$

$$= B\, O(0) + \alpha_s V\, O(0) + \alpha_s \int_0^1 dx\, \frac{R(x)\, O(x) - B\, O(0)}{x}$$

$$= \int_0^1 dx\, \left[O(0) \left(B + \alpha_s V - \alpha_s \frac{B}{x} \right) + O(x)\, \alpha_s \frac{R(x)}{x} \right]. \tag{2.273}$$

This subtraction procedure is a standard method to compute next–to–leading order corrections involving one-loop virtual contributions and the emission of one additional parton.

As a side remark, we can numerically improve this expression using a distribution relation

$$\int_0^1 dx\, \frac{f(x)}{x^{1-2\epsilon}} = \int_0^1 dx\, \frac{f(x) - \theta(x_c - x)\, f(0)}{x^{1-2\epsilon}} + f(0) \int_0^{x_c} dx\, x^{-1+2\epsilon}$$

$$= \int_0^1 dx\, \frac{f(x) - \theta(x_c - x)\, f(0)}{x} \left(1 + 2\epsilon \log x + \mathcal{O}(\epsilon^2) \right) + f(0) \frac{x_c^{2\epsilon}}{2\epsilon}$$

2.7 Next-to-Leading Orders and Parton Shower

$$= \int_0^1 dx \left(\frac{f(x) - \theta(x_c - x) f(0)}{x} + 2\epsilon \frac{f(x) - \theta(x_c - x) f(0)}{x} \log x \right.$$
$$\left. + \frac{x_c^{2\epsilon}}{2\epsilon} f(x)\delta(x) \right) \Leftrightarrow \frac{1}{x^{1-2\epsilon}} = \frac{x_c^{2\epsilon}}{2\epsilon} \delta(x) + \left(\frac{1}{x}\right)_c + 2\epsilon \left(\frac{\log x}{x}\right)_c ,$$
(2.274)

where the last line is a relation between appropriately defined distributions. This c-subtraction first introduced as part of the Frixione–Kunszt–Signer subtraction scheme is defined as

$$\int_0^1 dx \, f(x) \, g(x)_c = \int_0^1 dx \, [f(x)g(x) - f(0)g(x)\theta(x_c - x)] . \quad (2.275)$$

It is a generalization of the plus subtraction defined in Eq. (2.128) which we reproduce for $x_c = 1$. Linking the delta distribution to the divergent integral over $1/x$ it is also reminiscent of the principal value integration, but for an endpoint singularity and a dimensionally regularized phase space. Effectively combining phase space subtraction Eq. (2.272) and phase space slicing Eq. (2.270), we include a cutoff in the integrals holding the subtraction terms

$$\langle O \rangle_R = \alpha_s \int_0^1 dx \frac{R(x)O(x)}{x^{1-2\epsilon}}$$
$$= \alpha_s \int_0^1 dx \frac{R(x)O(x) - \theta(x_c - x)B \, O(0)}{x} (1 + 2\epsilon \log x)$$
$$+ \alpha_s BO(0) \frac{x_c^{2\epsilon}}{2\epsilon} + \mathcal{O}(\epsilon^2) . \quad (2.276)$$

The dependence on the cutoff parameter x_c drops out of the final result. The numerical behavior, however, should be improved if we subtract the infrared divergence only close to the actual pole where following Eq. (2.267) we understand the behavior of the real emission amplitude.

The formula Eq. (2.273) is, in fact, a little tricky: usually, the Born-type kinematics would come with an explicit factor $\delta(x)$, which in this special case we can omit because of the integration boundaries. We can re-write the same formula in a more appropriate way to compute distributions, possibly including experimental cuts

$$\boxed{\frac{d\sigma}{dO} = \int_0^1 dx \left[I(O)_{\text{LO}} \left(B + \alpha_s V - \alpha_s \frac{B}{x} \right) + I(O)_{\text{NLO}} \, \alpha_s \frac{R(x)}{x} \right]} . \quad (2.277)$$

The *transfer function* $I(O)$ is defined in a way that formally does precisely what we require: at leading order we evaluate $I(O)$ using the Born kinematics $x = 0$ while for the real emission kinematics it allows for general $x = 0 \cdots 1$.

2.7.2 MC@NLO Method

For example in Eq. (2.269) we integrate over the entire phase space of the additional parton. For a hard additional parton or jet the cross section looks well defined and finite, provided we fully combine real and virtual corrections. An infrared divergence appears after integrating over small but finite $x \to 0$ from real emission, and we cancel it with an infrared divergence in the virtual corrections proportional to a Born–type momentum configuration $\delta(x)$. In terms of a histogram in x we encounter the real emission divergence at small x, and this divergence is cancelled by a negative delta distribution at $x = 0$. Obviously, this will only give a well behaved distribution after integrating over at least a range of x values just above zero.

This soft and collinear subtraction scheme for next–to–leading order calculations leads us to the first method of combining or matching next–to–leading order calculations with a parton shower. Instead of the contribution from the virtual corrections contributing at $\delta(x)$ what we would rather want is a smeared virtual corrections pole which coincides with the justified collinear approximation and cancels the real emission over the entire low-x range. We can view this contribution as events with a negative weight or counter–events. Negative events trigger negative reactions with experimentalists, because they cause problems in a chain of probabilistic statements like a detector simulation. Fundamentally, there is no problem with them as long as any physical prediction we make after adding all leading order and next–to–leading order contributions gives a positive cross section.

Because we know they describe collinear jet radiation correctly such a modification will make use of Sudakov factors. We can write them as a function of the energy fraction z and define the associated probability as $d\mathscr{P} = \alpha_s P(z)/z \, dz$. Note that we avoid the complicated proper two-dimensional description of Eq. (2.166) in favor of the simpler picture just in terms of particle energy fractions as introduced in the last section.

Once we integrate over the entire phase space this modified subtraction scheme has to give the same result as the *next–to–leading order rate*. Smearing the integrated soft–collinear subtraction term using the splitting probabilities entering the parton shower means that the MC@NLO subtraction scheme has to be adjusted to the parton shower we use.

Let us consider the perturbatively critical but otherwise perfectly fine observable, the radiated *photon spectrum* as a function of the external energy scale z. We know what this spectrum looks like for the collinear and hard kinematic configurations

2.7 Next–to–Leading Orders and Parton Shower

$$\left.\frac{d\sigma}{dz}\right|_{\text{LO}} = \alpha_s \frac{BP(z)}{z} \qquad \left.\frac{d\sigma}{dz}\right|_{\text{NLO}} = \alpha_s \frac{R(z)}{z} \ . \tag{2.278}$$

The first term describes parton shower radiation from the Born diagram at order α_s, while the second term is the hard real emission defined in Eq. (2.266). According to Eq. (2.277) the transfer functions read

$$\left.I(z,1)\right|_{\text{LO}} = \alpha_s \frac{P(z)}{z}$$

$$\left.I(z,x_M)\right|_{\text{NLO}} = \delta(z-x) + \alpha_s \frac{P(z)}{z} \theta(x_M(x) - z) \ . \tag{2.279}$$

The second term in the real radiation transfer function arises because at the next order in perturbation theory the parton shower also acts on the real emission process. It requires that enough energy to radiate a photon with an energy z be available, where x_M is the energy available at the respective stage of showering, i.e. $z < x_M$.

We can include these transfer functions in Eq. (2.277) and obtain

$$\begin{aligned}\frac{d\sigma}{dz} &= \int_0^1 dx \left[I(z,1) \left(B + \alpha_s V - \alpha_s \frac{B}{x} \right) + I(z,x_M) \, \alpha_s \frac{R(x)}{x} \right] \\ &= \int_0^1 dx \left[\alpha_s \frac{P(z)}{z} \left(B + \alpha_s V - \alpha_s \frac{B}{x} \right) + (\delta(x-z) + \mathcal{O}(\alpha_s)) \, \alpha_s \frac{R(x)}{x} \right] \\ &= \int_0^1 dx \left[\alpha_s \frac{BP(z)}{z} + \alpha_s \frac{R(z)}{z} \right] + \mathcal{O}(\alpha_s^2) \\ &= \alpha_s \frac{BP(z) + R(z)}{z} + \mathcal{O}(\alpha_s^2) \ . \end{aligned} \tag{2.280}$$

All Born terms proportional to $\delta(z)$ vanish because their contributions would be unphysical. This already fulfills the first requirement for our scheme, without having done anything except for including a transfer function. Now, we can integrate over z and calculate the total cross section σ_{tot} with a cutoff $z_{\min}$ for consistency. However, Eq. (2.280) includes an additional term which spoils the result: the same kind of jet radiation is included twice, once through the matrix element and once through the shower. This is precisely the double counting which we avoid in the CKKW scheme. So we are still missing something.

We also knew we would fall short, because our strategy includes a smeared virtual subtraction term which for finite x should cancel the real emission. This subtraction is not yet included. Factorization tells us how to write such a subtraction term using the splitting function P as defined in Eq. (2.278), to turn the real emission term into a finite contribution

$$\frac{R(x)}{x} \longrightarrow \frac{R(x) - BP(x)}{x} \ . \tag{2.281}$$

This ad hoc subtraction term we have to add again to the Born–type contribution. This leads us to a modified version of Eq. (2.277), now written for general observables

$$\frac{d\sigma}{dO} = \int_0^1 dx \left[I(O,1) \left(B + \alpha_s V - \frac{\alpha_s B}{x} + \frac{\alpha_s BP(x)}{x} \right) + I(O, x_M) \, \alpha_s \frac{R(x) - BP(x)}{x} \right]. \tag{2.282}$$

Looking back at different methods of removing ultraviolet divergences this modification from the minimal soft and collinear subtraction in Eq. (2.277) to a physical subtraction term corresponding to the known radiation pattern reminds us of different renormalization schemes. The minimal $\overline{\text{MS}}$ scheme will always guarantee finite results, but for example the on–shell scheme with its additional finite terms has at least to a certain degree beneficial properties when it comes to understanding its physical meaning. This is the same for the MC@NLO method: we replace the minimal subtraction terms by physically motivated non–minimal subtraction terms such that the radiation pattern of the additional parton is described correctly.

When we use this form to compute the z spectrum to order α_s it will in addition to Eq. (2.280) include an integrated subtraction term contributing to the Born–type kinematics

$$\frac{d\sigma}{dz} \longrightarrow \int_0^1 dx \left[\alpha_s \frac{BP(z)}{z} + \alpha_s \, \delta(x-z) \left(\frac{R(x)}{x} - \frac{BP(x)}{x} \right) \right] + \mathcal{O}(\alpha_s^2)$$

$$= \int_0^1 dx \, \alpha_s \, \frac{BP(z) + R(z) - BP(z)}{z} + \mathcal{O}(\alpha_s^2)$$

$$= \alpha_s \frac{R(z)}{z} + \mathcal{O}(\alpha_s^2). \tag{2.283}$$

This is exactly the distribution we expect.

Following the above argument the subtraction scheme implemented in the MC@NLO Monte Carlo describes hard emission just like a next–to–leading order calculation. This includes the next–to–leading order normalization of the rate as well as the *next–to–leading order distributions* for those particles produced in the original hard process. For example for W+jets production such corrections to the W and leading jet distributions matter, while for the production of heavy new particles their distributions hardly change at next–to–leading order. The distribution of the first radiated parton is included at leading order, as we see in Eq. (2.283). Finally, additional collinear particle emissions is simulated using Sudakov factors, precisely like a parton shower.

Most importantly, this scheme avoids double counting between the first hard emission and the collinear jets, which means it describes the entire p_T range of jet emission for the *first and hardest* radiated jet consistently. Those additional jets, which do not feature in the next–to–leading order calculation, are added through the parton shower, i.e. in the collinear approximation. As usually, what looked fairly easy in our toy example is much harder in QCD reality, but the setup is the same.

2.7.3 POWHEG Method

As described in Sect. 2.7.2 the MC@NLO matching scheme for a next–to–leading order correction and the parton shower is based on an extended subtraction scheme. It starts from a given parton shower and avoids double counting by modifying the next–to–leading corrections. An interesting question is: can we also combine a next–to–leading order calculation by keeping the next–to–leading order structure and apply a modified parton shower? The main ingredient to this structure are Sudakov factors introduced in Sect. 2.5.1 and used for the CKKW merging scheme in Sect. 2.6.2.

In contrast to the MC@NLO scheme the POWHEG (Positive Weight Hardest Emission Generator) scheme does not introduce counter–events or subtraction terms. It considers the next–to–leading order calculation of a cross section a combination of an n-particle and an $(n + 1)$-particle process and attempts to adjust the parton shower attached to each of these two contributions such that there is no double counting.

Our starting point is the next–to–leading order computation of a cross section following Eq. (2.266). We can combine it with appropriate soft and collinear subtraction terms C in the factorized $(n + 1)$-particle phase space where for simplicity we assume that the integrated subtraction terms exactly cancel the divergences from the virtual corrections. In our simplified model where the extra radiation is only described by an integral over the energy fraction x we find

$$\begin{aligned}\frac{d\sigma}{dx} &= B\,\delta(x) + \alpha_s \left(-\frac{B}{2\epsilon} + V\right) \delta(x) + \alpha_s R \\ &= B\,\delta(x) + \alpha_s V\,\delta(x) + \alpha_s\,(R - C\mathbb{P}) \qquad \text{after soft--collinear subtraction} \\ &= B\left[\delta(x) + \frac{\alpha_s R}{B}(1 - \mathbb{P})\right] + \alpha_s\left[V\delta(x) + (R - C)\mathbb{P}\right]. \end{aligned} \qquad (2.284)$$

The projector $\mathbb{P}$ maps the nominal $(n + 1)$-particle phase space of the real emission onto the n-particle phase space of the leading order process. We keep it separate from the factor $\delta(x)$ and define $\mathbb{P}\delta(x) = \delta(x)$.

The first term in Eq. (2.284) consists of the Born contribution and the hard emission of one parton, so we have to avoid double counting when defining the appropriate Sudakov factors. The second term is suppressed by one power of α_s, so we can add a parton shower to it without any worry. A serious problem appears in Eq. (2.284) when we interpret it probabilistically: nothing forces the combination of virtual and subtracted real emission in the second bracket to be positive. To cure this shortcoming we can instead combine all n-particle contributions into one term

$$\frac{d\sigma}{dx} = \left[\delta(x) + \frac{\alpha_s R}{B}(1-\mathbb{P})\right][B + \alpha_s V + \alpha_s(R-C)\mathbb{P}] + \mathcal{O}(\alpha_s^2)$$

$$\equiv \overline{B}\left[\delta(x) + \frac{\alpha_s R}{B}(1-\mathbb{P})\right] + \mathcal{O}(\alpha_s^2)$$

$$= \overline{B}\left[\delta(x) + \frac{\alpha_s R}{B}\theta\left(p_T(x) - p_T^{\min}\right)\right] + \mathcal{O}(\alpha_s^2) \,. \quad (2.285)$$

Defined like this the combination $\overline{B}$ can only become negative if the regularized next–to–leading contribution over–compensates the Born term which would indicate a breakdown of perturbation theory. If we replace the symbolic projection $(1 - \mathbb{P})$ by a step function in terms of the transverse momentum of the radiated parton $p_T(x)$ we can ensure that it really only appears for hard radiation above $p_T^{\min}$ and at the same time keep the integral over the radiation phase space finite.

From CKKW jet merging we know what we have to do to combine an n-particle process with an $(n + 1)$-particle process, even in the presence of the parton shower: the n-particle process has to be exclusive, which means we need to attach a Sudakov factor Δ to veto additional jet radiation to the first term in the brackets of Eq. (2.285). In the CKKW scheme the factor in the front of the brackets would be B and not $\overline{B}$. The introduction of $\overline{B}$ is nothing but a re-weighting factor for the events contributing to the n-particle configuration which we need to maintain the next–to–leading order normalization of the combined n-particle and $(n + 1)$-particle rates. The second factor $\alpha_s R/B$ is essentially the multiplicative PYTHIA or HERWIG matrix element correction used for an improved simulation for example of W+jet events. The only technical issue with such a re-weighted shower is that the generating shower has to cover the entire radiation phase space. From Sect. 2.3.5 we know that for a proper resummation the collinear logarithms should only be integrated up to the combined renormalization and factorization scales, $p_T < \mu_R \equiv \mu_F$. This additional constraint needs to be addressed in the POWHEG approach.

The appropriate Sudakov factor for the real emission has to veto only hard jet radiation from an additional parton shower. This way we ensure that for the $(n+1)$-particle contribution the hardest jet radiation is given by the matrix element R, which means no splitting occurs in the hard regime $p_T > p_T^{\min}$. Such a *vetoed shower* we can define in analogy to the (diagonal) Sudakov survival probability Eq. (2.166) by adding a step function which limits the unwanted splittings to $p_T > p_T^{\min}$

$$\Delta(t, p_T^{\min}) = \exp\left(-\int_{t_0}^{t} \frac{dt'}{t'} \int_0^1 dz \frac{\alpha_s}{2\pi} \hat{P}(z)\, \theta\left(p_T(t',z) - p_T^{\min}\right)\right), \quad (2.286)$$

omitting the resolution t_0 in the argument and switching back to the proper real emission phase space in terms of z and t'. This modified Sudakov factor indicates that in contrast to the MC@NLO method we now modify the structure of the parton shower which we combine with the higher order matrix elements.

2.7 Next-to-Leading Orders and Parton Shower

For the vetoed Sudakov factors to make sense we need to show that they obey a DGLAP equation like Eq. (2.180), including the veto condition in the splitting kernel

$$f(x,t) = \Delta(t, p_T^{\min}) f(x, t_0) + \int_{t_0}^{t} \frac{dt'}{t'} \Delta(t, t', p_T^{\min}) \int_0^1 \frac{dz}{z} \frac{\alpha_s}{2\pi}$$

$$\hat{P}(z) \, \theta \left(p_T(t', z) - p_T^{\min} \right) f \left(\frac{x}{z}, t' \right) . \quad (2.287)$$

Again, we show the diagonal case to simplify the notation. The proof of this formula starts from Eq. (2.180) with the modification of an explicit veto. Using $1 = \theta(g) + (1 - \theta(g))$ we find Eq. (2.287) more or less straight away. The bottom line is that we can consistently write down vetoed Sudakov probabilities and build a parton shower out of them.

Inserting both Sudakov factors into Eq. (2.285) gives us for the combined next-to-leading order exclusive contributions

$$\boxed{\frac{d\sigma}{d\Phi_n} = \overline{B} \left[\Delta(t, 0) + \Delta(t', p_T^{\min}) \frac{\alpha_s R}{B} \theta \left(p_T(t', z) - p_T^{\min} \right) dt' dz \right] + \mathcal{O}(\alpha_s^2) .}$$

(2.288)

The first Sudakov factor is not vetoed which means it is evaluated at $p_T^{\min} = 0$.

Based on the next-to-leading order normalization of the integrated form of Eq. (2.288) we can determine the form of the splitting probability entering the Sudakov factor from the perturbative series: the term in brackets integrated over the entire phase space has to give unity. Starting from Eq. (2.287) we first compute the derivative of the Sudakov factor with respect to one of its integration boundaries, just like in Eq. (2.176)

$$\frac{d\Delta(t, p_T^{\min})}{dt} = \frac{d}{dt} \exp \left(-\int_{t_0}^{t} \frac{dt'}{t'} \int_0^1 dz \frac{\alpha_s}{2\pi} \hat{P}(z) \, \theta \left(p_T(t', z) - p_T^{\min} \right) \right)$$

$$= \Delta(t, p_T^{\min}) \frac{(-1)}{t} \int_0^1 dz \frac{\alpha_s}{2\pi} \hat{P}(z) \, \theta \left(p_T(t, z) - p_T^{\min} \right) . \quad (2.289)$$

Using this relation we indeed find for the integral over the second term in the brackets of Eq. (2.288)

$$\int_{t_0}^{t} dt' dz \, \Delta(t', p_T^{\min}) \frac{\alpha_s R}{B} \theta \left(p_T(t', z) - p_T^{\min} \right)$$

$$= -\int_{t_0}^{t} dt' \frac{d\Delta(t', p_T^{\min})}{dt'} \frac{\int dz \frac{\alpha_s R}{B} \theta \left(p_T(t', z) - p_T^{\min} \right)}{\int dz \frac{\alpha_s}{2\pi t'} \hat{P}(z) \, \theta \left(p_T(t', z) - p_T^{\min} \right)}$$

Table 2.3 Comparison of the MC@NLO and CKKW schemes combining collinear and hard jets

	MC@NLO/POWHEG matching	CKKW/MLM merging
Hard jets	First jet correct	All jets correct
Collinear jets	All jets correct, tuned	All jets correct, tuned
Normalization	Correct to NLO	Correct to LO plus real emission
Implementations	MC@NLO, POWHEG, SHERPA, HERWIG	SHERPA, Alpgen, Madgraph,...

$$= -\int_{t_0}^{t} dt' \, \frac{d\Delta(t', p_T^{\min})}{dt'}$$

$$= -\Delta(t, p_T^{\min})$$

$$\Leftrightarrow \quad \boxed{\frac{\alpha_s R}{B} = \frac{\alpha_s}{2\pi t'} \hat{P}(z)}. \tag{2.290}$$

Looking back at Eq. (2.99) this corresponds to identifying $B = \sigma_n$ and $\alpha_s R = \sigma_{n+1}$. In the POWHEG scheme the Sudakov factors are based on the simulated splitting probability $\alpha_s R/B$ instead of the splitting kernels. This replacement is nothing new, though. We can already read it off Eq. (2.99).

A technical detail which we have not mentioned yet is that the POWHEG scheme assumes that our Sudakov factors can be ordered in such a way that the hardest emission always occurs first. Following the discussion in Sect. 2.5.4 we expect any collinear transverse momentum ordering to be disrupted by soft radiation, ordered by the angle. The first emission of the parton shower might well appear at large angles but with small energy, which means it will not be particularly hard.

For the POWHEG shower this soft radiation has to be removed or moved to a lower place in the ordering of the splittings. The condition to treat soft emission separately we know from CKKW merging, namely Eq. (2.262): the scale at which we resolve a parton splitting does not have to identical with the lower boundary of simulated splittings. We can construct a parton shower taking into account such splitting kernels, defining a *truncated shower*. This modified shower is the big difference between the MC@NLO scheme and the POWHEG scheme in combining next–to–leading order corrections with a parton shower. In the MC@NLO scheme we modify the next–to–leading order correction for a given shower, but the shower stays the same. In the POWHEG scheme the events get re-weighted according to standard building blocks of a next–to–leading order calculation, but the shower has to be adapted.

In Sects. 2.6.2 and 2.7.2–2.7.3 we have introduced different ways to simulate jet radiation at the LHC. The main features and shortcomings of the matching and merging approaches we summarize in Table 2.3.

At this stage it is up to the competent user to pick the scheme which describes their analysis best. First of all, if there is a well defined and sufficiently hard scale in the process, the old-fashioned Monte Carlo with a tuned parton shower will be fine, and it is by far the fastest method. When for some reason we are mainly interested

2.7 Next–to–Leading Orders and Parton Shower

in one hard jet we can use MC@NLO or POWHEG and benefit from the next–to–leading order normalization. This is the case for example when a gluon splits into two bottoms in the initial state and we are interested in the radiated bottom jet and its kinematics. In cases where we really need a large number of jets correctly described we will end up with CKKW or MLM simulations. However, just like the old-fashioned parton shower Monte Carlo we need to include the normalization of the rate by hand. Or we are lucky and combined versions of CKKW and POWHEG, as currently developed by both groups, will be available.

I am not getting tired of emphasizing that the conceptual progress in QCD describing jet radiation for all transverse momenta is absolutely crucial for LHC analyses. If I were a string theorist I would definitely call this achievement a revolution or even two, like 1917 but with the trombones and cannons of Tchaikovsky's 1812. In contrast to a lot of other progress in theoretical physics jet merging solves a problem which would otherwise have limited our ability to understand LHC data, no matter what kind of Higgs or new physics we are looking for.

Further Reading

Just like the Higgs part, the QCD part of these lecture notes is something in between a text book chapter and a review of QCD and mostly focused on LHC searches. I cut some corners, in particular when calculations do not affect the main topic, namely the resummation of logarithms in QCD and the physical meaning of these logarithms. There is no point in giving a list of original references, but I will list a few books and review articles which should come in handy if you would like to know more:

- I started learning high energy theory including QCD from Otto Nachtmann's book. I still use his appendices for Feynman rules because I have not seen another book with as few (if not zero) typos [15].
- Similar, but maybe a little more modern is the Standard Model primer by Cliff Burgess and Guy Moore [2]. At the end of it you will find more literature.
- The best source to learn QCD at colliders is the pink book by Keith Ellis, James Stirling, and Bryan Webber [7]. It includes everything you ever wanted to know about QCD and more. This QCD section essentially follows its Chap. 5.
- A little more phenomenology you can find in Günther Dissertori, Ian Knowles and Michael Schmelling's book [6]. Again, I borrowed some of the discussions in the QCD section from there. In the same direction but more theory oriented is the QCD book by Ioffe, Fadin, and Lipatov [13].
- If you would like to learn how to for example compute higher order cross sections to Drell–Yan production, Rick Field works it all out [10].

- For those of you who are now hooked on QCD and jet physics at hadron colliders there are two comprehensive reviews by Steve Ellis et al. [8] and Gavin Salam [18].
- Aimed more at perturbative QCD at the LHC is the QCD primer by John Campbell, Joey Huston, and James Stirling [3].
- Coming to the usual brilliant TASI lectures, there are Dave Soper's [20] and George Sterman's [22] notes. Both of them do not exactly use my kind of notations and are comparably formal, but they are a great read if you know something about QCD already. More on the phenomenological side there are Mike Seymour's lecture notes [19].
- For a more complete discussion of the Catani–Seymour dipoles the very brief discussion in this writeup should allow you to follow the original long paper [4].
- The only review on leading order jet merging is by Michelangelo Mangano and Tim Stelzer [14]. The original CKKW paper beautifully explains the general idea for final state radiation, and I follow their analysis [5]. For other approaches there is a very concise discussion included with the comparison of the different models [1].
- To understand MC@NLO there is nothing like the original papers by Bryan Webber and Stefano Frixione [11].
- The POWHEG method is really nicely described in the original paper by Paolo Nason [16]. Different processes you can find discussed in detail in a later paper by Stefano Frixione, Paolo Nason, and Carlo Oleari [12].
- Even though they are just hand written and do not include a lot of text it might be useful to have a look at Michael Spira's QCD lecture notes [21] to view some of the topics from a different angle.

References

1. J. Alwall et al., Comparative study of various algorithms for the merging of parton showers and matrix elements in hadronic collisions. Eur. Phys. J. C **53**, 473 (2008)
2. C.P. Burgess, G.D. Moore, *The Standard Model: A Primer* (Cambridge University Press, Cambridge, 2007), 542 p.
3. J.M. Campbell, J.W. Huston, W.J. Stirling, Hard interactions of quarks and gluons: a primer for LHC physics. Rep. Prog. Phys. **70**, 89 (2007)
4. S. Catani, M.H. Seymour, A general algorithm for calculating jet cross-sections in NLO QCD. Nucl. Phys. B **485**, 291 (1997). (Erratum-ibid. B **510**, 503 (1998))
5. S. Catani, F. Krauss, R. Kuhn, B.R. Webber, QCD matrix elements + parton showers. JHEP **2001**(11), 063 (2001)
6. G. Dissertori, I.G. Knowles, M. Schmelling, *QCD—High Energy Experiments and Theory* (Clarendon, Oxford, 2003), 538 p.
7. R.K. Ellis, W.J. Stirling, B.R. Webber, *QCD and Collider Physics*. Cambridge Monographs on Particle Physics, Nuclear Physics, and Cosmology, vol. 8 (Cambridge University Press, Cambridge, 1996), p. 1
8. S.D. Ellis, J. Huston, K. Hatakeyama, P. Loch, M. Tonnesmann, Jets in hadron-hadron collisions. Prog. Part. Nucl. Phys. **60**, 484 (2008)

9. C. Englert, T. Plehn, P. Schichtel, S. Schumann, Jets plus missing energy with an autofocus. Phys. Rev. D **83**, 095009 (2011)
10. R.D. Field, *Applications of Perturbative QCD* (Addison-Wesley, Redwood City, 1989), 366 p. (Frontiers in Physics, **77**)
11. S. Frixione, B.R. Webber, Matching NLO QCD computations and parton shower simulations. JHEP **2002**(6), 029 (2002)
12. S. Frixione, P. Nason, C. Oleari, Matching NLO QCD computations with Parton Shower simulations: the POWHEG method. JHEP **2007**(11), 070 (2007)
13. B.L. Ioffe, V.S. Fadin, L.N. Lipatov, *Quantum Chromodynamics: Perturbative and Nonperturbative Aspects*. Cambridge Monographs on Particle Physics, Nuclear Physics, and Cosmology, vol. 30 (Cambridge University Press, Cambridge, 2010), p. 1
14. M.L. Mangano, T.J. Stelzer, Tools for the simulation of hard hadronic collisions. Ann. Rev. Nucl. Part. Sci. **55**, 555 (2005). CERN-PH-TH-2005-074
15. O. Nachtmann, *Elementary Particle Physics: Concepts and Phenomena* (Springer, Berlin, 1990), 559 p.
16. P. Nason, A new method for combining NLO QCD with shower Monte Carlo algorithms. JHEP **2004**(11), 040 (2004)
17. T. Plehn, T.M.P. Tait, Seeking sgluons. J. Phys. G **36**, 075001 (2009)
18. G.P. Salam, Towards Jetography. arXiv:0906.1833 [hep-ph]
19. M.H. Seymour, Quantum chromodynamics. arXiv:hep-ph/0505192
20. D.E. Soper, Basics of QCD perturbation theory. arXiv:hep-ph/0011256
21. M. Spira, QCD. people.web.psi.ch/spira/vorlesung/qcd
22. G. Sterman, QCD and jets. arXiv:hep-ph/0412013

Chapter 3
LHC Phenomenology

While the first two parts of these lecture notes focus on Higgs physics and on QCD, biased towards aspects relevant to the LHC, they hardly justify the title of the lecture notes. In addition, both introductions really are theoretical physics. The third section will introduce other aspects which theorists working on LHC topics need to know. It goes beyond what you find in theoretical physics text books and is usually referred to as 'phenomenology'.[1]

This terms indicates that these topics are not really theoretical physics in the sense that they rely on for example field theory. They are not experimental physics either, because they go beyond understanding the direct results of the LHC detectors. Instead, they lie in between the two fields and need to be well understood to allow theorists and experimentalists to interact with each other.

Sometimes, phenomenology has the reputation of not being proper theoretical physics. From these lecture notes it is clear that LHC physics is field theory, either electroweak symmetry breaking, QCD, or—not covered in these notes—physics models describing extensions of our Standard Model at the TeV scale. This chapter supplements the pure theory aspects and links them to experimental issues of the ATLAS and CMS experiments. In Sect. 3.1 we fill in some blanks from Sects. 1.5.4, 1.7, and 2.6.2. We first discuss jets and how to link the asymptotic final states of QCD amplitudes, partons, to experimentally observed QCD objects, jets. Then, we turn to a current field of research, so-called fat jets. In Sect. 3.2 we introduce a particularly efficient way of computing transition amplitudes from Feynman rules, the helicity method. Most professional tools for the computation of LHC cross sections or for simulating LHC events use this method instead of squaring amplitudes analytically. Section 3.3 discusses how to reconstruct particles which interact too weakly to be observed in LHC detectors. In the Standard Model

---

[1]The term 'phenomenology' is borrowed from philosophy where it means exactly the opposite from what it means in physics. Originally, phenomenology is a school based on Edmund Husserl, who were interested not in observations but the actual nature of things. Doing exactly the opposite, physicist phenomenologists are theorists who really care about measurements.

those would be neutrinos, but as part of the LHC program we hope to find dark matter candidates that way. Finally, in Sect. 3.4 we very briefly discuss LHC uncertainties from a theory point of view. In the public arXiv version more sections will follow, hopefully triggered by LHC measurements challenging theorists and their simulations.

3.1 Jets and Fat Jets

Throughout Sect. 2 we pretend that quarks and gluons produced at the LHC are what we observe in the LHC detectors. In perturbative QCD they are assumed to form the initial and final states, even though they cannot exist individually as long as QCD is asymptotically free. In Eq. (2.64) we even apply wave function renormalization factors to their quantum fields. On the other hand, in Sect. 2.2.2 we see that the strong coupling explodes at small energy scales around Λ_{QCD} which means that something has to happen with quarks and gluons on their way through the detectors. Indeed, the gluon and all quarks except for the top quark *hadronize* before they decay and form bunches of baryons and mesons which in turn decay in many stages. At the LHC these particles carry a lot of energy, typically around the electroweak scale. Relativistic kinematics then tells us that these baryons and mesons are strongly boosted together to form *jets*. Those jets we measure at hadron colliders and link to the partons produced in the hard interaction.

Consequently, in Sect. 2 we use the terms parton and jet synonymously, essentially assuming that each parton at the LHC turns into a jet and that the measured jet four-momentum can be linked to the parton four-momentum. The way we usually define jets is based on so-called *recombination algorithms*, including for example the Cambridge–Aachen or (anti-) k_T algorithms. Imagine we observe a large number of energy depositions in the ATLAS or CMS calorimeter which we would like to combine into jets. We know that they come from a small number of partons which originate in the hard QCD process and which since have undergone a sizeable number of splittings, hadronized and decayed to stable particles. Can we try to reconstruct the original partons?

The answer is yes, in the sense that we can combine a large number of subjets into smaller numbers, where unfortunately nothing tells us what the final number of jets should be. We know from Sect. 2 that in QCD we can produce an arbitrary number of hard jets in a hard matrix element and another arbitrary number of jets via soft or collinear radiation. Therefore, we need to tell the jet algorithm either how many jets it should arrive at or what the resolution of the smallest subjets we consider partons should be, whatever the measure for this resolution might be. Below we will therefore discuss what criteria exist for a subjet recombination to correspond to an assumed physical jet.

3.1.1 Jet Algorithms

The basic idea of recombination algorithms is to ask if a given subjet has a soft or collinear partner. This follows from Sect. 2: we know that partons produced in a hard process preferably turn into collinear pairs of partons as approximately described by the parton shower. To decide if two subjets have arisen from one parton leaving the hard process we have to define a collinearity measure. This measure will on the one hand include the distance in R space as introduced in Eq. (2.35) and on the other hand the transverse momentum of one subjet with respect to another or to the beam axis. Explicit measures weighted by the relative power of the two ingredients are

$$k_T \qquad y_{ij} = \frac{\Delta R_{ij}}{R} \min\left(p_{T,i}, p_{T,j}\right) \qquad y_{iB} = p_{T,i}$$

$$\text{C/A} \qquad y_{ij} = \frac{\Delta R_{ij}}{R} \qquad y_{iB} = 1$$

$$\text{anti-}k_T \qquad y_{ij} = \frac{\Delta R_{ij}}{R} \min\left(p_{T,i}^{-1}, p_{T,j}^{-1}\right) \qquad y_{iB} = p_{T,i}^{-1}. \qquad (3.1)$$

The parameter R balances the jet–jet and jet–beam criteria. In an exclusive jet algorithm we define two subjets as coming from one jet if $y_{ij} < y_{cut}$, where y_{cut} is a reference scale we give to the algorithm. Such an exclusive jet algorithm then proceeds as

(1) For all combinations of two subjets in the event find the minimum $y^{min} = \min_{ij}(y_{ij}, y_{iB})$
(2a) If $y^{min} = y_{ij} < y_{cut}$ merge subjets i and j and their momenta, keep only the new subjet i, go back to (1)
(2b) If $y^{min} = y_{iB} < y_{cut}$ remove subjet i, call it beam radiation, go back to (1)
(2c) If $y^{min} > y_{cut}$ keep all subjets, call them jets, done

The result of the algorithm will of course depend on the resolution y_{cut}. Alternatively, we can give the algorithm the minimum number of physical jets and stop there.

In an inclusive jet algorithm we do not introduce y_{cut}. We can postpone the decision if want to include a jet in our analysis to the point where all jets are defined. Instead, y_{iB} acts as the cutoff:

(1) For all combinations of two subjets in the event find the minimum $y^{min} = \min_{ij}(y_{ij}, y_{iB})$
(2a) If $y^{min} = y_{ij}$ merge subjets i and j and their momenta, keep only the new subjet i, go back to (1)
(2b) If $y^{min} = y_{iB}$ remove subjet i and call it a final state jet, go back to (1)

The algorithm ends when condition (2a) has left no particles or subjets in the event. Now, the smallest jet–beam distance y_{iB} sets the scale for all jet–jet separations. In the C/A example we immediately see that this translates into a geometric jet size given by R. For regular QCD jets we choose values of $R = 0.4 \ldots 0.7$. For the C/A and k_T cases we see that an inclusive jet algorithm produces jets arbitrarily close to the beam axis. Those are hard to observe and often not theoretically well defined, as we know from our discussion of collinear divergences. Therefore, inclusive jet algorithms have to include a final minimum cut on $p_{T,\text{jet}}$ which at the LHC can be anything from 20 GeV to more than 100 GeV, depending on the analysis.

A technical question is what 'combine jets' means in terms of the four-momentum of the new jet. The three-momentum vectors we simply add $\vec{k}_i + \vec{k}_j \to \vec{k}_j$. For the zero component we can assume that the new physical jet have zero invariant mass, which is inspired by the massless parton we are usually looking for. If instead we add the four-momenta we can compute the invariant mass of the jet constituents, the *jet mass*. As we will see in the next section this allows us to extend the concept of jet algorithms to massive particles like a W or Z boson, the Higgs boson, or the top quark.

All jet algorithms them have in common that they link physical objects, namely calorimeter towers, to other more or less physical objects, namely partons from the hard process. As we can see from the different choices in Eq. (3.1) we have all the freedom in the world to weight the angular and transverse momentum distances relative to each other. As determined by their power dependence on the transverse momenta, the three algorithms start with soft constituents (k_T), purely geometric (*Cambridge–Aachen C/A*) or hard constituents (anti-k_T) to form a jet. While for the k_T and the C/A algorithms it is fairly clear that the intermediate steps have a physical interpretation, this is not clear at all for the anti-k_T algorithm.

From Sect. 2 and the derivation of the collinear splitting kernels it is obvious why theorists working on perturbative QCD often prefer the k_T algorithm: we know that the showering probability or the collinear splitting probability is best described in terms of virtuality or transverse momentum. A transverse momentum distance between jets is therefore best suited to combine the correct subjets into the original parton from the hard interaction, following a series of actual physical intermediate splittings. Moreover, this transverse momentum measure is intrinsically infrared safe, which means the radiation of an additional soft parton cannot affect the global structure of the reconstructed jets. For other algorithms we have to ensure this property explicitly, and you can find examples in QCD lectures by Mike Seymour.

The problem of the k_T algorithm arises with pile–up or underlying event, i.e. very soft QCD activity entering the detectors undirectionally or from secondary partonic vertices. Such noise is easiest understood geometrically in a probabilistic picture. Basically, the low energy jet activity is constant all over the detector, so we *subtract it from each event*. How much energy deposition we have to subtract from a reconstructed jet depends on the area the jet covers in the detector. Therefore, it is a major step that even for the k_T algorithm we can compute an IR–safe geometric jet size. The C/A and anti-k_T algorithms are more compact and easier to interpret experimentally.

3.1.2 Fat Jets

Starting from the way the experiments at the Tevatron and the LHC search for bottom jets, including several detailed requirements on the content of such jets, the question arises if we can look for other *heavy objects* inside a jet. Such jets involving heavy particles and (usually) a large geometrical size are referred to as fat jets. For example, looking for boosted top quarks a fat jet algorithm will try to distinguish between two splitting histories, where we mark the massive splittings from boosted top decays:

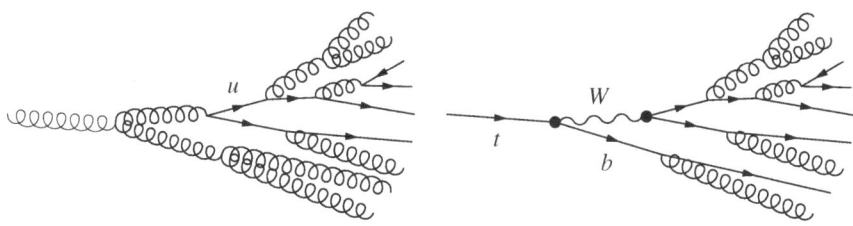

The splittings inside the light–flavor jet are predicted by the soft and collinear QCD structure. The splittings in the top decays differ because some of the particles involved have masses. This is the jet substructure pattern a fat jet algorithm looks for.

Three main motivations lead us into the direction of fat jets: first, dependent on our physics model heavy objects like W bosons or top quarks will be boosted enough to fit into a regular jet of angular size $R \lesssim 0.7$. Secondly, a jet algorithm might include hadronic decay products which we would not trust to include in a regular mass reconstruction based on reconstructed detector objects. And finally, even if only a fraction of the heavy particles we are searching for are sufficiently boosted such an algorithm automatically resolves signal combinatorics known to limit some LHC analyses.

At the LHC, we are guaranteed to encounter the experimental situation $p_T/m \gtrsim 1$ for electroweak gauge bosons, Higgs bosons, and top quarks. The more extreme case of $p_T \gg m$, for example searching for top quarks with a transverse momentum in excess of 1 TeV, is unlikely to appear in the Standard Model and will only become interesting if we encounter very heavy resonances decaying to a pair of top quarks. This is why we focus on the moderate scenario. Amusingly, the identification of W and top jets was part of the original paper studying the pattern of splittings y_{ij} defining the k_T algorithm. At the time this was mostly a gedankenexperiment to test the consistency of the general k_T algorithm approach. Only later reality caught up with it.

Historically, fat jet searches were first designed to look for strongly interacting W bosons. Based on the k_T algorithm they look for structures in the chain of y values introduced in Eq. (3.1), which define the kinematics of each jet. For such an analysis of y values it is helpful but not crucial that the intermediate steps of the jet algorithm have a physics interpretation. More recent fat jet algorithms looking for not too highly boosted heavy particles are based on the C/A algorithm which appears to be best suited to extract massive splittings inside the jet clustering history. A comparison of different jet algorithms can be found in the original paper on associated Higgs and gauge boson production. Using a C/A algorithm we can search for hadronically decaying boosted W and Z bosons. The problem is that for those we only have one hard criterion based on which we can reject QCD backgrounds: the mass of the W/Z resonance. Adding a second W/Z boson and possibly the mass of a resonance decaying to these two, like a heavy Higgs boson, adds to the necessary QCD rejection. For Higgs and top decays the situation is significantly more promising.

Starting with the *Higgs tagger* we search for jets which include two bottom quarks coming from a massive Higgs boson with $m_H \gtrsim 120\,\text{GeV}$. First, we run the C/A algorithm over the event, choosing a large geometric size $R = 1.2$ estimated to cover

$$R_{b\bar{b}} \sim \frac{1}{\sqrt{z(1-z)}} \frac{m_H}{p_{T,H}} > \frac{2m_H}{p_{T,H}}, \qquad (3.2)$$

in terms of the transverse momentum of the boosted Higgs and the momentum fractions z and $1-z$ of the two bottom jets.

We then uncluster again this *fat jet*, searching for a drop in jet mass indicating the decay of the massive Higgs to two essentially massless quarks. The iterative unclustering we start by undoing the last clustering of the jet j, giving us two subjets j_1, j_2 ordered such that $m_{j_1} > m_{j_2}$. If the mass drop between the original jet and its more massive splitting product is small, i.e. $m_{j_1} > 0.8\, m_j$, we conclude that j_2 is soft enough to come from the underlying event or soft–collinear QCD emission and discard j_2 while keeping j_1; otherwise we keep both j_1 and j_2; each surviving subjet j_i we further decompose recursively until it reaches some minimum value, $m_{j_i} < 30\,\text{GeV}$, ensuring it does not involve heavy states. This way we obtain a splitting pattern which should only include massive splittings and which for the Higgs jet uniquely identifies the $H \to b\bar{b}$ decay. Making use of the scalar nature of the Higgs boson we can add an additional requirement on the balance based on $\min(p_{Tj_1}^2, p_{Tj_2}^2) \Delta R_{j_1 j_2}^2$. Of course, all actual numbers in this selection are subject to experimental scrutiny and can only be determined after testing the algorithm on LHC data.

Experimentally, the goal of such a Higgs search is a distribution of the invariant mass of the bottom quarks which gives us a signal peak and side bins to estimate the background. However, applying jet algorithms with very large R size makes us

3.1 Jets and Fat Jets

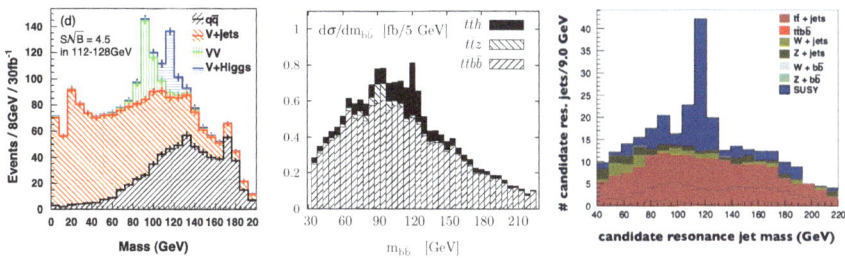

Fig. 3.1 Invariant mass distributions for Higgs searches using fat jets from $H \to b\bar{b}$ decays. For a Standard Model Higgs boson the production mechanisms are $pp \to WH/ZH$ (*left*) and $pp \to t\bar{t}H$ (*center*). In cascade decays of supersymmetric squarks and gluinos we can apply the same search for the light Higgs boson (*right*) (Figures from Refs. [5, 17] and [10] (left to right))

increasingly vulnerable to underlying event, pile–up, or even regular initial–state radiation as described in Sect. 2.3.2. Therefore, we cannot simply use the mass of a set of fat jet constituents. Instead, we apply a *filtering* procedure looking at the same constituent with a higher resolution which can for example be $R_{\text{filt}} = \min(0.3, R_{b\bar{b}}/2)$. This filtering significantly reduces the y-ϕ surface area of the relevant constituents and thereby the destructive power of the underlying event and pile–up. The invariant mass we include in the histogram is the mass of the three hardest filtered constituents, the two bottom quarks and possibly a radiated gluon.

In a *busy QCD environment* another problem arises: errand jets from initial–state radiation or other particles in the final state enter the fat jet algorithm and give us several mass drops in the fat jet history. To avoid shaping the background side bins we can include several (filtered) subjet combinations, ordered in the modified Jade distance $p_{T,1} p_{T,2} (\Delta R_{12})^4$—the original Jade distance is given by $p_{T,1} p_{T,2} (\Delta R_{12})^2$. The invariant mass distributions for different Higgs search channels in Fig. 3.1 include Standard Model Higgs searches in WH/ZH production, in $t\bar{t}H$ production, and in decays of squarks and gluinos.

From the above discussion we see that Higgs taggers rely only on *one kinematic criterion*, the mass of the $b\bar{b}$ pair. In terms of background rejection this is not much, so we usually add two bottom tags on the constituents which according to detector simulations can be very efficient. The two combined add to a QCD rejection of at least 10^{-4}, which might even allows us to run a Higgs tagger over any kind of event sample and see if we find any Higgs bosons for example in new physics decays.

While fat jet Higgs searches are targeted mostly at the Standard Model, looking for other boosted heavy particles is usually motivated by new physics scenarios. Looking for massive particles decaying to heavy quarks *top taggers* should be the next step. Starting from a C/A jet of size $R = 1.5 - 1.8$ we again search for mass drops, this time corresponding to the top and W masses. After appropriate filtering we apply two mass window conditions: first, the entire fat jet has to reproduce the top mass. Second, we require a mass drop corresponding to the W decay and effectively constrain a second combination of two decay jets evaluating the helicity angle of the left handed W decay. Instead of these two distinct steps we can also apply a two-dimensional condition on the kinematics of the three top decay products which avoids assigning the two W decay jets in cases where two combinations of decay jets give similar invariant masses. On the simulation level both methods give similar results.

Applying these three kinematic conditions for example in the HEPTopTagger implementation gives a QCD rejection of a few per-cent. If this should not be sufficient for a given analysis we can increase the rejection rate by requiring a bottom tag which as a bonus also tells us which of the three top decay jets should reconstruct the W mass. When we use top taggers to look for new particles decaying to top quarks we are not only interested in finding boosted top quarks, but we would like to know their invariant mass. This means we would like to reconstruct their direction and their energy. Such a reconstruction is possible in a reasonably clean sample, provided the top quarks have large enough energy to boost all three decay jets into a small enough cone.

While it seems like the C/A jet algorithm with its purely geometric selection has the best potential to search for massive particles in its jet history there exists a multitude of algorithms searching for boosted top quarks. Once the top quarks have very large transverse momenta the two-step mass drop criterion becomes less critical because the three decay jets are too close to be cleanly resolved. In this situation analyses based on the k_T or anti-k_T algorithms can be very promising, as could be event shapes which do not involve any jet algorithm.

3.2 Helicity Amplitudes

When we simulate LHC events we do not actually rely on the approach usually described in text books. This is most obvious when it comes to the computation of a transition matrix elements in modern LHC Monte Carlo tools, which you will not even recognize when looking at the codes. In Sect. 2.1 we compute the cross section for Z production by writing down all external spinors, external polarization vectors, interaction vertices and propagators and squaring the amplitude analytically. The amplitude itself inherits external indices for example from the polarization vectors, while $|\mathcal{M}|^2$ is a real positive number with a fixed mass dimension depending on the number of external particles.

For the LHC nobody calculates gamma matrix traces by hand anymore. Instead, we use powerful tools like FORM to compute traces of Dirac matrices in the calculation of $|\mathcal{M}|^2$. Nevertheless, a major problem with squaring Feynman diagrams and computing polarization sums and gamma matrix traces is that once we include more than three particles in the final state, the number of terms appearing in $|\mathcal{M}|^2$ soon becomes very large. Moreover, this approach requires symbolic computer manipulation instead of pure numerics. In this section we illustrate how we can transform the computation of $|\mathcal{M}|^2$ at the tree level into a purely numerical problem.

3.2 Helicity Amplitudes

As an example, we consider our usual toy process

$$u\bar{u} \to \gamma^* \to \mu^+\mu^- . \qquad (3.3)$$

The structure of the amplitude $\mathcal{M}$ with two internal Dirac indices μ and ν involves one vector current on each side $(\bar{u}_f \gamma_\mu u_f)$ where $f = u, \mu$ are to good approximation massless, so we do not have to be careful with the different spinors u and v. The entries in the external spinors are given by the spin of the massless fermions obeying the Dirac equation. For each value of $\mu = 0 \cdots 3$ each current is a complex number, computed from the four component of each spinor and the respective 4×4 gamma matrix γ^μ shown in Eq. (2.109). The intermediate photon propagator has the form $g_{\mu\nu}/s$, which is a real number for each value of $\mu = \nu$. Summing over μ and ν in both currents forms the matrix element. To square this matrix element we need to sum $\mathcal{M}^* \times \mathcal{M}$ over all possible spin directions of the external fermions.

Instead of squaring this amplitude symbolically we can follow exactly the steps described above and compute an array of numbers for different spin and helicity combinations numerically. Summing over the internal Dirac indices we compute the matrix element; however, to compute the matrix element squared we need to sum over external fermion spin directions or gauge boson polarizations. The helicity basis we have to specify externally. This is why this method is called helicity amplitude approach. To explain the way this method works, we illustrate it for muon pair production based on the implementation in the *Madgraph/Helas* package.

Madgraph is a tool to compute matrix elements this way. Other event generators have corresponding codes serving the same purposes. In our case, Madgraph5 automatically produces a Fortran routine which then calls functions to compute spinors, polarization vectors, currents of all kinds, etc. These functions are available as the so-called Helas library. For our toy process equation (3.3) the slightly shortened Madgraph5 output reads

```
          REAL*8 FUNCTION MATRIX1(P,NHEL,IC)
C
C         Generated by Madgraph 5
C
C         Returns amplitude squared summed/avg over colors
C         for the point with external lines W(0:6,NEXTERNAL)
C
C         Process: u u~ > mu+ mu- / z WEIGHTED=4 @1
C
          INTEGER     NGRAPHS, NWAVEFUNCS, NCOLOR
          PARAMETER (NGRAPHS=1, NWAVEFUNCS=5, NCOLOR=1)

          REAL*8 P(0:3,NEXTERNAL)
          INTEGER NHEL(NEXTERNAL), IC(NEXTERNAL)

          INCLUDE 'coupl.inc'

          DATA DENOM(1)/1/
          DATA (CF(I,  1),I=  1,  1) /    3/

          CALL IXXXXX(P(0,1),ZERO,NHEL(1),+1*IC(1),W(1,1))
          CALL OXXXXX(P(0,2),ZERO,NHEL(2),-1*IC(2),W(1,2))
          CALL IXXXXX(P(0,3),ZERO,NHEL(3),-1*IC(3),W(1,3))
```

```
          CALL OXXXXX(P(0,4),ZERO,NHEL(4),+1*IC(4),W(1,4))
          CALL FFV1_3(W(1,1),W(1,2),GC_2,ZERO, ZERO, W(1,5))
          CALL FFV1_0(W(1,3),W(1,4),W(1,5),GC_3,AMP(1))
          JAMP(1)=+AMP(1)

          DO I = 1, NCOLOR
            DO J = 1, NCOLOR
              ZTEMP = ZTEMP + CF(J,I)*JAMP(J)
            ENDDO
            MATRIX1 = MATRIX1 + ZTEMP*DCONJG(JAMP(I))/DENOM(I)
          ENDDO

          END
```

The input to this function are the external four-momenta $p(0:3,1:4)$ and the helicities of all fermions $n_{\text{hel}}(1:4)$ in the process. Remember that helicity and chirality are identical only for massless fermions because chirality is defined as the eigenvalue of the projectors $(\mathbb{1} \pm \gamma_5)/2$, while helicity is defined as the projection of the spin onto the momentum direction, i.e. as the left or right handedness. The entries of n_{hel} will be either $+1$ or -1. For each point in phase space and each helicity combination the Madgraph subroutine `MATRIX1` computes the matrix element using standard *Helas routines*.

- `IXXXXX`$(p, m, n_{\text{hel}}, n_{\text{sf}}, F)$ computes the wave function of a fermion with incoming fermion number, so either an incoming fermion or an outgoing anti–fermion. As input it requires the four-momentum, the mass and the helicity of this fermion. Moreover, $n_{\text{sf}} = +1$ marks the incoming fermion u and $n_{\text{sf}} = -1$ the outgoing anti–fermion μ^+, because by convention Madgraph defines its particles as u and μ^-.

  The fermion wave function output is a complex array $F(1:6)$. Its first two entries are the left–chiral part of the fermionic spinor, i.e. $F(1:2) = (\mathbb{1}-\gamma_5)/2\, u$ or $F(1:2) = (\mathbb{1}-\gamma_5)/2\, v$ for $n_{\text{sf}} = \pm 1$. The entries $F(3:4)$ are the right–chiral spinor. These four numbers can directly be computed from the four-momentum if we know the helicity. The four entries correspond to the size of one γ matrix, so we can compute the trace of the chain of gamma matrices. Because for massless particles helicity and chirality are identical, our quarks and leptons will only have finite entries $F(1:2)$ for $n_{\text{hel}} = -1$ and $F(3:4)$ for $n_{\text{hel}} = +1$.

  The last two entries of F contain the four-momentum in the direction of the fermion flow, namely $F(5) = n_{\text{sf}}(p(0)+ip(3))$ and $F(6) = n_{\text{sf}}(p(1)+ip(2))$.

- `OXXXXX`$(p, m, n_{\text{hel}}, n_{\text{sf}}, F)$ does the same for a fermion with outgoing fermion flow, i.e. our incoming $\bar{u}$ and our outgoing μ^-. The left–chiral and right–chiral components now read $F(1:2) = \bar{u}(\mathbb{1}-\gamma_5)/2$ and $F(3:4) = \bar{u}(\mathbb{1}+\gamma_5)/2$, and similarly for the spinor $\bar{v}$. The last two entries are $F(5) = n_{\text{sf}}(p(0)+ip(3))$ and $F(6) = n_{\text{sf}}(p(1)+ip(2))$.

- `FFV1_3`$(F_i, F_o, g, m, \Gamma, J_{io})$ computes the (off-shell) current for the vector boson attached to the two external fermions F_i and F_o. The coupling $g(1:2)$ is a complex array with the interaction of the left–chiral and right–chiral fermion in the upper and lower index. For a general Breit–Wigner propagator we need to know the mass m and the width Γ of the intermediate vector boson. The output array J_{io} again has six components which for the photon with momentum q are

3.2 Helicity Amplitudes

$$J_{io}(\mu+1) = -\frac{i}{q^2} F_o^T \gamma^\mu \left(g(1) \frac{\mathbb{1} - \gamma_5}{2} + g(2) \frac{\mathbb{1} + \gamma_5}{2} \right) F_i \qquad \mu = 0, 1, 2, 3$$

$$J_{io}(5) = -F_i(5) + F_o(5) \sim -p_i(0) + p_o(0) + i(-p_i(3) - p_o(3))$$

$$J_{io}(6) = -F_i(6) + F_o(6) \sim -p_i(1) + p_o(1) + i(-p_i(2) + p_o(2)) \ . \tag{3.4}$$

The first four entries in J_{io} correspond to the index μ or the dimensionality of the Dirac matrices in this vector current. The spinor index is contracted between F_o^T and F_i.

As two more arguments J_{io} includes the four-momentum flowing through the gauge boson propagator. They allow us to reconstruct q^μ from the last two entries

$$q^\mu = (\operatorname{Re} J_{io}(5), \operatorname{Re} J_{io}(6), \operatorname{Im} J_{io}(6), \operatorname{Im} J_{io}(5)) \ . \tag{3.5}$$

- FFV1_0(F_i, F_o, J, g, V) computes the amplitude of a fermion–fermion–vector coupling using the two external fermionic spinors F_i and F_o and an incoming vector current J which in our case comes from FFV1_3. Again, the coupling $g(1:2)$ is a complex array, so we numerically compute

$$F_o^T \, J \left(g(1) \frac{\mathbb{1} - \gamma_5}{2} + g(2) \frac{\mathbb{1} + \gamma_5}{2} \right) F_i \ . \tag{3.6}$$

All spinor and Dirac indices of the three input arguments are contracted in the final result. Momentum conservation is not enforced by FFV1_0, so we have to take care of it by hand.

Given the list above it is easy to follow how Madgraph computes the amplitude for $u\bar{u} \to \gamma^* \to \mu^+ \mu^-$. First, it calls wave functions for all external particles and puts them into the array $W(1:6, 1:4)$. The vectors $W(*, 1)$ and $W(*, 3)$ correspond to $F_i(u)$ and $F_i(\mu^+)$, while $W(*, 2)$ and $W(*, 4)$ mean $F_o(\bar{u})$ and $F_o(\mu^-)$.

The first vertex we evaluate is the incoming quark–photon vertex. Given the wave functions $F_i = W(*, 1)$ and $F_o = W(*, 2)$ FFV1_3 computes the vector current for the massless photon in the s-channel. Not much changes if we instead choose a massive Z boson, except for the arguments m and Γ in the FFV1_3 call. Its output is the photon current $J_{io} \equiv W(*, 5)$.

The last step combines this current with the two outgoing muons coupling to the photon. Since this number gives the final amplitude, it should return a complex number, not an array. Madgraph calls FFV1_0 with $F_i = W(*, 3)$ and $F_o = W(*, 4)$, combined with the photon current $J = W(*, 5)$. The result AMP is copied into JAMP without an additional sign which could have come from the relative ordering of external fermions in different Feynman diagrams contributing to the same process.

The only remaining sum left to compute before we square JAMP is the color structure, which in our simple case means one color structure with a color factor $N_c = 3$.

As an added bonus Madgraph produces a file with all Feynman diagrams in which the numbering of the external particles corresponds to the second argument of W and the numbering of the Feynman diagrams corresponds to the argument of AMP. This helps us identify intermediate results W, each of which is only computed once, even if is appears several times in the different Feynman diagrams.

As mentioned above, to calculate the transition amplitude Madgraph requires all masses and couplings. They are transferred through common blocks in the file coupl.inc and computed elsewhere. In general, Madgraph uses unitary gauge for all vector bosons, because in the helicity amplitude approach it is easy to accommodate complicated tensors, in exchange for a large number of Feynman diagrams.

The function MATRIX1 described above is not yet the full story. When we square $\mathcal{M}$ symbolically we need to sum over the spins of the outgoing states to transform a spinor product of the kind $u\bar{u}$ into the residue or numerator of a fermion propagator. To obtain the full transition amplitude numerically we correspondingly sum over all *helicity combinations* of the external fermions, in our case $2^4 = 16$ combinations.

```
      SUBROUTINE SMATRIX1(P,ANS)
C
C     Generated by Madgraph 5
C
C     Returns amplitude squared summed/avg over colors
C     and helicities for the point in phase space P(0:3,NEXTERNAL)
C
C     Process: u u~ > mu+ mu- / z
C
      INTEGER    NCOMB, NGRAPHS, NDIAGS, THEL
      PARAMETER (NCOMB=16, NGRAPHS=1, NDIAGS=1, THEL=2*NCOMB)

      REAL*8 P(0:3,NEXTERNAL)

      INTEGER I,J,IDEN
      INTEGER NHEL(NEXTERNAL,NCOMB),NTRY(2),ISHEL(2),JHEL(2)
      INTEGER JC(NEXTERNAL),NGOOD(2), IGOOD(NCOMB,2)
      REAL*8 T,MATRIX1
      LOGICAL GOODHEL(NCOMB,2)

      DATA NGOOD /0,0/
      DATA ISHEL/0,0/
      DATA GOODHEL/THEL*.FALSE./

      DATA (NHEL(I,   1),I=1,4) /-1,-1,-1,-1/
      DATA (NHEL(I,   2),I=1,4) /-1,-1,-1, 1/
      DATA (NHEL(I,   3),I=1,4) /-1,-1, 1,-1/
      DATA (NHEL(I,   4),I=1,4) /-1,-1, 1, 1/
      DATA (NHEL(I,   5),I=1,4) /-1, 1,-1,-1/
      DATA (NHEL(I,   6),I=1,4) /-1, 1,-1, 1/
      DATA (NHEL(I,   7),I=1,4) /-1, 1, 1,-1/
      DATA (NHEL(I,   8),I=1,4) /-1, 1, 1, 1/
      DATA (NHEL(I,   9),I=1,4) / 1,-1,-1,-1/
      DATA (NHEL(I,  10),I=1,4) / 1,-1,-1, 1/
      DATA (NHEL(I,  11),I=1,4) / 1,-1, 1,-1/
      DATA (NHEL(I,  12),I=1,4) / 1,-1, 1, 1/
      DATA (NHEL(I,  13),I=1,4) / 1, 1,-1,-1/
      DATA (NHEL(I,  14),I=1,4) / 1, 1,-1, 1/
      DATA (NHEL(I,  15),I=1,4) / 1, 1, 1,-1/
      DATA (NHEL(I,  16),I=1,4) / 1, 1, 1, 1/
```

```
DATA IDEN/36/

DO I=1,NEXTERNAL
  JC(I) = +1
ENDDO

DO I=1,NCOMB
  IF (GOODHEL(I,IMIRROR) .OR. NTRY(IMIRROR).LE.MAXTRIES) THEN
    T = MATRIX1(P ,NHEL(1,I),JC(1))
    ANS = ANS+T
  ENDIF
ENDDO

ANS = ANS/DBLE(IDEN)
END
```

The important part of this subroutine is the list of possible helicity combinations stored in the array $n_{\text{hel}}(1:4, 1:16)$. Adding all different helicity combinations means a loop over the second argument and a call of MATRIX1 with the respective helicity combination. Because of the naive helicity combinations many are not allowed the array GOODHEL keeps track of valid combinations. After an initialization to all 'false' this array is only switched to 'true' if MATRIX1 returns a finite value, otherwise Madgraph does not waste time to compute the matrix element. At the very end, a complete spin–color averaging factor is included as IDEN and in our case given by $2 \times 2 \times N_c^2 = 36$.

Altogether, Madgraph provides us with the subroutine SMATRIX1 and the function MATRIX1 which together compute $\overline{|\mathcal{M}|^2}$ for each phase space point given as an external momentum configuration. This helicity method might not seem particularly appealing for a simple $(2 \to 2)$ process, but it makes it possible to compute processes with many particles in the final state and typically up to 10,000 Feynman diagrams which we could never square symbolically, no matter how many graduate students' live times we throw in.

3.3 Missing Transverse Energy

Some of the most interesting signatures at the LHC involve dark matter particles. From cosmological constraints we know that dark matter definitely interacts gravitationally and that it cannot carry electromagnetic or color charges. Weak interactions are allowed because of their limited reach. It turns out that a weakly interacting particle with a mass around the electroweak scale typically gives the observed relic density in the universe. This is called the *WIMP miracle*. It it the reason why in modern TeV-scale model building every model (and its dog) predict a stable WIMP. From supersymmetry we know that this is not hard to achieve: all we need is a $\mathbb{Z}_2$ symmetry to induce a *multiplicative quantum number* for a sector of newly introduced particles. In supersymmetry this is called R parity, in little-Higgs models T parity, and in extra-dimensional models Kaluza–Klein parity.

At the LHC we typically produce strongly interacting new particles, provided they exist. In the presence of a conserved dark matter quantum number exists we will always produce them in pairs. Each of them decays to the weakly interacting sector which includes a stable dark matter agent. On the way, the originally produced particles have to radiate quarks or gluons to shed their color charge. If in some kind of cascade decays they also radiate leptons or photons those can be very useful to trigger on and to reduce QCD backgrounds, but this depends on the details of the weakly interacting new physics sector. The decay steps ideally are two body decays from on–shell particle to on–shell particle, but they do not have to be. What we therefore look for is jets in association with pairs of only weakly interacting, hence invisible particles in the ATLAS and CMS detectors.

From Eq. (2.28) and the discussion of parton densities we remember that at hadron colliders we do not know the kinematics of the initial state. While in the transverse plane by definition the incoming partons have zero momentum, in beam direction we only know its boost statistically. The way to look for invisible particles therefore is a mis-balance of three-momentum in the transverse plane. The actual observable is the *missing transverse momentum* defined as the vector sum of the transverse momenta of all invisible particles. We can convert it into a missing transverse energy which in the absence of any information on particle masses is defined as the absolute value of the two-dimensional missing momentum vector. LHC events including dark matter are characterized by a high jet multiplicity and large missing transverse energy.

At the end of Sect. 2.6.2 we focus on the proper simulation of W+jets and Z+jets samples, which are the Standard Model backgrounds to such signals. It will turn out that jet merging is needed to reliably predict the missing transverse momentum distributions in Standard Model processes. After all our studies in Sect. 2 we are at least theoretically on safe ground. However, this is not the whole story of missing transverse momentum.

3.3.1 Measuring Missing Energy

The left panel of Fig. 3.2 is a historic distribution of missing transverse energy from DZero. It nicely illustrates that by just measuring missing transverse energy, Tevatron would have discovered supersymmetry based on two beautiful peaks around 150 GeV and around 350 GeV. However, this preliminary experimental result has nothing to do with physics, it is purely a detector effect.

We can illustrate the problem of missing energy using a simple number: to identify and measure a lepton we need around 500 out of 200,000 calorimeter cells in an experiment like ATLAS, while for missing transverse energy we need all of them. To cut on a variable like missing transverse momentum we need to understand our detectors really well, and this level of understanding needs a lot of time and effort.

There are several sources of missing energy which we have to understand before we get to search for new physics:

3.3 Missing Transverse Energy

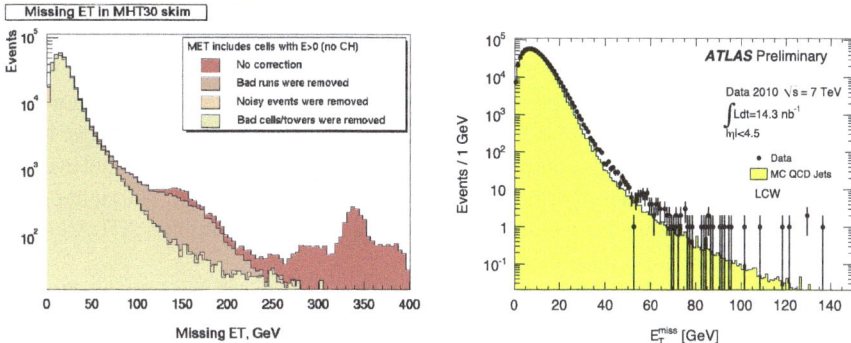

Fig. 3.2 *Left*: missing energy distribution from the early running phase of the DZero experiment at the Tevatron (Figure from Beate Heinemann). *Right*: corrected missing energy distribution in QCD events at ATLAS using only data from April/May 2010 at 7 TeV collider energy (Figure from Ref. [1])

- First, we have to subtract bad runs. They happen if for a few hours parts of the detector do not work properly. We can identify them by looking at the detector response and its correlation. One example is a so-called ring of fire where we see coherent effects in detector modules of circular shape around the beam axis.
- Next, there will be coherent noise in the calorimeter. With 200,000 cells we know that some of them will individually fail or produce noise. Some sources of noise, like leaking voltage or other electronic noise can be correlated geometrically and lead to beautiful missing momentum signals. The way to get rid of such noise event by event is to again look for usual detector response. Combined with bad runs such events can constitute $\mathcal{O}(0.1\%)$ of all events and get removed from the data sample.
- In addition, there might be particles crossing our detector, but not coming from the interaction point. They can be cosmic rays and lead to unbalanced energy deposition as well. Such events will have reconstructed particle tracks which are not compatible with the measured primary vertex.
- Another source of fake missing energy is failing calorimeter cells, like continuously hot cells or dead cells. ATLAS for example has developed such a hole by 2010. Events where missing energy points into such a region can often be removed once we understand the detector.
- While not really a detector fake the main source of missing energy at hadron colliders are mis-measured QCD jets. If parts of jets point into regions with poor calorimetry, like support structures, the jet energy will be wrongly measured, and the corresponding QCD event will show missing transverse energy. One way to tackle this problem is to require a geometric separation of the missing momentum vector and hard jets in the event. ATLAS detector studies indicate that up to $\mathcal{O}(0.1-1\%)$ of all hard QCD events at the LHC lead to more than 100 GeV of well separated fake missing transverse energy. Figure 1.9 in Sect. 1.5.4 shows that this is not at all a negligible number of events.

Once we understand all sources of fake missing momentum we can focus on real missing momentum. This missing transverse momentum we compute from the momenta of all tracks seen in the detector. This means that any uncertainty on these measurements, like the jet or lepton energy scale will smear the missing momentum. Moreover, we know that there is for example dead matter in the detector, so we have to compensate for this. This compensation is a global correction to individual events, which means it will generally smear the missing energy distribution. The right panel of Fig. 3.2 shows a very early missing transverse energy distribution of ATLAS after some of the corrections described above.

To simulate a realistic missing transverse momentum distribution at the LHC we have to smear all jet and lepton momenta, and in addition apply a Gaussian *smearing* of the order

$$\frac{\Delta E_T}{\text{GeV}} \sim \frac{1}{2} \sqrt{\frac{\sum E_T}{\text{GeV}}} \gtrsim 20 \; . \tag{3.7}$$

While this sounds like a trivial piece of information it is impossible to count the number of papers where people forget this smearing and discover great channels to look for Higgs bosons or new physics. They fall apart when experimentalists take a careful look. The simple rule is: phenomenological studies are right or wrong based on if they can be reproduced by real experimentalists and real detectors or not.

3.3.2 Missing Energy in the Standard Model

In the Standard Model there exists a particle which only interacts weakly: the neutrino. At colliders we produce them in reasonably large numbers in W decays. This means that in W + jets production we can learn how to reconstruct the mass of a leptonically decaying W from one observed and one missing particle. We construct a *transverse mass* in analogy to an invariant mass, but neglecting the longitudinal momenta of the decay products

$$\begin{aligned} m_T^2 &= (E_{T,\text{miss}} + E_{T,\ell})^2 - (\vec{p}_{T,\text{miss}} + \vec{p}_{T,\ell})^2 \\ &= m_\ell^2 + m_{\text{miss}}^2 + 2 \left(E_{T,\ell} E_{T,\text{miss}} - \vec{p}_{T,\ell} \cdot \vec{p}_{T,\text{miss}} \right) \; , \end{aligned} \tag{3.8}$$

in terms of a transverse energy $E_T^2 = \vec{p}_T^2 + m^2$. Since the transverse mass is always smaller than the actual W mass and reaches this limit for realistic phase space regions we can extract m_W from the upper edge in the $m_{T,W}$ distribution. Obviously, we can define the transverse mass in many different reference frames. However, its value is invariant under—or better independent of—longitudinal boosts. Moreover, given that we construct it as the transverse projection of an invariant mass it is also invariant under transverse boosts. By construction we cannot analyze the transverse

3.3 Missing Transverse Energy

mass event by event, so this W mass measurement only uses the fraction of events which populate the upper end of the transverse mass distribution.

Alternatively, from single top production and the production of mixed leptonically and hadronically decaying top pairs we know another method to conditionally reconstruct masses and momenta involving one invisible particle: from a leptonically decaying top quark we only miss the longitudinal momentum of the neutrino. On the other hand, we know at least for the signal events that the neutrino and the lepton come from an on–shell W boson, so we can use this *on–shell condition* to reconstruct the longitudinal neutrino momentum under the assumption that the neutrino has zero mass. Recently, we have seen that sufficiently boosted top quarks with leptonic decays can be fully reconstructed even without using the measured missing energy vector. Instead, we rely on the W and t on–shell conditions and on an assumption about the neutrino momentum in relation to the bottom-lepton decay plane.

From Higgs searches we know how to extend the transverse mass to two leptonic W decays with two neutrinos in the final state. The definition of this transverse mass

$$\begin{aligned} m_{T,WW}^2 &= (E_{T,\text{miss}} + E_{T,\ell\ell})^2 - (\vec{p}_{T,\text{miss}} + \vec{p}_{T,\ell\ell})^2 \\ &= m_{\ell\ell}^2 + m_{\text{miss}}^2 + 2\left(E_{T,\ell\ell} E_{T,\text{miss}} - \vec{p}_{T,\ell\ell} \cdot \vec{p}_{T,\text{miss}}\right) \end{aligned} \quad (3.9)$$

is not unique because it is not clear how to define m_{miss}, which also enters the definition of $E_{T,\text{miss}}$. From Monte Carlo studies it seems that identifying $m_{\text{miss}} \equiv m_{\ell\ell}$, which is correct at threshold, is most strongly peaked. On the other hand, setting $m_{\text{miss}} = 0$ to define a proper bounded–from–above transverse mass variable seems to improve the Higgs mass extraction.

For an unspecified number of visible and invisible particles in the final state there also exist global observables we can rely on. The *visible mass* is based on the assumption that we are looking for the decay of two heavy new states where the parton densities will ensure that these two particles are non–relativistic. We can then approximate the partonic energy $\sqrt{\hat{s}} \sim m_1 + m_2$ by some kind of visible energy. If the heavy states are produced with little energy, boost invariance is not required for this construction. Without taking into account missing energy and adding leptons ℓ and jets j the visible mass looks like

$$m_{\text{visible}}^2 = \left[\sum_{\ell,j} E\right]^2 - \left[\sum_{\ell,j} \vec{p}\right]^2. \quad (3.10)$$

Similarly, the Tevatron experiments have for a long time used an effective transverse mass scale which is usually evaluated for jets only, but can trivially be extended to leptons:

$$H_T = \sum_{\ell,j} E_T = \sum_{\ell,j} p_T, \quad (3.11)$$

where the last step assumes that all final-state particles are massless. In an alternative definition of H_T we sum over a number of jets plus the missing energy and skip the hardest jet in this sum.

When combining such a measure with missing transverse momentum the question arises: do we want to pair up the missing transverse momentum with the visible transverse momenta or with the complete visible momenta? For example, we can use the scalar sum of all transverse momenta in the event, now including the missing transverse momentum

$$m_{\text{eff}} = \sum_{\ell,j,\text{miss}} E_T = \sum_{\ell,j,\text{miss}} p_T \,. \quad (3.12)$$

This effective mass is known to trace the mass of the heavy new particles decaying for example to jets and missing energy. This interpretation relies on the non-relativistic nature of the production process and our confidence that all jets included are really decay jets.

3.3.3 Missing Energy and New Physics

The methods described in the last section are well studied for different Standard Model processes and can be applied in new physics searches for various lengths of decay chains. However, there is need for one significant modification, namely to account for a finite *unknown mass* of the missing energy particle. This is a problem of relativistic kinematics and at leading order does not require any knowledge of QCD or new physics models.

The chain of three successive three-body decays shown in Fig. 3.3 is the typical left handed *squark cascade decay* in supersymmetry. The same topology we can interpret in extra-dimensional models (universal extra dimensions or UED) as the decay of a Kaluza–Klein quark excitation

$$\tilde{q}_L \to \tilde{\chi}_2^0 q \to \tilde{\ell}^\pm \ell^\mp q \to \tilde{\chi}_1^0 \ell^+ \ell^- q \qquad Q_L^{(1)} \to Z^{(1)} q \to \ell^{(1)\pm} \ell^\mp q \to \gamma^{(1)} \ell^+ \ell^- q. \quad (3.13)$$

In both cases the last particle in the chain, the lightest neutralino or the Kaluza–Klein photon excitation pass the detectors unobserved. The branching ratio for such decays might not be particularly large; for example in the supersymmetric parameter point SPS1a with a mass spectrum we will discuss later in Fig. 3.7 the long squark decay ranges around 4 %. On the other hand, strongly interacting new particles should in principle be generously produced at the LHC, so we usually assume that there will be enough events to study. The question is how we can then extract the four masses of the new particles appearing in this decay from the three observed external momenta.

3.3 Missing Transverse Energy

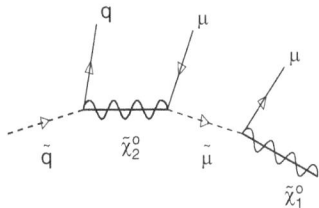

Fig. 3.3 Feynman diagram for the long decay chain shown in Eq. (3.13)

The proposals to solve this problem can be broadly classified into four classes. While all of them should in principle work and would then differ mostly by statistics, we only know how QCD and detector smearing affect the first strategy.

1. *Endpoint methods* extract masses from lower (threshold) and upper (edge) kinematic endpoints of invariant mass distributions of visible decay products. This method is best suited for long decay chains, where the number of independent endpoint measurements in one leg at least matches the number of unknown masses in the cascade. An implicit assumption of these endpoint techniques is that the form of the matrix element populates the phase space close to the endpoint well. Otherwise, the endpoint will be soft and difficult to identify on top of the continuum background.

    The squark decay equation (3.13) has a particular kinematic feature: the invariant mass distributions of the two leptons $m_{\ell\ell}$. Looked at in the rest frame of the intermediate slepton it is a current–current interaction, similar to the Drell–Yan process computed in Eq. (2.11). Because in the s-channel there now appears a scalar particle there cannot be any angular correlations between the two currents, which means the $m_{\ell\ell}$ distribution will have a triangular shape. We can compute its upper limit, called the *dilepton edge*: in the rest frame of the scalar lepton the three-momenta of the incoming and outgoing pair of particles have the absolute values $|\vec{p}| = |m^2_{\tilde{\chi}^0_{1,2}} - m^2_{\tilde{\ell}}|/(2m_{\tilde{\ell}})$. The lepton mass we set to zero. The invariant mass of the two lepton reaches its maximum if the two leptons are back–to–back and the scattering angle is $\cos\theta = -1$

$$\begin{aligned}
m^2_{\ell\ell} &= (p_{\ell^+} + p_{\ell^-})^2 \\
&= 2\left(E_{\ell^+}E_{\ell^-} - |\vec{p}_{\ell^+}||\vec{p}_{\ell^-}|\cos\theta\right) \\
&< 2\left(E_{\ell^+}E_{\ell^-} + |\vec{p}_{\ell^+}||\vec{p}_{\ell^-}|\right) \\
&= 4\,\frac{m^2_{\tilde{\chi}^0_2} - m^2_{\tilde{\ell}}}{2m_{\tilde{\ell}}}\,\frac{m^2_{\tilde{\ell}} - m^2_{\tilde{\chi}^0_1}}{2m_{\tilde{\ell}}} \qquad \text{using}\quad E^2_{\ell^\pm} = \vec{p}^{\,2}_{\ell^\pm}\,.
\end{aligned} \qquad (3.14)$$

This kinematic statement is independent of the shape of the $m_{\ell\ell}$ distribution. For the particle assignments shown in Eq. (3.13) the kinematic endpoints are given by

$$0 < m_{\ell\ell}^2 < \frac{(m_{\tilde{\chi}_2^0}^2 - m_{\tilde{\ell}}^2)(m_{\tilde{\ell}}^2 - m_{\tilde{\chi}_1^0}^2)}{m_{\tilde{\ell}}^2} \qquad 0 < m_{\ell\ell}^2 < \frac{(m_{Z^{(1)}}^2 - m_{\ell^{(1)}}^2)(m_{\ell^{(1)}}^2 - m_{\gamma^{(1)}}^2)}{m_{\ell^{(1)}}^2}.$$
(3.15)

A problem in realistic applications of endpoint methods is combinatorics. We need to either clearly separate the decays of two heavy new states, or we need to combine a short decay chain on one side with a long chain on the other side. In supersymmetry this is naturally the case for associated squark and gluino production. A right handed squark often decays directly to the lightest neutralino which is the dark matter candidate in the model. The gluino has to radiate two quark jets to reach the weakly interacting sector of the model and can then further decay in many different ways. In other models this feature is less generic. The impressive potential of endpoint methods in the case of supersymmetry we will illustrate later in this section.

When looking at long cascade decays for example with two leptons we usually cannot tell which of the two leptons is radiated first. Therefore, endpoint techniques will always be plagued with *combinatorial background* from the mapping of the particle momenta on the decay topology. The same applies to QCD jet radiation vs decay jets. In this situation it is useful to consider the correlation of different invariant masses and their endpoints. The endpoint method can be extended to use invariant mass distributions from both sides of the event (hidden threshold techniques), and correlations between the distributions from each leg (wedgebox techniques).

2. *Mass relation methods* generalize the single top example in Sect. 3.3.2 and completely reconstruct the kinematics event by event. For each event this construction provides a set of kinematic constraints. While for one event the number of parameters can be larger than the number of measurements, adding signal events increases the number of measurements while keeping the number of unknowns constant. Eventually, the system of equations will solve, provided all events are really signal events. Implicitly, we always assume that all decaying particles are on–shell.

In Fig. 3.4 we show the general topology of a three-step cascade decay on each side of the event, like we expect it for a pair of left handed squarks following Eq. (3.13). To extract the masses of the new particles we need to solve the system of equations

$$(p_1 + p_2 + p_3 + p_4)^2 = m_Z^2$$
$$(p_2 + p_3 + p_4)^2 = m_Y^2$$
$$(p_3 + p_4)^2 = m_X^2$$
$$p_4^2 = m_N^2,$$
(3.16)

3.3 Missing Transverse Energy

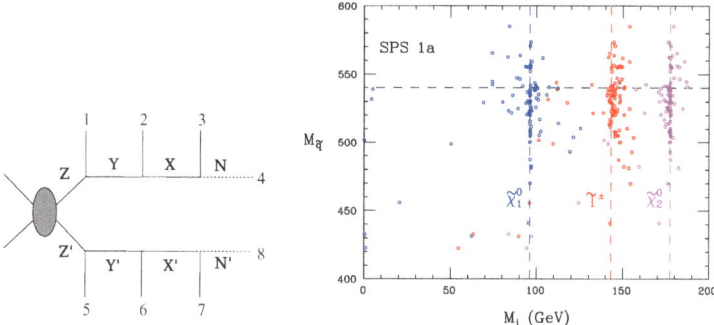

Fig. 3.4 Example for the mass relation method using three successive two-body decays on both sides of the events (*left*). After detector smearing we can reconstruct the masses for the supersymmetric parameter point SPS1a with the squark decay chain shown in Eq. (3.13) (*right*) (Figure from Ref. [21])

for each side of the event. For each event there are eight unknown masses and six unknown missing momentum components of which we measure two combinations as the missing transverse momentum. All of these 12 unknowns we can determine if we add a sufficiently large number of events.

One strategy to solve this problem is to assume eight test masses $m = (m_Z^2, m_Y^2, m_X^2, m_N^2, \ldots)$, use the three first equations in Eq. (3.16) for each event plus the two missing transverse momentum components to determine both missing four-momenta, and test the consistency of this solution using the last line of Eq. (3.16) for each of the two legs. In this consistency test we combine the information from several events.

We can conveniently solve the first three lines in Eq. (3.16) for the missing momentum p_4

$$-2(p_1 p_4) \equiv s_1 = m_Y^2 - m_Z^2 + 2(p_1 p_2) + 2(p_1 p_3)$$
$$-2(p_2 p_4) \equiv s_2 = m_X^2 - m_Y^2 + 2(p_2 p_3)$$
$$-2(p_3 p_4) \equiv s_3 = m_N^2 - m_X^2 , \qquad (3.17)$$

for simplicity assuming massless Standard Model decay products. Similarly, we define the measured combinations $s_{5,6,7}$ from the opposite chain. In addition, we measure the two-dimensional missing transverse momentum, so we can collect the two missing four-momenta into $p_{\text{miss}} = (\vec{p}_4, E_4, \vec{p}_8, E_8)$ and define two additional entries of the vector s in terms of measured quantities and masses like

$$(\hat{x} p_4) + (\hat{x} p_8) = s_4$$
$$(\hat{y} p_4) + (\hat{y} p_8) = s_8 . \qquad (3.18)$$

Combining the first equal signs of Eqs. (3.17) and (3.18) for both halves of the events reads $A \cdot p_{\text{miss}} = s$, where the matrix A includes only components of measured momenta and is almost block diagonal, so it can be inverted. Following the second equal sign in Eq. (3.17) we can then write $s = B \cdot m + c$, where the matrix B only contains non-zero entries ± 1 and the vector c consists of measured quantities. Together, this gives us

$$p_{\text{miss}} = A^{-1}s = A^{-1}Bm + A^{-1}c. \tag{3.19}$$

We show the result for all masses in the decay chain using 25 events per set and including all combinatorics in Fig. 3.4. The mass relation method has also been successfully applied to single legs as well as in combination with kinematic endpoints.

3. *MT2 methods* are based on a global variable m_{T2}. It generalizes the transverse mass known from W decays to the case of two massive invisible particles, one from each leg of the event. The observed missing energy in the event we can divide into two scalar fractions $p_{T,\text{miss}} = q_1 + q_2$. Given the two fractions q_j we can construct a transverse mass for each side of the event, assuming we know the invisible particle's mass $m_{T,j}(q_j; \hat{m}_{\text{miss}})$; the second argument is an external assumption, so $\hat{m}_{\text{miss}}$ is an assumed value for m_{miss}.

Inspired by the usual transverse mass we are interested in a mass variable with a well–defined upper edge, so we need to construct some kind of minimum of $m_{T,j}$ as a function of the splitting of $p_{T,\text{miss}}$. Naively, this minimum will simply be the zero transverse momentum limit of m_T on one leg, which is not very interesting. On the other hand, in this case the transverse mass from the other leg reaches a maximum, so we can instead define

$$m_{T2}(\hat{m}_{\text{miss}}) = \min_{p_{T,\text{miss}} = q_1 + q_2} \left[\max_j m_{T,j}(q_j; \hat{m}_{\text{miss}}) \right] \tag{3.20}$$

as a function of the unknown missing particle mass. There are two properties we know by construction

$$m_{\text{light}} + \hat{m}_{\text{miss}} < m_{T2}(\hat{m}_{\text{miss}})$$
$$m_{\text{light}} + m_{\text{miss}} < m_{T2}(m_{\text{miss}}) < m_{\text{heavy}}. \tag{3.21}$$

The first line means that each of the $m_{T,j}$ lie between the sum of the two decay products' masses and the mass of the decaying particle, so for massless Standard Model decay products there will be a global m_{T2} threshold at the missing particle's mass.

Moreover, for the correct value of m_{miss} the m_{T2} distribution has a sharp edge at the mass of the parent particle. In favorable cases m_{T2} may allow the measurement of both the parent particle and the LSP based on a single–step decay chain. These two aspects for the correct value $\hat{m}_{\text{miss}} = m_{\text{miss}}$ we can see

3.3 Missing Transverse Energy

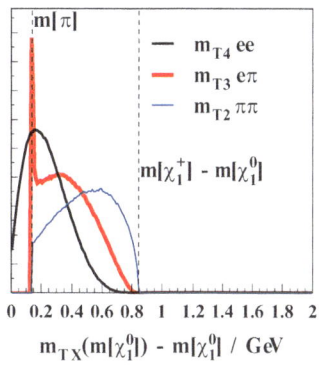

Fig. 3.5 Simulations for different m_{TX}, for the decay $\tilde{\chi}_1^+ \to \tilde{\chi}_1^0 \pi$ or $\tilde{\chi}_1^+ \to \tilde{\chi}_1^0 e^+ \nu$. The *blue m_{T2} line* applies to the two-body decay (Figure from Ref. [2])

in Fig. 3.5: the lower threshold is indeed given by $m_{T2} - m_{\tilde{\chi}_1^0} = m_\pi$. while the upper edge of $m_{T2} - m_{\tilde{\chi}_1^0}$ coincides with the dashed line for $m_{\tilde{\chi}_1^+} - m_{\tilde{\chi}_1^0}$.

An interesting aspect of m_{T2} is that it is boost invariant if and only if $\hat{m}_{\text{miss}} = m_{\text{miss}}$. For a wrong assignment of m_{miss} it has nothing to do with the actual kinematics and hence with any kind of invariant, and house numbers are not boost invariant. We can exploit this aspect by scanning over m_{miss} and looking for so-called kinks, defined as points where different events kinematics all return the same value for m_{T2}.

Similar to the more global m_{eff} variable we can generalize m_{T2} to the case where we do not have a clear assignment of the two decay chains involved. This modification $M_{T\text{Gen}}$ again has an upper edge, which unfortunately is not as sharp as the one in m_{T2}. Similarly, the procedure can be generalized to any one-step decay, for example a three-body decay with either one or two missing particles on each side of the event. Such M_{TX} distributions are useful as long as they have a sharp enough edge, as illustrated in Fig. 3.5.

4. *Extreme kinematics* can also give us a handle on mass reconstruction from an incomplete set of observables. One such phase space region are points close kinematic endpoints where particles are produced at rest. Other examples are the approximate collinear Higgs mass reconstruction in a decay to boosted tau pairs described in Sect. 1.6.3 or the boosted leptonic top decays mentioned before.

The way mass measurements can lead to proper *model reconstruction* we sketch for one scenario. The classic example for the endpoint method is the long supersymmetric left handed squark decay chain shown in Eq. (3.13) and in Fig. 3.3. The quoted supersymmetric partner masses are by now ruled out, but in the absence of more recent studies we stick to their historic values. When we use such kinematic endpoints or other methods to extract mass parameters it is crucial to start from a signal–rich sample to avoid combinatorics and washed out endpoints vanishing in a fluctuating or even sculptured background. For jets, leptons and missing energy a major background will be top pairs. The key observation is that in long cascade decays the leptons are flavor–locked, which means the combination $e^+e^- + \mu^+\mu^- - e^-\mu^+ - e^+\mu^-$ becomes roughly twice $\mu^+\mu^-$ for the signal, while

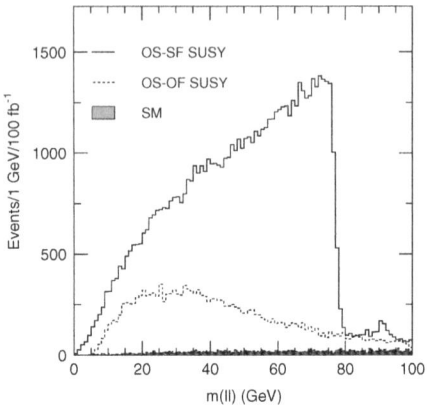

Fig. 3.6 Invariant mass of two leptons after selection cuts for the SPS1a parameter point: SUSY signal Opposite Sign Same Flavor (OS-SF): *full line*; SUSY signal Opposite Sign Opposite Flavor (OS-OF): *dotted line*; Standard Model background: *grey* (Figure from Giacomo Polesello (ATLAS))

it cancels for top pairs. Using such a combination for the endpoint analysis means the top background is subtracted purely from data, as illustrated in Fig. 3.6.

The long squark decay in by now ruled out SPS1a-like parameter points with squark masses in the 500–600 GeV range has an important advantage: for a large *mass hierarchy* we should be able to isolate the one decay jet just based on its energy. In complete analogy to the dilepton edge shown in Eq. (3.15), but with somewhat reduced elegance we can measure the threshold and edge of the $\ell^+\ell^-q$ distribution and the edges of the two $\ell^\pm q$ combinations. Then, we solve the system for the intermediate masses without any model assumption, which allows us to even measure the dark matter mass to $\mathcal{O}(10\%)$. The limiting factors will likely be our chances to observe enough endpoints in addition to $m_{\ell\ell}^{\max}$ and the jet energy scale uncertainty. An interesting question is how well we will do with tau leptons, where the edge is softened by neutrinos from tau decays.

Provided the gluino or heavy gluon is heavier than the squarks or heavy quarks we can measure its mass by extending the squark chain by one step: $\tilde{g} \to q\tilde{q}$. This measurement is hard if one of the two jets from the gluino decay is not very hard, because its information will be buried by the combinatorial error due to QCD jet radiation. The way around is to ask for two bottom jets from the strongly interacting decay: $\tilde{g} \to b\tilde{b}^*$ or $G^{(1)} \to b\bar{B}^{(1)}$. The summary of all measurements in Fig. 3.7 shows that we can extract for example the gluino mass at the per-cent level, a point at which we might have to start thinking about off–shell propagators and at some point even define what exactly we mean by 'masses as appearing in cascade decays'.

A generic feature or all methods relying on decay kinematics is that it is easier to constrain the differences of squared masses than the absolute mass scale. This is also visible in Fig. 3.7. It is due to the form of the endpoint formulas which involve the difference of mass squares $m_1^2 - m_2^2 = (m_1+m_2)(m_1-m_2)$. This combination is much more sensitive to (m_1-m_2) than it is to (m_1+m_2). Experimentally, correlated jet and lepton energy scale uncertainties do not make life easier either. Nevertheless, the common lore that kinematics only constrain mass differences is obviously not true for two body decays.

3.3 Missing Transverse Energy

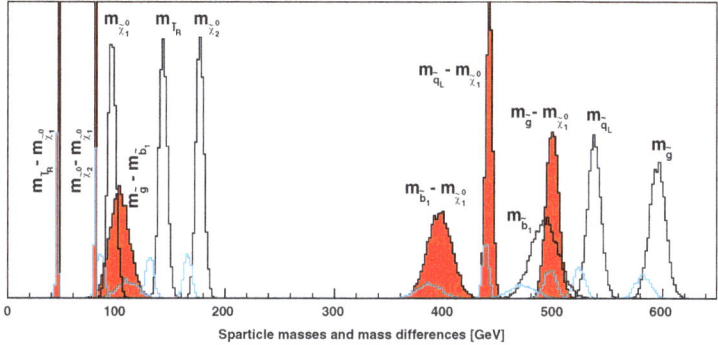

Fig. 3.7 Masses extracted from the gluino-sbottom decay chain, including estimated errors. The *faint blue lines* indicate wrong solutions when inverting the endpoint–mass relations. The supersymmetric mass spectrum is given by the SPS1a parameter point (Figure from Ref. [9])

Alternatively, we can use the same gluino decay to first reconstruct the intermediate neutralino or Kaluza–Klein Z momentum for lepton pairs residing near the $m_{\ell\ell}$ edge. In that case the invisible heavy state is produced approximately at rest, and the momenta are correlated as

$$\vec{p}_{\tilde{\chi}_2^0} = \left(1 - \frac{m_{\tilde{\chi}_1^0}}{m_{\ell\ell}}\right) \vec{p}_{\ell\ell} \qquad \vec{p}_{Z^{(1)}} = \left(1 - \frac{m_{\gamma^{(1)}}}{m_{\ell\ell}}\right) \vec{p}_{\ell\ell} \qquad (3.22)$$

If both neutralino masses (or the Kaluza–Klein photon and Z masses) are known, we can extract the sbottom (Kaluza–Klein bottom) and gluino (Kaluza–Klein gluon) masses by adding the measured bottom momenta to this neutralino (Kaluza–Klein photon) momentum. Again, for the mass spectrum shown in Fig. 3.7 we can measure the gluino mass to few per-cent, depending on the systematic errors.

For a complete analysis, kinematic endpoints can be supplemented by any other method to measure new physics masses. For short decay chains m_{T2} is best suited to measure the masses of particles decaying directly to the dark matter agent. In supersymmetry, this happens for right handed sleptons or right handed squarks. The issue with short decay chains is that they often require on some kind of jet veto, which following Sects. 2.6.2 and 1.6.2 is problematic for low-p_T jets.

Keeping in mind that endpoint analyses only use a small fraction of the events, namely those with extreme kinematics, an obvious way to improve their precision is to include the complete shape of the invariant mass distributions. However, this strategy bears a serious danger. Invariant masses are just an invariant way of writing angular correlations between outgoing particles. Those depend on the *spin and quantum numbers* of all particles involved. For example, in the case of the $m_{\ell\ell}$ endpoint the triangular shape implies the absence of angular correlations, because the intermediate particle is a scalar. This means that we should be careful when extracting information for example from kinematic endpoints we do not observe. Depending on the quantum numbers and mixing angles in the new physics scenario, kinematic endpoints can for example be softened, so they vanish in the background noise.

This argument we can turn around. Measuring discrete quantum numbers, like the spin of new particles, is hard in the absence of fully reconstructed events. The usual threshold behavior is not observable at hadron colliders, in particular when the final state includes missing transverse energy. Instead, we rely on angular correlation in decays. For the squark decay chain given in Eq. (3.13) there exists a promising method to simultaneously determine the spin of all new particles in the chain:

1. Instead of trying to measure spins in a general parameterization we start from the observation that cascade decays radiate particles with known spins. This is most obvious for long gluino decays where we know that the radiated bottom quarks as well as muons are fermions. The spins inside the decay chain can only alternate between fermions and bosons. Supersymmetry switches this fermion/boson nature compared to the corresponding Standard Model particle, so we can contrast it with another hypothesis where the spins in the decay chain follow the Standard Model assignments. Such a model are Universal Extra Dimensions, where each Standard Model particle acquires a Kaluza–Klein partner from the propagation in the bulk of the additional dimensions.
2. Thresholds and edges of all invariant masses of the radiated fermions are completely determined by the masses inside the decays chain. Kinematic endpoints cannot distinguish between supersymmetry and universal extra dimensions. In contrast, the shape of the distribution between the endpoints is nothing but an angular correlation in some reference frame. For example, the $m_{j\ell}$ distribution in principle allows us to analyze spin correlations in squark decays in a Lorentz invariant manner. The only problem is the link between $\ell^{\pm}$ and their ordering in decay chain.
3. A proton–proton collider like the LHC produces considerably more squarks than antisquarks in the squark–gluino associated channel. For the SPS1a spectrum at 14 TeV collider energy their ratio ranges around 2:1. A decaying squark radiates a quark while an antisquark radiates an antiquark, which means that we can define a non-zero production-side asymmetry between $m_{j\ell^+}$ and $m_{j\ell^-}$. Such an asymmetry we show in Fig. 3.8, for the SUSY and for the UED hypotheses. Provided the masses in the decay chain are not too degenerate we can indeed distinguish the two hypotheses.

This basic idea has since been applied to many similar situations, like decays including gauge bosons, three-body decays, gluino decays with decay–side asymmetries, cascades including charginos, weak boson fusion signatures, etc. They show that the LHC can do much more than just discover some kind of particle beyond the Standard Model. It actually allows us to study underlying models and symmetries.

3.4 Uncertainties   317

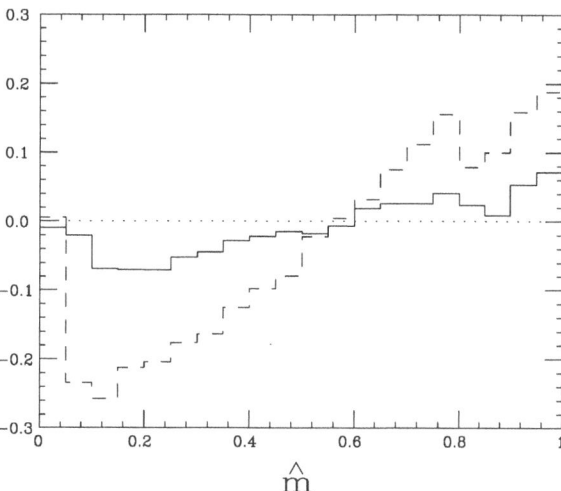

Fig. 3.8 Asymmetry in $m_{j\ell}/m_{j\ell}^{\max}$ for supersymmetry (*dashed*) and universal extra dimensions (*solid*). The spectrum is assumed to be hierarchical, typical for supersymmetric theories (Figure taken from Ref. [19])

3.4 Uncertainties

As we argue in the very beginning of the lecture, LHC physics always means extracting signals from often large backgrounds. This means, a correct error estimate is crucial. For LHC calculations we are usually confronted with three types of errors.

1. The first and easiest one are the *statistical errors*. For small numbers of events these experimental errors are described by Poisson statistics,

$$f(N; N_{\text{theo}}) = \frac{e^{-N_{\text{theo}}} N_{\text{theo}}^N}{\Gamma(N+1)}, \qquad (3.23)$$

where N_{theo} is the expected number of events and N the observed numbers. The mean value as well as the variance of this distribution is N_{theo}. For large event numbers N' they converge to the Gaussian limit, Eq. (1.182). This is an example of the central limit theorem which says that a sufficiently large number of independent random variables will eventually follow a Gaussian shape. In that limit the number of standard deviations in terms of the number of signal and background events is $S/\sqrt{B}$. The two event numbers are proportional to the integrated luminosity $\mathscr{L}$ which means that the statistical significance in the Gaussian limit increases with $\sqrt{\mathscr{L}}$. In high energy physics five standard deviations above a known background we call a discovery, three sigma is often referred to as an evidence. The Poisson region is the only complication we encounter for statistical errors. It means that for small number of signal and background events we need more luminosity than the Gaussian limit suggests.

2. The second set of errors are *systematic errors*, like the calibration of the jet and lepton energy scales, the measurements of the luminosity, or the efficiencies to for instance identify a muon as a muon. Some readers might remember a bunch of theorists mistaking a forward pion for an electron—that happened right around my TASI in 1997 and people not only discovered supersymmetry but also identified its breaking mechanism. Of course, our experimentalist CDF lecturer told us that the whole thing was a problem of identifying a particle in the detector with an efficiency which does not have to be zero or one.

    Naively, we would not assume that systematic errors follow a Gaussian distribution, but experimentally we determine efficiencies and scaling factors largely from well understood background processes. Such counting experiments in background channels like $Z \to$ leptons and their extracted parameters also follow a Gaussian distribution. The only caveat is the shape of far-away tails, which often turn out to be higher than the exponentially suppressed Gaussian tails.

    Systematic errors which do not follow a Gaussian distribution can scale like S/B, which means they do not improve with increasing luminosity. Again, five standard deviations are required to claim a discovery, and once we are systematics dominated waiting for more data does not help.

3. The third source of errors are *theoretical errors*. They are hardest to model because they are often dominated by higher–order QCD effects in fixed order perturbation theory. From Sect. 2 we know that higher order corrections for example to LHC production rates do not follow a naive power counting in α_s but are enhanced by large logarithms. If we can compute or reliably estimate some higher order terms in the perturbative QCD series we call this a prediction. In other words, once we consider a statement about perturbative QCD a statement about its uncertainty we are probably only giving a wild guess.

    To model theoretical uncertainties it is crucial to realize that higher order effects are not any more likely to give a K factor of 1.0 than 0.9 or 1.1. In other words, likelihood distributions accounting for theoretical errors do not have a peak and are definitely not Gaussian. Strictly speaking, all we know is a range of theoretically acceptable values. There is a good reason to choose the Gaussian short cut, which is that folding three Gaussian shapes for statistical, systematic and theoretical errors gives us a Gaussian distribution altogether, which makes things numerically much easier. But this approach assumes that we know much more about QCD than we actually do which means it is not conservative at all.

    On the other hand, we also know that theoretical errors cannot be arbitrarily large. Unless there is a very good reason, a K factor for a total LHC cross section should not be larger than something like two or three. If that happens we need to conclude that perturbative QCD breaks down, and the proper description of error bars is our smallest problem. In other words, the centrally flat theory probability distribution for an LHC observable has to go to zero for very large deviations from the currently best value. Strictly speaking, even this minimalist distribution is not well defined, because there is not frequentist interpretation of a range of theory uncertainties which could be used to define or test such a distribution.

3.4 Uncertainties

For example in the case if Higgs coupling measurements these different sources of errors pose two problems: first, we need to construct a probability measure combining all these three sources. Second, this exclusive probability or likelihood has to be reduced in dimensionality such that we can show an error bar on one of the Higgs couplings in a well defined manner. Both of these problems lead us to the same objects, likelihoods vs probabilities:

A *likelihood* is defined as a probability to obtain an experimental outcome $\vec{x}_{\text{meas}}$ given model predictions $\vec{x}_{\text{mod}}$, varied over the model space. We can link it to the corresponding probability as

$$P(\vec{x}_{\text{meas}}|\vec{x}_{\text{mod}}) = \mathscr{L}(\vec{x}_{\text{mod}}|\vec{x}_{\text{meas}}) ,\qquad(3.24)$$

where in both expressions are meant to be evaluated over model parameter space. In Sect. 1.8.1 we introduce the logarithm of a likelihood equation (1.239) as a generalization of the usual χ^2 distribution

$$\chi^2(\vec{m}) = \sum_{j=1}^{n_{\text{meas}}} \frac{|\vec{x}_{\text{meas}} - \vec{x}_{\text{mod}}(\vec{m})|_j^2}{\sigma_j^2} ,\qquad(3.25)$$

which is expressed in terms of the measurements $\vec{x}_{\text{meas}}$, the model predictions $\vec{x}_{\text{mod}}$, and the variance σ^2. This definition is really only useful in the Gaussian limit where we know what the variance is.

According to the definition equation (3.24) we can replace the Gaussian form of χ^2 by any other estimated shape for the statistical distribution of $\vec{x}_{\text{meas}}$. This includes the Poisson, Gaussian, and box shapes discussed above, as well as any combination of the three. Before we discuss in detail how to construct a likelihood for example for a Higgs couplings measurement we should link this likelihood to a mathematically properly defined probability.

Bayes' theorem tells us how to convert the likelihood equation (3.24) into the *probability* that a choice of model parameters $\vec{x}_{\text{mod}}$ is true given the experimental data, $\vec{x}_{\text{meas}}$. This is what we are actually interested in when we measure for example Higgs couplings

$$P(\vec{x}_{\text{mod}}|\vec{x}_{\text{meas}}) = P(\vec{x}_{\text{meas}}|\vec{x}_{\text{mod}}) \, \frac{P(\vec{x}_{\text{mod}})}{P(\vec{x}_{\text{meas}})} \equiv \mathscr{L}(\vec{x}_{\text{mod}}|\vec{x}_{\text{meas}}) \, \frac{P(\vec{x}_{\text{mod}})}{P(\vec{x}_{\text{meas}})} .\qquad(3.26)$$

In this relation $P(\vec{x}_{\text{meas}})$ is a normalization constant which might be hard to evaluate but which ensures that the probability $P(\vec{x}_{\text{mod}}|\vec{x}_{\text{meas}})$ summed over all possible experimental outcomes is normalized to unity. The problem is the prior $P(\vec{x}_{\text{mod}})$ which is a statement about the model or the model parameter choice and which obviously cannot be determined from experiment. If we bring it to the other side of Eq. (3.26) it ensures that the conditional probability $P(\vec{x}_{\text{meas}}|\vec{x}_{\text{mod}})$ integrated over model space is unity. This implies some kind of measure in model space or model parameter space. As an example, if we want to measure the mass of a particle we can integrate m over the entire allowed or interesting range, but we can also integrate $\log m$ instead. In an ideal world of perfect measurement the difference between these two measures will not affect the final answer for $P(\vec{x}_{\text{mod}}|\vec{x}_{\text{meas}})$. The problem is that Higgs coupling measurements are far from ideal, so we have to decide on a measure in Higgs couplings space to compute a probability for a set of couplings to be true.

One aspect we can immediately learn from this Bayesian argument is how to combine different uncertainties, i.e. statistical, systematic, and theoretical uncertainties for the same observable: we introduce one so-called nuisance parameter for each of these errors, describing the deviation of the measured value from the expected value for the given observable. All three nuisance parameters combined correspond to an actual observable, individually they are not interesting. This means that we want to remove them as dimensions or degrees of freedom from our big exclusive likelihood map. If the integral over model parameter space, including nuisance parameters, is well defined we simply integrate them out, leaving for example the normalization of our probability intact. If we write out this integration it turns out to be a *convolution*.

As is well known, the convolution of several Gaussians is again a Gaussian. The convolution of a Gaussian experimental error with a flat theory error returns a two-sided error distribution which has a peak again. While we started with the assumption that theory errors should not give a preferred value within the error band, the measure in model space after convolution again returns such a maximum.

The frequentist construction to reduce the number of model parameters avoids any measure in model parameter space, but leads to mathematical problem: to keep the mathematical properties of a likelihood as a probability measure, including the normalization, we would indeed prefer to integrate over unwanted directions. In the frequentist approach such a measure is not justified. An alternative solution which is defined to keep track of the best–fitting points in model space is the *profile likelihood*. It projects the best fitting point of the unwanted direction onto the new parameter space; for each binned parameter point in the $(n-1)$-dimensional space we explore the nth direction which is to be removed $\mathscr{L}(x_{1,\ldots,n-1}, x_n)$. Along this direction we pick the best value and identify it with the lower–dimensional parameter point $\mathscr{L}(x_{1,\ldots,n-1}) \equiv \mathscr{L}^{\max(n)}(x_{1,\ldots,n-1}, x_n)$. Such a projection avoids defining a measure but it does not maintain for example the normalization of the likelihood distribution.

We first compute the profile likelihood for two one-dimensional Gaussians affecting the same measurement x, removing the nuisance parameter y, and ignoring the normalization. The form of the combined likelihood is the same as for a convolution, except that the integral over y is replaced by the maximization,

$$\mathscr{L}(x) \sim \max_{y}\ e^{-y^2/(2\sigma_1^2)}\ e^{-(x-y)^2/(2\sigma_2^2)}$$

$$= \max_{y}\ \exp\left[-\frac{\sigma_1^2 + \sigma_2^2}{2\sigma_1^2 \sigma_2^2}\left(y^2 - 2xy\frac{\sigma_1^2}{\sigma_1^2+\sigma_2^2} + x^2\frac{\sigma_1^2}{\sigma_1^2+\sigma_2^2}\right)\right]$$

3.4 Uncertainties

$$= \max_y \exp\left[-\frac{\sigma^2}{2\sigma_1^2\sigma_2^2}\left(\left(y - x\frac{\sigma_1^2}{\sigma^2}\right)^2 - x^2\frac{\sigma_1^4}{\sigma^4} + x^2\frac{\sigma_1^2}{\sigma^2}\right)\right] \quad \text{with } \sigma^2 = \sigma_1^2 + \sigma_2^2$$

$$= \max_{y'} \exp\left[-\frac{\sigma^2}{2\sigma_1^2\sigma_2^2}\left(y'^2 - x^2\frac{\sigma_1^4}{\sigma^4} + x^2\frac{\sigma_1^2}{\sigma^2}\right)\right] \quad \text{with } y' = y - x\frac{\sigma_1^2}{\sigma^2}$$

$$= \max_{y'} \exp\left[-\frac{\sigma^2}{2\sigma_1^2\sigma_2^2}\left(y'^2 + x^2\frac{\sigma_1^2\sigma_2^2}{\sigma^4}\right)\right]$$

$$= e^{-x^2/(2\sigma^2)} \max_{y'} e^{-y'^2\sigma^2/(2\sigma_1^2\sigma_2^2)} = e^{-x^2/(2\sigma^2)} . \quad (3.27)$$

We use that the profile likelihood over y and y' is the same after the linear transformation. Just like in the case of the convolution, the profile likelihood of two Gaussian is again a Gaussian with $\sigma^2 = \sigma_1^2 + \sigma_2^2$. Next, we use the same reasoning to see what happens if we combine two sources of flat errors with identical widths,

$$\mathscr{L}(x) = \max_y \Theta(x_{\max} - y) \Theta(y - x_{\min}) \Theta(x_{\max} - x + y) \Theta(x - y - x_{\min})$$

$$= \max_{y \in [x_{\min}, x_{\max}]} \Theta((x_{\max} + y) - x) \Theta(x - (x_{\min} + y))$$

$$= \Theta(2x_{\max} - x) \Theta(x - 2x_{\min}) . \quad (3.28)$$

Each of the original boxes starts with a of $x_{\max} - x_{\min}$. The width of the box covering the allowed values for x after computing the profile likelihood is $2(x_{\max} - x_{\min})$, so unlike for the Gaussian case the two flat errors get *added linearly*, even though they are assumed to be uncorrelated. We can follow the same kind of calculation for the combination of a Gaussian and a flat box-shaped distribution,

$$\mathscr{L}(x) = \max_y \Theta(x_{\max} - y) \Theta(y - x_{\min}) e^{-(x-y)^2/(2\sigma^2)}$$

$$= \max_{y \in [x_{\min}, x_{\max}]} e^{-(x-y)^2/(2\sigma^2)}$$

$$= \begin{cases} e^{-(x-x_{\min})^2/(2\sigma^2)} & x < x_{\min} \\ 1 & x \in [x_{\min}, x_{\max}] \\ e^{-(x-x_{\max}^2)/(2\sigma^2)} & x > x_{\max} . \end{cases} \quad (3.29)$$

This profile likelihood construction is called *Rfit scheme* and is used for example by CKMfitter or SFitter. We obtain the combined distribution by cutting open the experimental Gaussian distribution and inserting a flat theory piece. Exactly the same happens for the profile likelihood combination of a Poisson distribution and a flat box. The last combination we need to compute is a Gaussian with widths σ with a Poisson with expectation value N. This projection is not trivial to compute,

$$\mathscr{L}(x) = \max_y \frac{e^{-N} N^y}{y!} e^{-(x-y)^2/(2\sigma^2)}$$

$$= \max_y \exp\left[-N + y \log N - \log y! - \frac{(x-y)^2}{2\sigma^2}\right]$$

$$= \max_y \exp\left[-N + y \log N - \frac{1}{2}\log\frac{2\pi}{y+1} - (y+1)\log\frac{y+1}{e} - \frac{(x-y)^2}{2\sigma^2}\right]. \quad (3.30)$$

However, we can approximate the result numerically as

$$\frac{1}{\log \mathscr{L}(x)} = \frac{1}{\log \mathscr{L}_{\text{Poisson}}} + \frac{1}{\log \mathscr{L}_{\text{Gauss}}} = \frac{1}{\log \frac{e^{-N} N^x}{x!}} + \frac{1}{-\frac{x^2}{2\sigma^2}}. \quad (3.31)$$

We can check this formula for the case of two Gaussians

$$\frac{1}{\log \mathscr{L}(x)} = -\frac{2\sigma_1^2}{x^2} - \frac{2\sigma_2^2}{x^2} = -\frac{2\sigma^2}{x^2} \quad \Leftrightarrow \quad \mathscr{L} = e^{-x^2/(2\sigma^2)}, \quad (3.32)$$

with $\sigma^2 = \sigma_1^2 + \sigma_2^2$. This is precisely the result of Eq. (3.27). Another sanity check is that if one of the likelihoods becomes very large it decouples from the final results and the combined likelihood is dominated by the bigger deviation. We can test that Eq. (3.32) reproduces the full result to a few per-cent.

Numerically, we usually compute the logarithm of the likelihood instead of the likelihood itself. The reason is that for many channels we need to multiply all individual likelihoods, leading to a vast numerical range of our likelihood map. It is numerically much more stable to use the logarithm instead and add the *log-likelihoods* instead. In the Gaussian limit this is related to the χ^2 value via $\chi^2 = -2 \log \mathscr{L}$. If we allow for a general correlation matrix C between the entries in the measurements vector $\vec{x}_{\text{meas}}$ and a symmetric theory error $x \pm \sigma^{(\text{theo})}$ we find the RFit expression

$$-2 \log \mathscr{L} = \vec{x}^T C^{-1} \vec{x}$$

$$x_i = \begin{cases} \dfrac{x_{\text{meas},i} - x_{\text{mod},i} + \sigma_i^{(\text{theo})}}{\sigma_i^{(\text{exp})}} & x_{\text{meas},i} < x_{\text{mod},i} - \sigma_i^{(\text{theo})} \\ 0 & |x_{\text{meas},i} - x_{\text{mod},i}| < \sigma_i^{(\text{theo})} \\ \dfrac{x_{\text{meas},i} - x_{\text{mod},i} - \sigma_i^{(\text{theo})}}{\sigma_i^{(\text{exp})}} & x_{\text{meas},i} > x_{\text{mod},i} + \sigma_i^{(\text{theo})} \end{cases} \quad (3.33)$$

This distribution implies that for very large deviations there will always be tails from the experimental errors, so we can neglect the impact of the theoretical errors on this range. In the center the distribution is flat, reflecting our ignorance of the theory prediction. The impact of the size of the flat box we need to test.

3.4 Uncertainties

This concludes our construction of the multi-dimensional correlated likelihood map with different types of errors, which we can apply for example in the Higgs couplings analysis introduced in Sect. 1.8.1. In principle, it is possible to compute an exclusive likelihood map even more generally by keeping all the nuisance parameters, avoiding any of the profile constructions described below, and then removing the nuisance parameter alongside the unwanted couplings at the end. However, this hugely increases the number of dimensions we initially encounter, so it is numerically more economical to first apply analytical profiling as done in SFitter.

Further Reading

Again, there exist several good review articles with more in-depth discussions of different aspects touched in this section:

- As mentioned in Sect. 2, two very useful reviews of jet physics are available by Steve Ellis and collaborators [8] and by Gavin Salam [18].
- If you are interested in top identification using fat jet techniques we wrote a short pedagogical review article illustrating the different techniques and tools available [16].
- For the general phenomenology of the heaviest Standard Model particles, the top quark, have a look at Sally Dawson's TASI lectures [7].
- If you use Madgraph/HELAS to compute helicity amplitudes there is the original documentation which describes every routine [15].
- A lot of experimental knowledge on new physics searches well described and theoretically sound you can find in the CMS technical design report. Some key analyses are described in detail while most of the documentation focuses on the physics expectations [4].
- More on the magical variable m_{T2} you can find in an article by Alan Barr, Chris Lester and Phil Stephens [2]. Chris Lester's thesis [13] is a good point to start with. Recently, Alan Barr and Chris Lester published a broad review on techniques to measure masses in models with dark matter particles [3].
- As mentioned in the introduction, there is our more advanced review on new physics at the LHC which includes an extensive chapter on LHC signatures [14].
- A lot of insight into new physics searches at the LHC and at a linear collider you can find in a huge review article collected by Georg Weiglein [22].
- The pivotal work on determining spins in cascade decays is Jennie Smillie's PhD thesis [19]. On the same topic there exists a nicely written review by Liantao Wang and Itay Yavin [20].
- Many useful pieces of information on mass extraction, errors, and the statistical treatment of new-physics parameter spaces you can find in the big SFitter publication [11]. The SFitter analysis of the Higgs sector [12] is very similar in structure, but different in the physics application.

- If you are interested in a recent discussion of experimental and theoretical errors and how to factorize them, you can try a recent paper we wrote with Kyle Cranmer, Sven Kreiss, and David Lopez–Val [6].

References

1. ATLAS Collaboration, Performance of the missing transverse energy reconstruction and calibration in proton-proton collisions at a center-of-mass energy of $\sqrt{s} = 7$ TeV with the ATLAS detector. ATLAS-CONF-2010-057
2. A. Barr, C. Lester, P. Stephens, m(T2): the truth behind the glamour. J. Phys. G **29**, 2343 (2003)
3. A.J. Barr, C.G. Lester, A review of the mass measurement techniques proposed for the large Hadron collider. J. Phys. G **37**, 123001 (2010)
4. G.L. Bayatian et al. (CMS Collaboration), CMS technical design report, volume II: physics performance. J. Phys. G **34**, 995 (2007)
5. J.M. Butterworth, A.R. Davison, M. Rubin, G.P. Salam, Jet substructure as a new Higgs search channel at the LHC. Phys. Rev. Lett. **100**, 242001 (2008)
6. K. Cranmer, S. Kreiss, D. Lopez-Val, T. Plehn, A novel approach to Higgs coupling measurements. arXiv:1401.0080 [hep-ph]
7. S. Dawson, The Top quark, QCD, and new physics. arXiv:hep-ph/0303191.
8. S.D. Ellis, J. Huston, K. Hatakeyama, P. Loch, M. Tonnesmann, Jets in hadron-hadron collisions. Prog. Part. Nucl. Phys. **60**, 484 (2008)
9. B.K. Gjelsten, D.J. Miller, P. Osland, Measurement of the gluino mass via cascade decays for SPS 1a. JHEP **0506**, 015 (2005)
10. G.D. Kribs, A. Martin, T.S. Roy, M. Spannowsky, Discovering the Higgs boson in new physics events using jet substructure. Phys. Rev. D **81**, 111501 (2010)
11. R. Lafaye, T. Plehn, M. Rauch, D. Zerwas, Measuring supersymmetry. Eur. Phys. J. C **54**, 617 (2008)
12. R. Lafaye, T. Plehn, M. Rauch, D. Zerwas, M. Dührssen, Measuring the Higgs sector. JHEP **0908**, 009 (2009)
13. C. Lester, Model independent sparticle mass measurement at ATLAS. CERN-THESIS-2004–003. www.hep.phy.cam.ac.uk/lester/thesis/index.html
14. D.E. Morrissey, T. Plehn, T.M.P. Tait, Physics searches at the LHC. Phys. Rep. **515**, 1 (2012)
15. H. Murayama, I. Watanabe, K. Hagiwara, HELAS: HELicity amplitude subroutines for Feynman diagram evaluations. KEK-91-11.
16. T. Plehn, M. Spannowsky, Top Tagging. J. Phys. G **39**, 083001 (2012)
17. T. Plehn, G.P. Salam, M. Spannowsky, Fat jets for a light Higgs. Phys. Rev. Lett. **104**, 111801 (2010)
18. G.P. Salam, Towards Jetography. arXiv:0906.1833 [hep-ph].
19. J.M. Smillie, B.R. Webber, Distinguishing spins in supersymmetric and universal extra dimension models at the large Hadron collider. JHEP **0510**, 069 (2005)
20. L.T. Wang, I. Yavin, A review of spin determination at the LHC. Int. J. Mod. Phys. A **23**, 4647 (2008)
21. B. Webber, Mass determination in sequential particle decay chains. JHEP **0909**, 124 (2009)
22. G. Weiglein et al. (LHC/LC Study Group), Physics interplay of the LHC and the ILC. Phys. Rep. **426**, 47 (2006)

Index

Absorptive integral, 167
Angular ordering, 252, 286
Anomalous dimension, 221
Asymptotic freedom, 188
Axial gauge, 78, 239
Azimuthal angle, 120, 173, 200, 250

Bethe–Salpeter equation, 235
Breit frame, 119

Callan–Symanzik equation, 192
Cascade decay, 304, 308, 309, 313, 315
Cauchy distribution, 169
Cauchy integral, 81, 165
Chiral projectors, 10, 158
Collinear divergence, 195
Collinear limit, 196, 227
Collinear radiation, 265
Combinatorial background, 266, 272, 295
Convolution, 212
Covariant derivative, 8
Cross section
    exclusive rate, 266
    hadronic, 170
    inclusive rate, 266
    Tevatron and LHC processes, 89
    total $2 \to 1$, 160
    total $2 \to 2$, 162
Cutkosky cut rules, 81

Dark matter, WIMP miracle, 90, 303
Derivative interaction, 7
Detector smearing, 306
DGLAP equation, 215, 216, 218, 285
    parton shower, 235
    solutions, 218
Dimensional regularization, 181, 213
Dimensional transmutation, 65, 190
Dipole radiation, 240
Dirac delta distribution, 169
Dirac matrices, 204

Effective W approximation, 96
Eikonal approximation, 238, 244, 246, 277
Electric charge, 11
Electroweak precision measurements
    Higgs mass, 27
    ρ parameter, 22, 23
Equivalence theorem, 39
Error
    RFit scheme, 321
    statistical error, 317
    systematic error, 318
    theoretical error, 229, 318
Event generation
    negative weights, 280
    re-weighting, 284
    unweighted events, 177
    weighted events, 177
Event generators
    ALPGEN, 271
    HERWIG, 267, 284
    Madgraph, 271, 299, 302
    PYTHIA, 267, 284
    SHERPA, 244, 271, 273

Factorization, 201, 210, 228
Fierz transformation, 130
FORM, 79, 298

Gaussian distribution, 70, 169
Gell–Mann matrices, 135
Goldstone boson, 7, 17, 28
    Feynman rules, 39
    linear representation, 20
    linear representation, 17
    non–linear representation, 17
Goldstone's theorem, 7, 28, 135, 148

Helicity amplitudes, HELAS, 299, 300, 302
Hierarchy problem, 134
Higgs boson, 7
    branching ratios, 68
    collinear decay, 103
    LHC cross sections, 78
Higgs coupling, 43
    dimension-6 CP basis, 117
    form factor, 79
    loop–induced, 78
    self coupling, 33, 122
Higgs field, 19
    quantum fluctuations, 19
Higgs mass
    experimental upper bound, 27
    quadratic divergence, 133, 135, 142
    stability bound, 45
    triviality bound, 45
Higgs potential, 14, 32, 59, 142
    dimension-6 operators, 30
Histogram, 176
Hypercharge, 11

Infrared safety, 276

Jet, 195, 292
    Cambridge-Aachen algorithm, 294
    fat jet, 296
    Higgs tagger, 106, 296
    jet mass, 294
    top tagger, 297

K factor, 274
Kinematic endpoint, 309
Klein–Gordon equation, 164

Landau pole, 45, 124, 190
Levi–Civita tensor, 120
Likelihood, 319
Luminosity, 88, 317, 318

Mandelstam variable, 160
Markov chain, 112
Mass
    factorization, 197
    fermion mass, Dirac mass, 10–12
    gauge boson mass, 13
    MT2 construction, 312
    transverse mass, 306
    visible mass, 307
Massive photon, 6
Mellin transform, 219
Minuit, 111
Monte Carlo
    generator, 234
    importance sampling, 178
    integration, 178
    Markov process, 233
MSSM, 51, 67

Next–to–leading order corrections, 275, 280, 282

Optical theorem, 41

Particle width, 166
    Breit–Wigner propagator, 68, 74, 168, 179, 300
    narrow width approximation, 168
Parton densities, 169
Parton shower
    backwards evolution, 236
    truncated shower, 286
    vetoed shower, 268, 284, 285
Pauli matrices, 8, 9, 17, 126, 143
Pauli–Villars regularization, 133, 180
Perturbative unitarity, 42, 68
Phase space
    generator, 177
    mapping, 179
    slicing, 276
    subtraction, 277
    Sudakov decomposition, 198
Plus subtraction, 213, 279
Poisson scaling, 100, 260, 264
Propagator
    Breit–Wigner propagator, 169
    cutting, 81
    Feynman $i\epsilon$, 164
    gauge boson, 18
    gluon, 185
    Goldstone boson, 18

Index 327

residue, 167
spinor, 207
Pseudo-rapidity, 173

QCD field strength, 79
QCD perturbative series, 186
QCD sum rules, 170

Rapidity, 172
Renormalizable operators, 29
Renormalization
  mass, 167
  $\overline{MS}$ scheme, 184
  squark mass, 183
  strong coupling, 181
  Thomson limit, 184
  top quark mass, 182
  wave function, 167
Renormalization group equation
  Higgs self coupling, 44
  strong coupling, 189
Resummation
  collinear logarithms, 222
  scaling logarithms, 194
R ratio, 163, 169, 191
Running coupling, 185, 229

Scalar–photon mixing, 5
Scale(s)
  factorization scale, 197, 218, 220, 226
  renormalization scale, 44, 181, 220
Scale artifact, 227
Simulated annealing, 113
Soft gluon emission, 237
Splitting

no-splitting probability, 233
phase space, 199
space–like branching, 210
time–like branching, 198
Splitting kernel, 197, 201, 211, 222, 231, 285
  subtracted $P_{g \leftarrow g}$, 218
  subtracted $P_{g \leftarrow q}$, 215, 216
  subtracted $P_{q \leftarrow g}$, 215
  subtracted $P_{q \leftarrow q}$, 215
  unsubtracted $\hat{P}_{g \leftarrow g}$, 203
  unsubtracted $\hat{P}_{g \leftarrow q}$, 209
  unsubtracted $\hat{P}_{q \leftarrow g}$, 207
  unsubtracted $\hat{P}_{q \leftarrow q}$, 209
Spontaneous symmetry breaking, 7, 18
Staircase scaling, 101, 264, 273
Sudakov factor, 231, 255, 267, 280
Supersymmetry, 50, 89, 104, 180, 273, 303, 308, 316

Tagging jet, 95–97, 118, 120
Transfer function, 280
Transverse momentum ordering, 223
Transverse momentum size, 227
Transverse tensor, 79, 119
Trigger, 90, 304
Two Higgs doublet model, 51

Unitary gauge, 15, 17, 18, 23, 302

Virtuality, 211

Weak interaction, 8, 22, 303
  charged current, 16
  neutral current, 16

The manufacturer's authorised representative in the EU is Springer Nature Customer Service Centre GmbH, Europaplatz 3, 69115 Heidelberg, Germany. If you have any concerns regarding our products, please contact ProductSafety@springernature.com

Printed and bound by CPI Group (UK) Ltd, Croydon, CR0 4YY

23/03/2026

02076686-0001